Editorial Board Members' Collection Series: Integrated Surface Water and Groundwater Resources Management

Editorial Board Members' Collection Series: Integrated Surface Water and Groundwater Resources Management

Guest Editors

Kristine Walraevens
Marco Delle Rose

MDPI

Basel • Beijing • Wuhan • Barcelona • Belgrade • Novi Sad • Cluj • Manchester

Guest Editors
Kristine Walraevens
Department of Geology
Ghent University
Gent
Belgium

Marco Delle Rose
Institute of Atmospheric Sciences
and Climate
National Research Council of Italy
Lecce
Italy

Editorial Office
MDPI AG
Grosspeteranlage 5
4052 Basel, Switzerland

This is a reprint of the Special Issue, published open access by the journal *Hydrology* (ISSN 2306-5338), freely accessible at: https://www.mdpi.com/journal/hydrology/special_issues/L7E6890066.

For citation purposes, cite each article independently as indicated on the article page online and as indicated below:

Lastname, A.A.; Lastname, B.B. Article Title. *Journal Name* **Year**, *Volume Number*, Page Range.

ISBN 978-3-7258-4861-4 (Hbk)
ISBN 978-3-7258-4862-1 (PDF)
https://doi.org/10.3390/books978-3-7258-4862-1

Cover image courtesy of Adugnaw Birhanu Zegeye

Contents

hydrology MDPI

Article

Investigating Nitrate with Other Constituents in Groundwater in Two Contrasting Tropical Highland Watersheds

Feleke K. Sishu [1], Seifu A. Tilahun [1,2], Petra Schmitter [3] and Tammo S. Steenhuis [1,4,*]

1 Faculty of Civil and Water Resources Engineering, Bahir Dar Institute of Technology, Bahir Dar University, Bahir Dar P.O. Box 26, Ethiopia; felek2004@gmail.com (F.K.S.); satadm86@gmail.com (S.A.T.)
2 International Water Management Institute, Accra PMB CT 112, Ghana
3 International Water Management Institute, 127 Sunil Mawatha, Battaramulla, Colombo P.O. Box 2075, Sri Lanka
4 Department of Biological and Environmental Engineering, Cornell University, 206 Riley Robb Hall, Ithaca, NY 14853, USA
* Correspondence: tss1@cornell.edu

Abstract: Nitrate is globally the most widespread and widely studied groundwater contaminant. However, few studies have been conducted in sub-Saharan Africa, where the leaching potential is enhanced during the rainy monsoon phase. The few monitoring studies found concentrations over drinking water standards of 10 mg N-NO_3^- L^{-1} in the groundwater, the primary water supply in rural communities. Studies on nitrate movement are limited to the volcanic Ethiopian highlands. Therefore, this study aimed to evaluate the transport and fate of nitrate in groundwater and identify processes that control the concentrations. Water table height, nitrate, chloride, ammonium, reduced iron, and three other groundwater constituents were determined monthly in the groundwater in over 30 wells in two contrasting volcanic watersheds over two years in the Ethiopian highlands. The first watershed was Dangishta, with lava intrusion dikes that blocked the subsurface flow in the valley bottom. The water table remained within 3 m of the surface. The second watershed without volcanic barriers was Robit Bata. The water table dropped rapidly within three months of the end of the rain phase and disappeared except near faults. The average nitrate concentration in both watersheds was between 4 and 5 mg N-NO_3^- L^{-1}. Hydrogeology influenced the transport and fate of nitrogen. In Dangishta, water was blocked by volcanic lava intrusion dikes, and residence time in the aquifer was larger than in Robit Bata. Consequently, nitrate remained high (in several wells, 10 mg N-NO_3^- L^{-1}) and decreased slowly due to denitrification. In Robit Bata, the water residence time was lower, and peak concentrations were only observed in the month after fertilizer application; otherwise, it was near an average of 4 mg N-NO_3^- L^{-1}. Nitrate concentrations were predicted using a multiple linear regression model. Hydrology explained the nitrate concentrations in Robit Bata. In Dangishta, biogeochemistry was also significant.

Keywords: Ethiopia; groundwater; highlands; nitrate; Sub-Saharan Africa; volcanic

1. Introduction

Nitrate is the most extensive groundwater contaminant globally [1–3]. Diffuse sources such as excess nitrogen fertilizer in intensively cropped land and urban areas without sanitation facilities are significant sources of nitrate [4–7]. Shallow aquifers are often the only safe drinking water source for rural communities [8,9]. For instance, more than 90% of the public drinking water in Ethiopia is supplied from groundwater wells [10], most of which are shallow [11]. High nitrate concentration in drinking water causes diseases such as thyroid and methemoglobinemia in infants [1,12]. The World Health Organization [13] has established 10 mg N-NO_3^- L^{-1} as a maximum allowable level of concentration for drinking water.

Nitrate concentration in groundwater has been well studied in the past 100 years. A search with "groundwater and nitrate" resulted in 12,500 citations in the Science Citation Index. However, only 1% of these studies were conducted in Africa and less than 0.1% in Ethiopia. These studies in Ethiopia were mainly focused on the evaluation of concentration levels in groundwater [14–17]. Only one study looked at the mechanisms of nitrate transport in groundwater [8]. Thus, despite the overabundance of global studies on nitrate pollution in groundwater, investigations on nitrate transport and its fate in groundwater are still needed in sub-Saharan Africa and Ethiopia because of the complex hydrologic pathways in volcanic soils. Moreover, nitrate leaching is faster in the sub-humid and humid tropical monsoon climate than in temperate climates because of more than 1200 mm of rainfall within four to five months [18]. On the other hand, the temperature is higher in the tropics, and watersheds are wetter, with many valley bottoms being saturated, enhancing denitrification and lowering nitrate concentrations.

The hydrology in the Ethiopian highlands is affected by the volcanic past. Despite their loam or clay texture, volcanic soils have high infiltration rates that, in many cases, are greater than 100 mm h^{-1} [19,20]. Only when the soils are saturated does overland flow occur. In addition, water transport through fissures and faults, and blockage by volcanic dikes, in Ethiopia's volcanic landscape significantly affect hydrology [14]. Subsurface water transfer between watersheds and basins is commonplace [21,22]. Additionally, land-use alteration due to agricultural practice for many years has modified the hydrology [23].

To feed the increasing population in sub-Sahara Africa, nitrogen fertilizer use increased by 6 Gg over the last 20 years to intensify agriculture [24]. In Ethiopia, nitrogen fertilizer use increased twofold within the past two decades, and the current input is 43 kg N ha^{-1} [25]. Therefore, to understand the dangers of groundwater pollution in the highlands with a monsoon climate such as Ethiopia, this study aims to contribute to understanding the transport and fate of nitrate in the volcanic soils and groundwater systems in the sub-humid and humid tropical highlands with a monsoon climate. To meet the objective, groundwater nitrate concentration and other geological, hydrological, and chemical parameters were evaluated in two geologically contrasting watersheds over two years. The watersheds were different in hydrogeology, farming practices, and extent of irrigation.

2. Materials and Methods

2.1. Description of the Two Watersheds

The 57 km^2 Dangishta and 9 km^2 Robit Bata watersheds are part of the Lake Tana basin in the headwaters of the Blue Nile. The Dangishta watershed is south of Lake Tana in the upper Gilgel Abay catchment. Robit Bata is at the southeastern edge of Lake Tana (Figure 1). Dangishta has a tropical highland monsoon climate rain phase between May and October. Over the past ten years, the annual average rainfall in Dangishta was 1745 mm a^{-1}, with temperatures varying between 10 °C and 25 °C. In Robit Bata, the average annual rainfall was 1420 mm a^{-1}, of which most falls between June and September. The temperature was between 11 °C and 27 °C.

Quaternary olivine volcanic rock is the geologic unit in the Dangishta watershed. The unit consists of porphyritic basalts, basaltic breccias, minor tuff, and scoriaceous lava flows with young cinder cones and vesicular basalts [26]. The transmissivity of Quaternary basalts is high and ranges from 100 to 200 m^2 day^{-1} [27]. Quaternary lacustrine and alluvial deposits are found in the lower part of the Robit Bata watershed, where the aquifer was located. Transmissivity of the Quaternary alluvial sediment is around 700 m^2 day^{-1} [27].

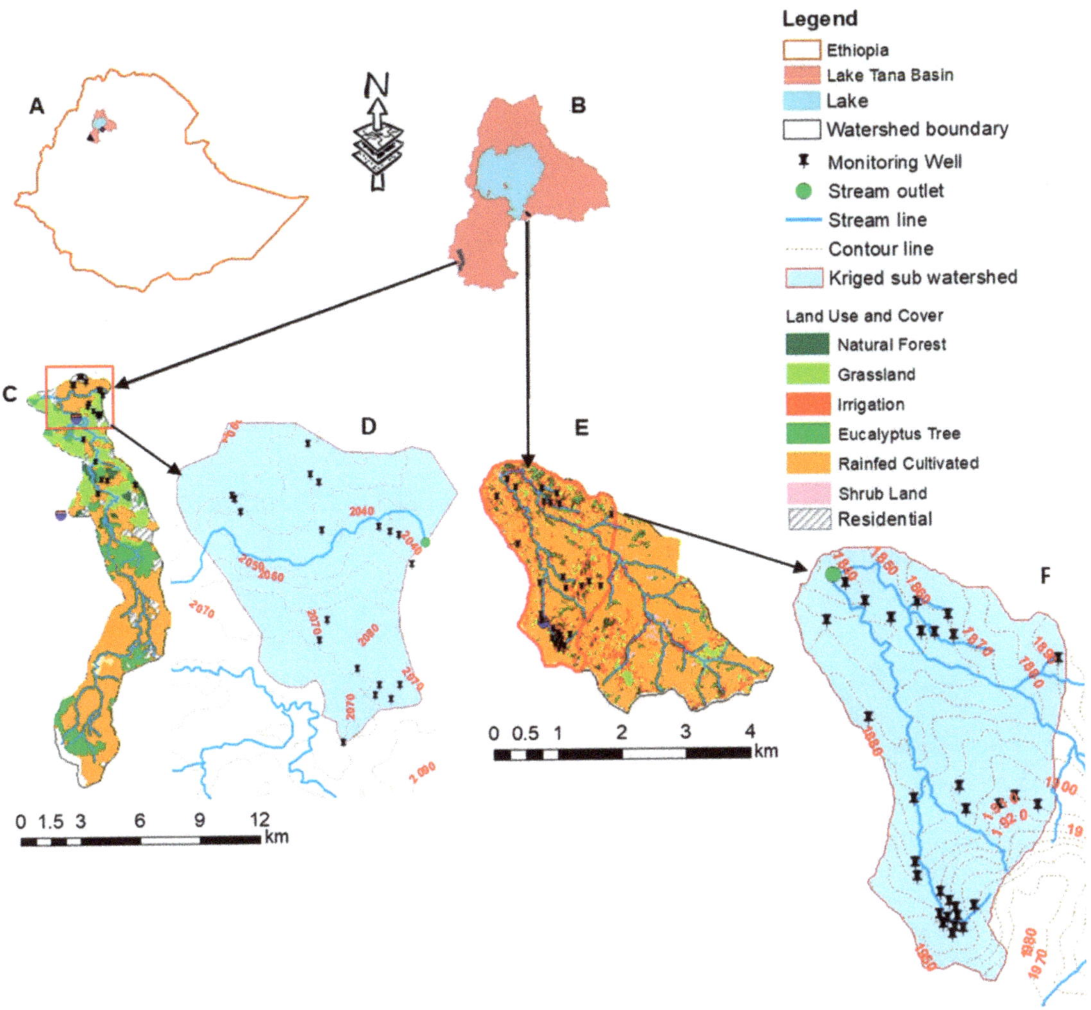

Figure 1. Location of Dangishta and Robit Bata watersheds in Lake Tana basin. (**A**) Lake Tana basin and (**B**) Lake Tana basin with Dangishta and Robit Bata watershed. Land-use maps, monitoring wells, and rain gauges for (**C**) Dangishta and (**E**) Robit Bata watersheds. (**D**) The topographic map of Dangishta and (**F**) Robit Bata, the area used for nitrate kriging.

In Dangishta, the surface soil consists of Regosols, Alisols, Nitosols, and Vertisols [19]. The soil depth varies from 4 to 21 m (Figure 2C), with the deepest soils in the uplands. In Robit Bata, the upper part of the watershed is severely degraded (Figure 2D), with shallow Eutric Cambisols and Luvisols soils over bedrock [28]. Deep Alisols are underlain by a conductive regolith layer above the bedrock in the mid and lower portion of the watershed (Figure 2E,F).

Figure 2. Maps and photos depicting features in the Dangishta and Robit Bata watersheds: (**A**) map of Dangishta showing seasonally saturated floodplain (taken from Google Earth image on 6 June 2018); yellow line is the boundary of the floodplain, and arrows indicate surface water flow directions; (**B**) photo taken in the rain phase of 2021 that shows subsurface flow is surfaced and appears as springs in a floodplain in Dangishta; (**C**) riverbank in Dangishta shows that the valley bottom has deep soils (except for areas with volcanic dikes); (**D**) a thin weathered topsoil with gravel at steeply sloping areas in Robit Bata; (**E**) excavated well materials from a 12 m deep hand-dug well; and (**F**) photograph of bedrock geology with visible outcrops in riverbeds of Robit Bata. (**C–E**) were photographed in 2018 and (**F**) in 2017.

The volcanic surface soils are highly conductive. In both watersheds, the infiltration rate exceeded the rainfall intensity less than 2.5% of the time [19,20]. Otherwise, the two watersheds are hydrologically different. In Dangishta, 87% of the area is underlain by the aquifer [29,30]. About 90% of the excess rainfall flows laterally through fissures and faults to the main watercourse. It is forced to the surface by volcanic dikes nearly perpendicular to the stream and appears as springs in the valley bottom (Figure 2B). It results in a grass-covered floodplain with a high groundwater table and is unsuitable as cropland (Figure 2A). A hillslope aquifer underlies Robit Bata's middle and lower portion of the watershed. The upper part with thin soils becomes saturated during rainfall, generating saturation excess overland flow. Unlike Dangishta, interflow does not resurface but directly discharges to the stream. Hence, a minimum grassland along the creeks periodically saturates, as seen in Figure 2F.

Rainfed cultivated land is practiced on 65% of the land in Dangishta and 74% in Robit Bata. The combined acreage of grassland and eucalyptus is 32% of the land in Dangishta but under 10% in Robit Bata [29]. Thirteen percent of the land is irrigated in Robit Bata during the dry monsoon phase. Irrigation in Dangishta is negligible at less than 1%.

Crop and livestock-integrated agriculture practices are followed [31,32]. Cereal crops, including 'tef' (*Eragrostis tef*), maize (*Zea mays*), millet (*Eleusine coracana*), and barley (*Hordeum vulgare*) are grown on rainfed land. Khat, a mild narcotic perennial small bush, is irrigated during the dry phase in Robit Bata [33]. According to the local agricultural office, livestock densities are 975 animal units per square kilometer in Robit Bata and 1500 in Dangishta.

Fertilizer use in the two watersheds was documented in 2017 by the Bahir Dar University in collaboration with Feed the Future Innovation Lab for Small-Scale Irrigation (ILSSI). The results are summarized in Table 1. Di-ammonium phosphate (DAP), urea, and organic compost fertilizers were applied in split doses. The first application consisting of DAP and compost occurred when crops were planted at the beginning of the rainy phase in May or early June in Dangishta and the end of June in Robit Bata. The second application of urea was in July in both watersheds. In Robit Bata, urea was applied to irrigated khat in the dry phase from November to March.

Table 1. Summary of fertilizers used and average N and P application rate in Dangishta and Robit Bata.

Fertilizers	Ingredients	Application Rate (kg N or P ha^{-1})						National-Level Application (CSA 2021)	
		Dangishta		Robit Bata					
		Rainfed		Rainfed		Irrigation			
		N	P	N	P	N	P	N	P
UREA $CO(NH_2)_2$	N.P.K (46:0:0)	48	0	45	0	5	0		
DAP $(NH_4)_2HPO_4$	18% N, and 20% P	24	27	23	25	2	3		
Compost	1.8% N, 0.25% P	3	0.4	1	0.1	22	3		
Total	Total applied (kg ha^{-1})	75	27	69	25	29	6	43	17

2.2. Data Collection

Meteorological, hydrological, and water quality data were collected from January 2017 to December 2018. These include groundwater table depth and precipitation. In addition, pH, electric conductivity, NO_3^-, NH_4^+, K^+, Mg^{2+}, Ca^{2+}, Cl^-, and Fe^{2+} were analyzed in groundwater samples. Additionally, chloride in precipitation and nitrate concentration in streams were measured.

2.2.1. Precipitation

In Dangishta, the rainfall depth was recorded every 10 min with a tipping bucket located in the middle of the catchment (Figure 1C). In addition, a bucket-type rainwater gauge collected water samples for chloride concentration. Missing precipitation data were filled in by linear regressing of rainfall recorded at the National Meteorological Agency (ENMA) weather station 5.3 km away [29].

For Robit Bata, in 2017, precipitation data were obtained from the National Meteorological Agency (NMA) Zenzelima weather station located at a 3.4 km distance away. Precipitation was measured on-site using a tipping bucket in 2018.

2.2.2. Data Collected at the Monitored Shallow Wells

One-meter-diameter hand-dug wells used for domestic water supply and irrigation were monitored. The wells were dug until bedrock along fault lines where water is available most of the year. In Dangishta, 32 wells were monitored. The well depth varied with topography and was around 21 m upslope and 4 m in the valley bottom. In Robit Bata, 35 wells were observed (Figure 1F). Well depths varied from 6 to 18 m. Wells with the shallowest depths were close to the river. Tilahun et al. (2020) [20] used some of these wells to determine the groundwater potential for irrigation beginning in 2015.

Groundwater levels were measured monthly using a tape measure in the dry phase and weekly in the rain phase. The pH and electrical conductivity (EC) were measured on-site using Yellow Springs Instrument (YSI) ProDSS (digital sampling system) multi-parameter. Monthly groundwater samples were collected: 748 samples from 32 wells in Dangishta and 792 from 35 wells in Robit Bata. The samples were filtered through Whatman filter paper with a 0.45 μm pore size filter and stored below 4 °C until analyzed within a week. The standard Palin test method was used to analyze all chemical parameters [34,35] using photometer 7100. The detailed analysis methods are described in the Palintest Operation Manual [36].

2.2.3. Analysis of Chloride in Precipitation and Nitrate in Streams

The chloride concentration in the rain was measured in Dangishta in 45 rain events in the rain phase from April to October 2017. The rainwater was collected in a bucket 2 m above the ground surface.

The nitrate and ammonium concentrations in the stream at the outlet in the rain phase were measured for 34 runoff events in Dangishta and 47 in Robit Bata. Between 4 and 15 water samples were collected per storm event. On days without a runoff event, the baseflow was sampled once. In the dry phase, the concentrations in baseflow were measured once a month. The stream in Robit Bata did not have water from January to May, and concentrations were not recorded for these months. Based on these data, average monthly flow-weighted concentrations were calculated. The detail on the streamflow measurements and sampling handling are published in [29].

2.2.4. Land-Use and Cover Data Acquisition

The land use for Dangishta was obtained from the Amhara Design and Supervision Works Enterprise [37]. Land use was classified as natural forests, grassland, eucalyptus, shrubs, homestead (residential), and rainfed cropland. For Robit Bata, land use was summarized by Takele [37] and consisted of natural forests, grassland, irrigated land, eucalyptus, shrubs, homestead, and rainfed cropland.

2.3. *Data Analysis*

2.3.1. Surface and Groundwater Interaction

The effect of fertilizer application on groundwater nitrate concentrations level was investigated by cataloging the land use in a sphere around each well with GIS (Figure S1). Spheres with radii of 25, 50, 100, and 150 m were examined based on the work of [38–40]. Spheres with a 100 m radius in Dangishta and 50 m in Robit Bata were selected because nitrate concentrations in wells were most closely related to land-use-associated fertilizer applications shown in Supplementary Materials A (Tables S1 and S2). Nitrogen fertilizer applications based on farmers' survey data are shown in Supplementary Materials A (Table S3).

2.3.2. Groundwater Redox Potential

The redox framework of McMahon et al. [41], which is based on the dissolved concentrations of Fe^{2+} and NH_4^+, was used to assess the redox condition of the groundwater. Concentrations for oxic and anoxic conditions are listed in Table S4.

2.3.3. Mapping of Groundwater Constituents

The ordinary kriging option in the Spatial Analyst Tool of ArcGIS 10.8 was used [42,43] to spatially interpolate the monthly collected concentration data in portions of watersheds with a sufficient number of wells (Figure 1D,F). The semi-variogram spherical model [44] had the smallest root mean square error and was used for interpolation. The kriged areas were 346 ha in the eastern part of the Dangishta area and 413 ha in the northern part of Robit Bata. Kriged maps were prepared for nitrate and those parameters that affected the nitrate distribution in the watershed, such as water table depth, chloride nitrate ratio, and Fe^{2+}.

2.3.4. Statistical Data Analysis

Statistical testing was performed to identify processes that influence temporal nitrate in shallow groundwater. Using the monthly measurements, we first employed the Pearson correlation test to evaluate the relationship between nitrate and groundwater constituents and the groundwater table. Then, the multiple linear regression model test was implemented to explain the combined effects of biogeochemical and physical factors on nitrate concentrations. Parameters included were precipitation (P), groundwater depth (GWD), and all groundwater constituents including pH, EC, NH_4^+, K^+, Mg^{2+}, Ca^{2+}, Cl^-, and Fe^{2+}. A stepwise regression procedure was employed to remove variables in the model by checking their significance ($a > 0.05$ for removing a parameter from the model). Seventy percent of the data were used to train the model and 30% for validation. The goodness of fit of the regression models was judged by maximizing the adjusted R-squared values. Histogram and Q–Q plots were used to verify the normality assumption in the residuals. The R programming statistical software package was used.

3. Results

3.1. Precipitation

Figure 3 shows rainfall amounts over the two years. It rained more in 2017 than in 2018 with a difference of 620 mm in Dangishta and 480 mm in Robit Bata (Supplementary Materials A Table S5). July and August were the wettest months. The highest recorded daily rainfall was 123 mm on 21 July 2017 in Robit Bata and 61 mm on 28 August 2017 in Dangishta (Figure 3A,C). The rainfall intensities are less than soil infiltration rates in unsaturated soils [20].

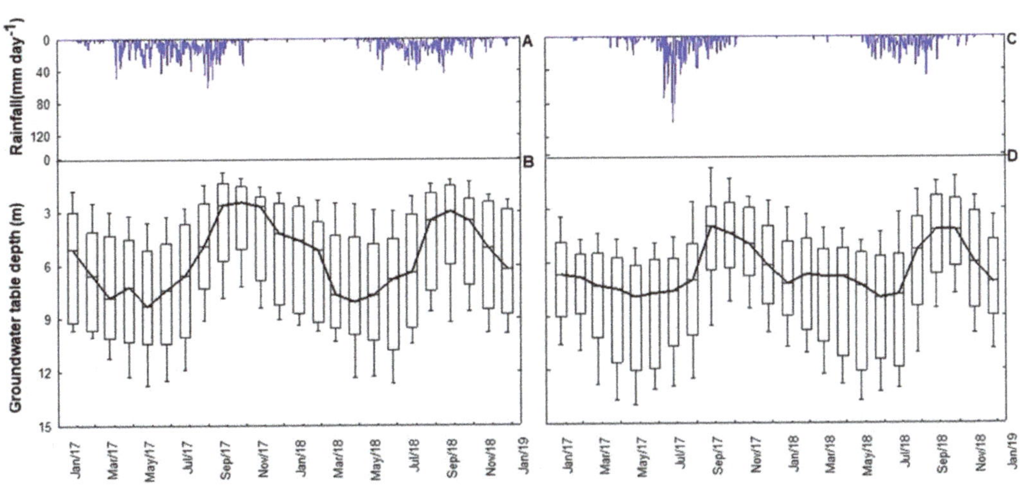

Figure 3. Boxplots of water table depth of 2017 and 2018. (**A**,**B**) is for Dangishta, and (**C**,**D**) is for Robit Bata watershed. Each box plot is based on data from 32 wells in Dangishta and 35 in Robit Bata, with 25th and 75th percentiles lower and upper ends. The black line passes through the median.

3.2. Chloride Concentration of the Precipitation

Average chloride concentrations in rainwater measured in Dangishta early in the 2017 rain phase were 10 mg L^{-1} from April to June (Figure 4). Concentrations decreased to an average of 5 mg L^{-1} in the remaining months. The same trend of Cl^- concentration in rainfall in Ethiopia's highlands was reported by Banks et al. [8], but concentrations were slightly lower compared to this study.

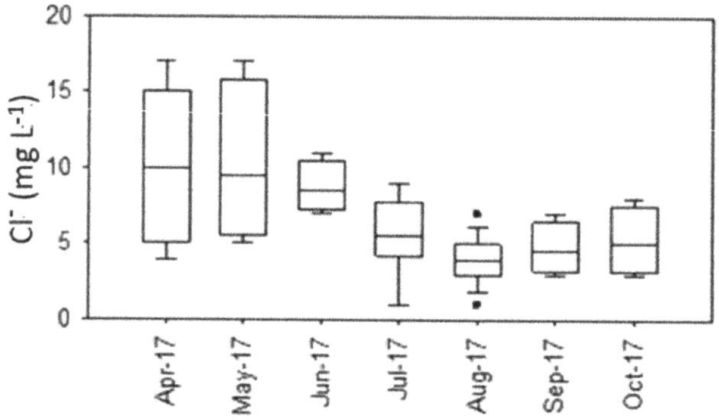

Figure 4. Boxplots of month chloride concentrations in precipitation in Dangishta in the 2017 rain phase. The dots placed above and below whiskers indicate outliers.

3.3. *Groundwater*

The water table measurements from 1 January 2017 to 31 December 2018 are summarized in Figure 3B,D and Figure 5 in Dangishta and Robit Bata.

3.3.1. Temporal Variation in Groundwater Levels

The temporal characteristics of the rise in the groundwater table at the beginning of the rain phase were the same in the two watersheds (Figure 3B,D). The rains and runoff began in May in Dangishta (Figure 3A) and in June in Robit Bata (Figure 3C). Otherwise, the watersheds behaved differently. The groundwater table reached the maximum level in September for Dangishta and in August for Robit Bata. After September, when the rains had stopped, water tables decreased. In Robit Bata, the rivers dried up in December, after which the groundwater was only found near the faults [20]. In Dangishta, volcanic dikes across the valley prevented subsurface flow from leaving [21,45]. The water table was maintained year-round in the lower part of the valley bottom, and the river flowed all year long in Dangishta.

3.3.2. Spatial Occurrence of the Groundwater Table

The monthly water table depths averaged over both years are shown in Figure 5. As expected, groundwater tables were the shallowest at the lowest elevations in Dangishta (west) and Robit Bata (north). At the end of the rain phase (August and September), the water tables were near the surface in both watersheds (i.e., the blue-colored areas in Figure 5 with depths less than 3 m). It remained at less than 4 m during the dry phase in the flood plain near the outlet in Dangishta (Figures 2A and 5A), where volcanic dikes blocked the subsurface flow, forcing the water to the surface and to appear as springs (Figure 2B). This also kept the water table elevated during the dry phase. Volcanic dikes are absent in Robit Bata. The groundwater depth near the faults varied from 6 to 8 m during the dry phase.

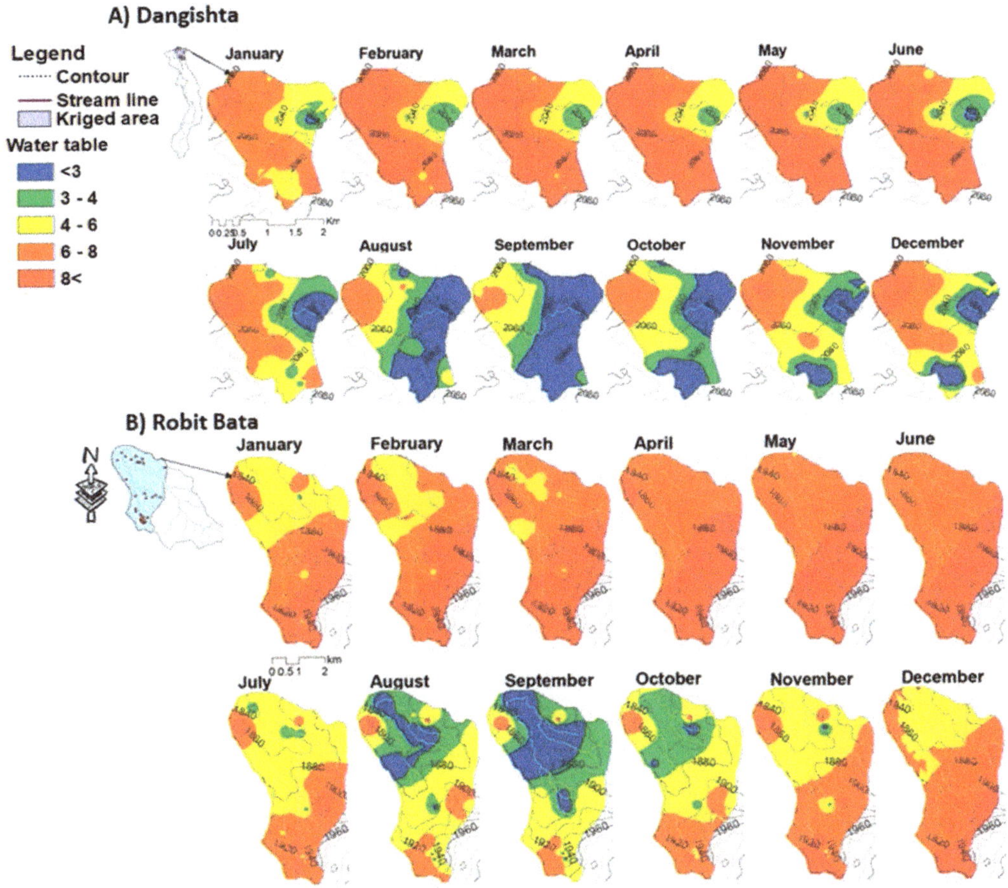

Figure 5. Kriged groundwater table level averaged over 2017 and 2018 for January to December (**A**) Dangishta and (**B**) Robit Bata.

Close inspection of Figure 5A shows that in northwest Dangishta at the watershed boundary, the water table remained below 6 m throughout the year. In contrast, the water tables increased during the rain phase to near the surface in the southeast. The land surface is less sloping and has a larger contributing area in the southeast than in the northwest (Figure 1D). Since the land slope and the water table gradient nearly coincide, the downslope velocity is smaller in the southeast, and the subsurface flow is greater. Consequently, the water builds up. In the northeast, where the steep slopes border the watershed divide, the subsurface flow is faster, and the water table does not increase as much [20]. In Robit Bata, the upper slopes are steeper in the south of the aquifer area than at the outlet (Figure 1F). In addition, the flow increases going downhill. It results in a water table closer to the surface during the rain phase near the outlet in the north.

3.4. Groundwater Constituents

3.4.1. pH

In Dangishta, the pH of the groundwater ranges from 5.1 to 7.1, with an average of 5.8. The surface soil pH was slightly more acidic, with a pH of 5.0 [29]. In Robit Bata, the groundwater pH ranged from 5.6 to 7.5, with an average of 6.7 (Table 2). The average surface soil pH was 5.6. The pH in the humid Dangishta with high rainfall was lower than in the sub-humid Robit Bata because more cations such as Ca^{2+}, Mg^{2+}, and K^+ were leached out of the topsoil with the higher rainfall. In the southern part of Ethiopia in the Dawa River basin, Woldemariyam and Ayenew [46] similarly found that groundwater in volcanic formations in the upper reach of a basin with more than 1500 mm rainfall had a pH below 7. The lower reach that received 200 mm of rain had a pH greater than 7.

Table 2. Maximum, minimum, and average EC (μS cm^{-1}), pH, and constituent concentrations (mg L^{-1}) in the groundwater of Dangishta and Robit Bata.

Watershed	Statistic	pH	EC	N-NO_3^-	N-NH_4^+	Cl^-	K^+	Mg^{2+}	Ca^{2+}	Fe^{2+}
Dangishta	Min.	5.1	44	0.1	<0.01	<0.01	<0.01	2	1	<0.01
	Max.	7.1	296	20.2	4.4	82	31	41	44	0.7
	mean	5.8	152	4.6	0.1	6.9	1.8	9	12	0.2
	Sd.	0.4	56	3.9	0.3	9.1	2	8	12	0.1
	Med.	6.1	150	4.0	0.1	4	1.5	6	7	0.3
	%Cv.	6.7	37	84.4	289	132.1	90.4	71	56	83.3
Robit Bata	Min.	5.6	43	0.1	<0.01	<0.01	<0.01	4	1	<0.01
	Max.	7.5	646	19	2.4	38	10.5	42	48	0.8
	Mean	6.7	144	4.0	0.1	4.1	1.3	10	13	0.1
	Sd.	0.3	59	2.6	0.2	4.6	1	9	12	0.1
	Med.	6.5	138	4.1	0.1	3	1.2	5.5	6.8	0.11
	%Cv.	4.8	41	65.3	199	112	80.2	57.7	51.3	89.9

3.4.2. Calcium, Magnesium, and Potassium

The yearly averaged cations (Ca^{2+}, Mg^{2+}, and K^+) were insignificantly different between the two watersheds (Table 2). However, the monthly concentrations were different between the watersheds. Ca^{2+} was high from April to June and less in September (Figure 6B) in Dangishta. In Robit Bata, Ca^{2+} and Mg^{2+} were high from January to June, when the groundwater was only available near the faults and evaporation was much greater than the rainfall. The cation concentrations were low from July to September when groundwater was recharged [47].

Potassium concentrations in both watersheds hardly varied throughout the year, except some wells recorded high concentrations in Robit Bata following high rainfall (Figure 6D,I). The order of cation abundance was ranked (i.e., $Ca^{2+} > Mg^{2+} > K^+$), and was the same as reported by Fenta et al. [14] in groundwater around Dangila town near the Dangishta watershed.

3.4.3. Chloride

The average chloride (Cl^-) concentration in Dangishta was nearly 7 mg L^{-1}. It was only 4 mg L^{-1} in Robit Bata (Table 2). The higher Cl^- concentrations in the groundwater of Dangishta were caused by the high chloride concentration in the early rainfalls in April, May, and partially in June (Figures 4 and 7D). These early rains with high Cl^- did not fall in Robit Bata (Figure 7F), and consequently, the chloride contribution of the rainfall to the groundwater was less. The monthly Cl^- concentration in the groundwater in Dangishta followed the precipitation with a high concentration early in the rain phase and gradually declined (Figure 7D). The groundwater chloride concentration in Robit Bata was less in the dry phase than in the rain phase. In volcanic highlands, Cl^- geological formation for halide dissolution is negligible [8,14,48]. In contrast, adsorption in the soil is expected [49], and it is likely the reason that the chloride concentration declined in the dry phase during transient unsaturated flow.

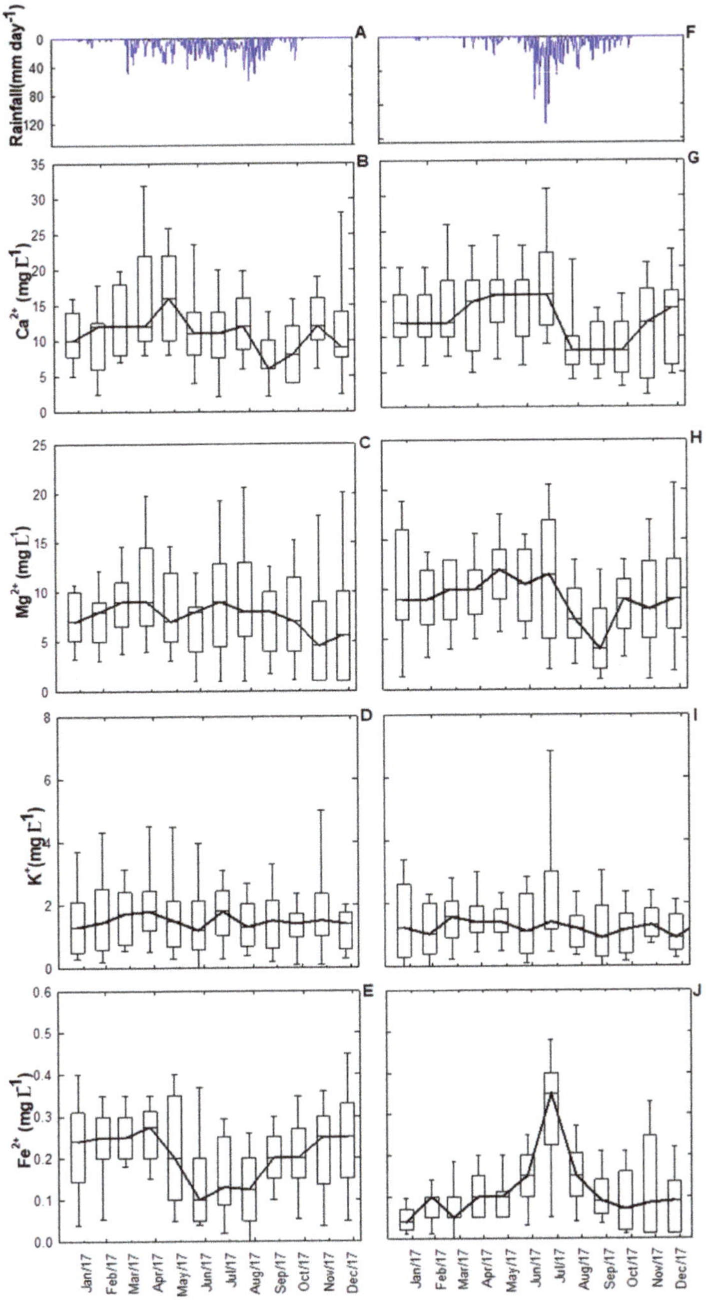

Figure 6. The concentration of major cations (Ca^{2+}, Mg^{2+}, and K^{+}) and iron (Fe^{2+}) in groundwater in 2017. (**A–E**) is for Dangishta, and (**F–J**) is for Robit Bata watershed. The box plots are based on data from 32 wells in Dangishta and 35 wells in Robit Bata.

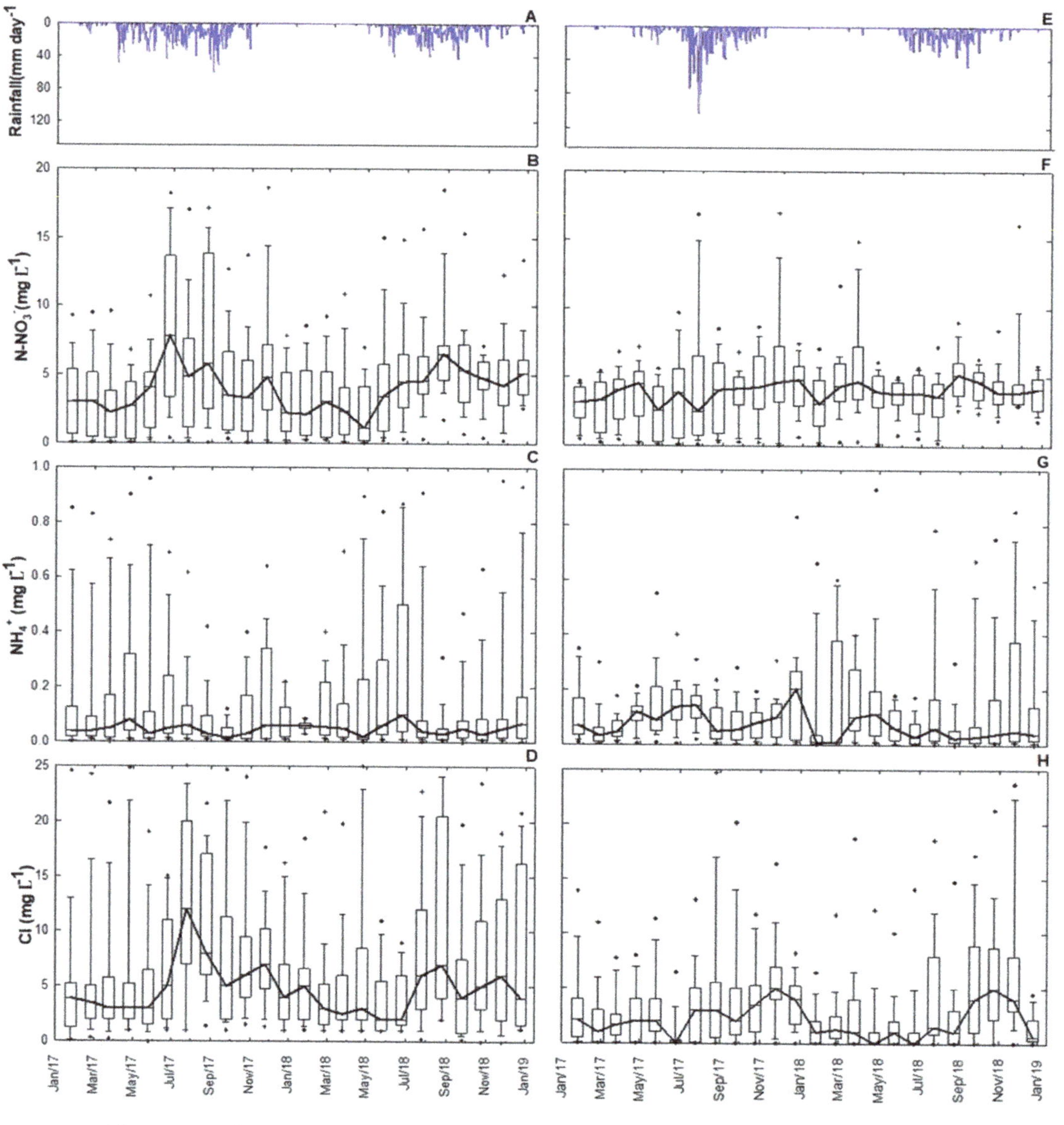

Figure 7. Box plots indicating perched groundwater nitrate ammonium and chloride concentrations in 2017 and 2018. (**A–D**) is for Dangishta, and (**E–H**) is for Robit Bata. Each box plot is based on data from 32 wells in Dangishta and 35 in Robit Bata. The dots placed above and below whiskers indicate outliers.

3.4.4. Nitrate

Nitrate concentrations (expressed as N-NO_3^-) averaged 4.6 mg L^{-1} in Dangishta and 4.0 mg L^{-1} in Robit Bata over the study period (Table 2). They were not significantly different between the two watersheds. Despite this, the monthly concentrations throughout the year differed between the watersheds.

Concentrations were unusually high in Dangishta from June to August in the cropped part of the watersheds on the well-drained soils (Figure 7B), with concentrations of over

7 mg L^{-1} and in portions of the aquifer over the drinking water standard of 10 mg L^{-1} (Figures 7A and 8A). In June, the crop growing area had a higher nitrate concentration than the old forest in the west and grassland (pasture) and eucalyptus (plantation forest) in the periodically saturated valley bottoms in the east near the outlet, which were too wet for cropland [29] (Figure 8A). The groundwater moves from the croplands to the grass and eucalyptus, causing the concentrations to rise in August to above 7 mg L^{-1} in the valley bottoms (Figure 8A). In August, the concentrations under cropland were nearly the same as the valley bottom after fertilizer application in July. In September, concentrations in the valley bottom started declining to less than 5 mg L^{-1} (Figure 8A). The cropland sections remained at 7 mg L^{-1} despite the water moving downhill. Akale et al. [50] also noted that nitrate concentrations in other parts of the Lake Tana basin were relatively higher in cropland areas.

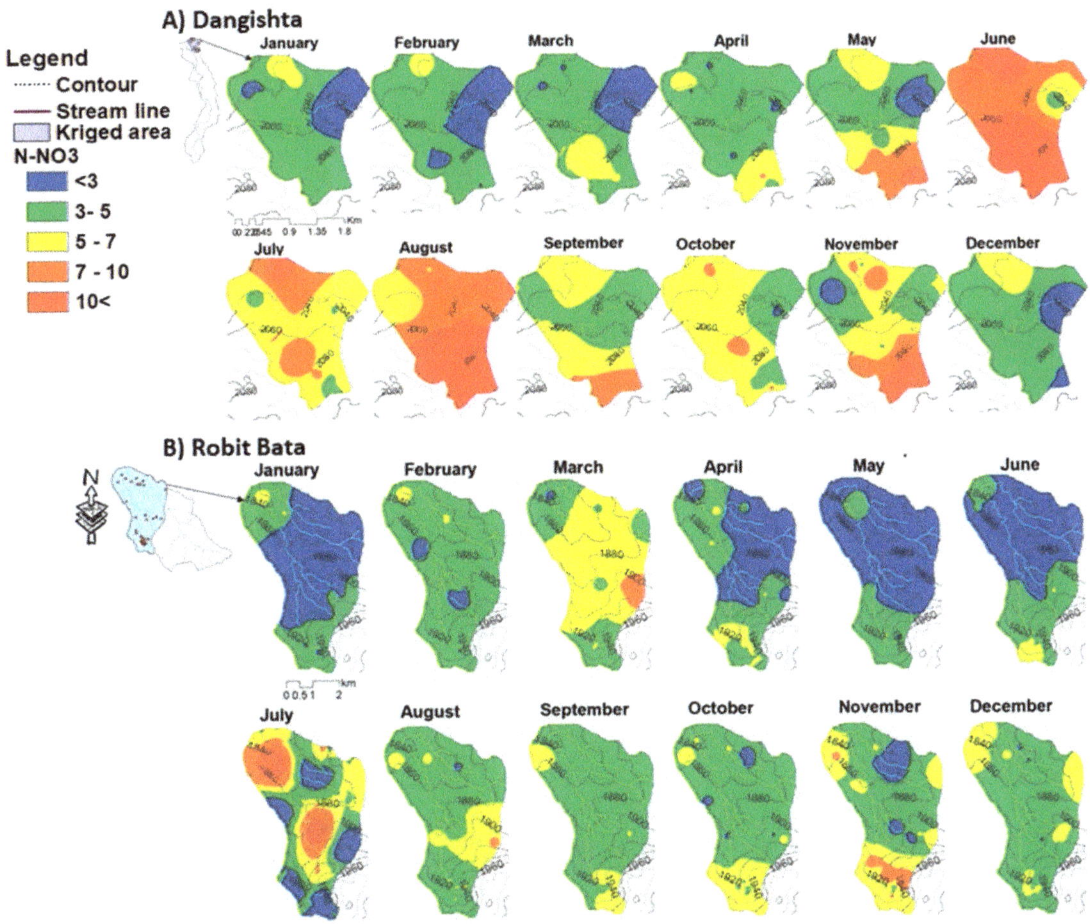

Figure 8. Monthly, two-year-averaged kriged nitrate concentration (2017 to 2018) in mg N-NO_3^- L^{-1} in the aquifers of (**A**) Dangishta and (**B**) Robit Bata watersheds.

In Robit Bata, median nitrate concentrations were nearly steady, around 4 mg N-NO_3^- L^{-1} throughout the two years, except with slight increases in July after fertilizer application on rainfed crops at the end of June (Figures 7F and 8B). In addition, nitrate was slightly elevated in November and March when the khat was irrigated on plots close to wells. Nitrate, derived from applied fertilizers, leached with excess irrigation water in the groundwater from fields close to the wells. Similar observations were reported by Lentz and Lehrsch [51].

3.4.5. Reduced Iron and Ammonia (Fe^{2+} and NH_4^+) Concentration

Reduced iron concentrations were greater in Dangishta than in Robit Bata (Table 2 and Figure 9). The pattern of monthly concentrations of reduced iron in the two watersheds was opposite (Figure 6E,J). During the rain phase, from June to September, Fe^{2+} concentrations were between 0.1 to 0.2 mg L^{-1} in Dangishta (Figure 9). In Robit Bata, the peak was 0.4 mg L^{-1} (Figure 6E,J). The reverse was true during the dry phase in Robit Bata. The mean Fe^{2+} concentration was less than 0.1 mg L^{-1}; in Dangishta, the concentrations range between 0.2 to 0.3 mg L^{-1} (Figure 9). The reasons are discussed in Section 4.1.

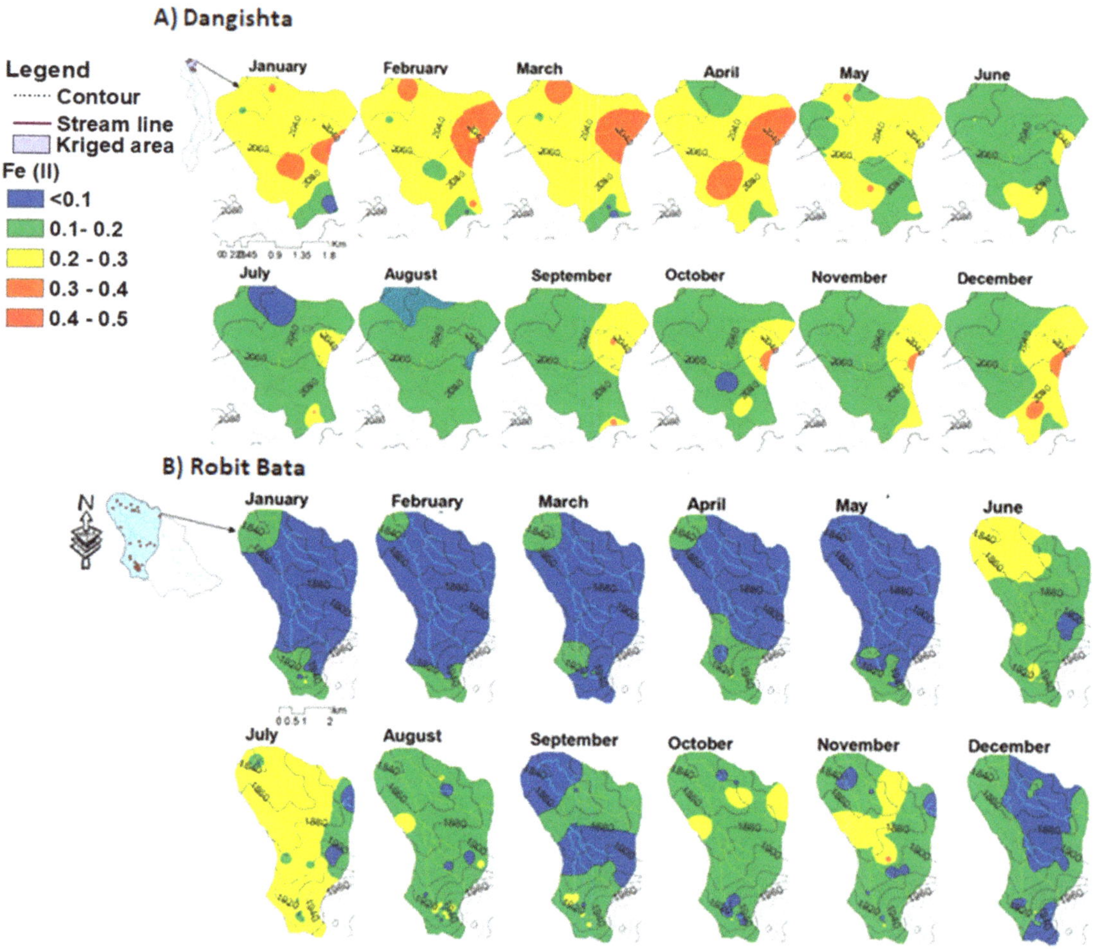

Figure 9. Monthly, kriged Fe^{2+} concentration (2017 to 2018) in mg L^{-1} in the aquifers of (**A**) Dangishta and (**B**) Robit Bata watersheds.

The averaged ammonium concentrations were around 0.1 mg L^{-1} in both watersheds (Table 2). In some wells, the ammonia concentrations were elevated. In the dry period, consistent with the iron level, the ammonium concentration increased when nitrate was reduced in Dangishta and decreased in the rainy period (Figure 7C,G) [52]. The coefficient of variation of NH_4^+ was the largest of all groundwater constituents, indicating leaching preferentially of NH_4^+ from the surface (Figure 7C,G). The matrix flow of NH_4^+ is hampered by the adsorption of NH_4^+ to negatively charged soil particles.

3.5. Surface Runoff Nitrate and Ammonia Concentrations

The monthly averaged nitrate and ammonia concentrations in the stream differ between the two watersheds (Figure 10). In Dangishta, the nitrate concentrations at the outlet follow the same pattern as the nitrate concentration in the groundwater. The concentrations are elevated during the rain phase when the crops are fertilized, and the groundwater is recharged. Nitrate concentrations decrease in the dry phase to the lowest levels (Figures 7C and 10A). Monthly nitrate concentrations in the stream were approximately half of the groundwater concentrations due to denitrification in the wetlands around the streams (Figure 2C, Figure 7C, and Figure 10A). The ammonia mobility pattern was the same with the phosphorus reported by [29]. We hypothesize runoff fertilizer transport and also that the wetlands are likely the cause of the high ammonia concentrations due to the cattle grazing the wetlands. The urea in the urine of grazing cattle is transported directly to the stream by the overland flow in the saturated area. Urea is rapidly converted to ammonia with the prevailing temperatures in the highlands.

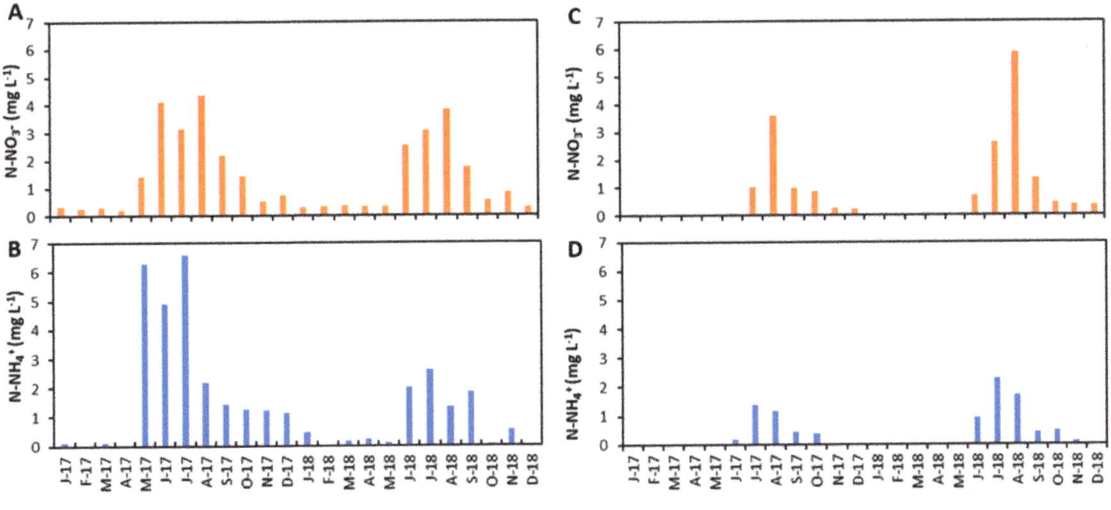

Figure 10. Monthly stream nitrate and ammonium concentrations (2017 to 2018). (**A**,**B**) are for Dangishta, and (**C**,**D**) are for Robit Bata watershed.

In Robit Bata, the average monthly nitrate concentrations in groundwater (Figure 8) and surface runoff (Figure 10C) were elevated after the urea fertilizer was applied. The greatest concentration occurred in July (Figure 8) and in August in the surface runoff (Figure 10). The average nitrate concentration in groundwater was 4 mg $N\text{-}NO_3^-\ L^{-1}$ (Table 2, Figure 7F). The nitrate concentration in the stream comprises the nitrate transported in the overland flow due to saturated excess runoff from the upper area and the baseflow from the middle and lower part. The concentration in the baseflow is diluted by the water coming from the upper part. Hence the lower concentration of nitrate in the stream compared with the groundwater.

The ammonia transport in a stream had the same pattern as the dissolved phosphorus reported by [29]. It transports with the saturation excess surface runoff from the upper part of the watershed and valley bottom.

3.6. *Relating Nitrate with Groundwater Constituents and Hydrology*

This section investigates the spatial and temporal interactions of nitrate concentrations in groundwater with water table depth, precipitation, and other groundwater constituents. It illustrates the different processes in the two watersheds that govern the fate of nitrate.

3.6.1. Relationships between Nitrate, Hydrology, and Groundwater Constituents

The results of the Pearson correlation test showed that precipitation, groundwater table depth, and chloride concentration were significantly and positively correlated with nitrate concentrations in both watersheds (Tables 3 and 4). The pH, Fe^{2+}, and N-NH_4^+ were negatively correlated with nitrate in Dangishta but not in Robit Bata (Table 3A,B). Magnesium and calcium were not correlated with nitrate levels in groundwater in either of the two watersheds.

Table 3. (**A**) Pearson correlation coefficients between hydrological and chemical parameters for Dangishta; (**B**) Pearson correlation coefficients between hydrological and chemical parameters for Dangishta.

(A)	**P**	**GWTD**	**pH**	**EC**	**Mg^{2+}**	**Ca^{2+}**	**K^+**	**Fe^{2+}**	**N-NH_4^+**	**Cl^-**	**N-NO_3^-**
P	1										
GWTD	−0.191 **	1									
pH	0.044	−0.105	1								
Ec	−0.361 **	0.029	0.323 **	1							
Mg^{2+}	0.038	−0.040	0.007	0.320 **	1						
Ca^{2+}	0.009	−0.020	0.008	0.357 **	0.981 **	1					
K^+	−0.002	0.156 **	0.313 **	0.235 **	−0.007	−0.015	1				
Fe^{2+}	−0.256 **	0.106	0.064	0.030	−0.064	−0.064	0.162 **	1			
N-NH_4^+	−0.04	0.095	0.088	0.098	−0.081	−0.088	0.420 **	0.136	1		
Cl^-	0.196 **	−0.110 *	−0.162 *	−0.096	0.242 **	0.235 **	0.092	−0.168 **	−0.012	1	
N-NO_3^-	0.127 **	0.197 **	−0.248 **	−0.128 **	0.089	0.099	−0.060	−0.551 **	−0.136 **	0.511 **	1
(B)	**P**	**GWTD**	**pH**	**EC**	**Mg^{2+}**	**Ca^{2+}**	**K^+**	**Fe^{2+}**	**N-NH_4^+**	**Cl^-**	**N-NO_3^-**
P	1										
GWTD	−0.216 **	1									
pH	−0.156 *	−0.190 **	1								
Ec	−0.295 **	0.259 **	0.010	1							
Mg^{2+}	−0.153 *	0.077	0.232 **	0.265 **	1						
Ca^{2+}	−0.113	0.347 **	0.117 *	0.412 **	0.286 **	1					
K^+	0.148 *	0.222 **	0.165 **	0.131 *	0.228 **	0.124 *	1				
Fe^{2+}	0.236 **	−0.077	0.174 **	−0.037	−0.015	−0.015	0.022	1			
N-NH_4^+	−0.005	−0.018	0.138 *	0.074	0.094	0.196 **	0.288 **	0.083	1		
Cl^-	0.129 *	−0.170 **	0.091	−0.122	0.025	−0.104	−0.032	−0.019	0.024	1	
N-NO_3^-	0.162 *	0.144 *	−0.008	−0.053	0.106	−0.044	0.291 **	−0.059	−0.035	0.672 **	1

** Correlation is significant at the 0.01 level, and * correlation is significant at the 0.05 level (two-tailed).

Table 4. Coefficient of the regression equations to estimate nitrate concentration.

Watersheds	**Name**	**Coefficient (β)**	***t*-Test**	***p***	**Adjusted R^2**	**Relative Importance**
Dangishta	Intercept	8.6	3.6	0.02	0.70	
	Cl^-	0.31	13	<0.001		0.52
	Fe^{2+}	12.6	−6.7	<0.001		0.39
	GWD	0.25	5.5	<0.001		0.08
Robit Bata	Intercept	0.55	1.3	0.09	0.63	
	Cl^-	0.62	15.1	<0.001		0.84
	Fe^{2+}	−1.85	−2.0	0.042		0.01
	GWD	0.14	3.5	<0.001		0.07
	P	0.002	3.2	0.002		0.06

3.6.2. Multiple Linear Regression Model for Nitrate Concentrations

The monthly nitrate concentration can be expressed by following multiple linear regression models. The multiple regression models also fulfilled all the assumptions, such as normality, equal variance, and linearity indicates.

$$\text{Dangishta}: \ \text{N-NO}_3^- = 8.6 + 0.31\text{Cl}^- + 0.25\text{GWT} - 12.3\text{Fe}^{2+} \quad (1)$$

$$\text{Robit Bata}: \ \text{N-NO}_3^- = 0.55 + 0.62\text{Cl}^- + 0.002\text{P} + 0.14\text{GWT} - 1.85\text{Fe}^{2+} \quad (2)$$

where the concentrations are expressed in (mg L^{-1}) and groundwater table (GWT) in (m).

The variance explained by the model was 70% in Dangishta and 63% in Robit Bata (Figure 11). Ranking the variables by their relative importance (Table 4) shows that Fe^{2+} made a 1% contribution to the overall adjusted R^2 in Robit Bata and 40% in Dangishta. Thus, nitrate concentration in Dangishta is affected by the redox potential of the groundwater, while in Robit Bata, the transport-related parameters, chloride, and precipitation together explain 90% of the variation (Table 4). In Dangishta, 50% of the variance was transport-related.

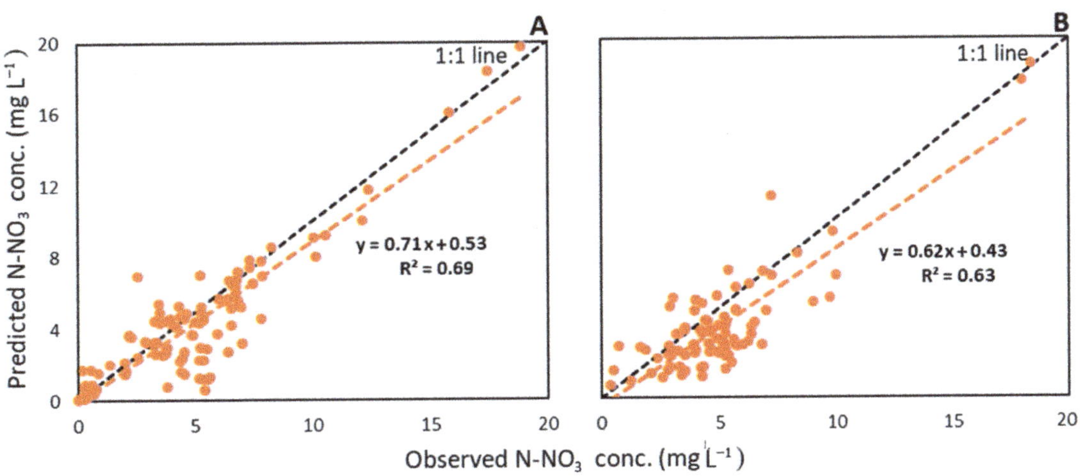

Figure 11. Predicted vs. observed nitrate concentration for the multiple linear regression model. (**A**) Dangishta and (**B**) Robit Bata watershed.

The model was validated using an independent data set collected from nine wells among monitored wells in the study (i.e., 30% of the total data set). A plot of the model predicted vs. observed fit (Figure 11) for validation showed that the analytical models satisfactorily predicted the monthly nitrate concentrations. The observed versus predicted values have an R^2 of 0.69 in Dangishta and 0.63 in Robit Bata, indicating that the model fit is acceptable (Figure 10). The fit line is very close to the 1:1 line.

4. Discussion

In this discussion section the reasons for the differences in groundwater constituent concentrations between the two watersheds are first discussed. Next, the fate and transport of nitrate in sloping hillside aquifers are discussed in relation to the watershed properties.

4.1. Constituent Concentrations in Sloping Hillside Aquifers

The groundwater table depth and nitrate concentrations in both watersheds were correlated (Table 3). At the same time, the temporal and spatial changes in groundwater depth and nitrate concentrations were different in the two watersheds (Figures 4, 5, 8 and 9). To explain these almost contrary findings, we will first discuss geology and hydrology.

The geology in Dangishta influenced water residence time. The decrease in groundwater level was slower in Dangishta than in Robit Bata. Hence, the residence time was greater in Dangishta (Figure 5A). The volcanic dikes nearly perpendicular to the streams in the valley bottom formed a barrier for the interflow. It prevented incision of the streams (Figure 2C) and resulted in a 400 m-wide strip along the river covered by grass during the rain phase because it was too wet to grow crops (Figure 2A). It is clearly shown in the kriged area (Figure 5A) near the outlet in the east part. The water tables remained within 3 m from the surface from June to January. The rivers also flow throughout the year in Dangishta. Thus, the water was held back.

In Robit Bata, volcanic dikes were absent, and the stream formed a gully at the outlet, enhancing the groundwater drainage in the watershed [29]. As a result, the water table at the outlet dropped rapidly from the surface to 4 to 8 m depth after September, when the rains stopped (Figure 5B). The water table disappeared around December in wells away from fault lines [20]. The river stopped flowing at this time as well. The groundwater is relatively fast-flowing in this aquifer, with an approximate slope of 8% and a high transmissivity.

Tilahun et al. [20] confirmed high groundwater flux in Robit Bata. They found that approximately two-thirds of the recharge became interflow by September, and by the end of the rain phase, only 10% of the water recharged to the aquifer was still left as storage. Setargie et al. [53] showed using isotopes that in Robit Bata, after the aquifer reached its maximum level, 90% of all discharge at the outlet was pre-event water, indicating a significant subsurface flow component.

The chloride concentrations in groundwater confirm these differences in hydrology as well. During the rain phase, when rain moves preferentially to the groundwater and has a short residence time, the median chloride concentration in the groundwater is in the same order as in Robit Bata (see Section 3.3.2 and Figure 7H). Concentrations remained rising after the rains ended in October until December. Starting from January, the groundwater dropped below 6 m (Figures 3D and 5B) and remained low until June. In this period, groundwater near the faults (tapped by the wells) originated from the unsaturated soils and has a median Cl^- concentration below 2 mg L^{-1}. The unsaturated flow in volcanic soil is subject to the adsorption of negative ions [49] and decreasing chloride concentration to below that of the rainfall. In Dangishta, the same adsorption of ions occurs in unsaturated soil. Still, the water in the wells is a mixture of groundwater from the summer stored behind the volcanic dikes and the unsaturated flow. It resulted in a greater concentration in Dangishta than in Robit Bata in the dry monsoon phase. In the volcanic terrain of Ethiopia, groundwater Cl from halide dissociation is not significant; rainfall is the source of chloride [8,46].

On average, the two watersheds had similar groundwater cation concentrations of Ca^{2+}, Mg^{2+}, and K^+, but concentration for Ca^{2+} and Mg^{2+} varied during monthly measurement. These indicate similarities of rock and soil minerals, but differences in hydrogeology between the watersheds. High concentrations in the dry period and low concentrations in the rainy period indicate evaporation and dilution processes controlling Ca^{2+} and Mg^{2+} in watersheds [14]. The K^+ in both watersheds remained nearly the same throughout the year except for the peak concentration observed in July in Robit Bata (Figure 6D,I).

The concentrations of Fe^{2+} were greater in Dangishta than in Robit Bata, and thus groundwater in Dangishta is more anoxic (Table 2; Figure 6E,J and Figure 10 [51]. In Dangishta, one possible reason for the low concentration of Fe^{2+} in the rain phase from June to October (Figure 7E) is that the nitrate was reduced instead of the iron [40,41] when the nitrate concentrations were high. In Robit Bata, Fe concentration rose in the summer monsoon period for a few months but declined fast. Perhaps the groundwater reached more organic-rich layers, the water fill-up soil matric that blocks oxygen transfer and increases iron reduction [52]. From January to May, when the groundwater was concentrated near the faults, nearly the groundwater was oxic, i.e., less than 0.1 mg L^{-1} of Fe^{2+} (Figure 9B).

4.2. Fate and Transport of Nitrate in Sloping Hillside Aquifers

Understanding the fate and transport of nitrate in watersheds requires knowing the sources of nitrate and factors that control nitrate mobility through the hydrologic pathways [7,39,40]. Applied N fertilizer is a primary source of nitrate concentrations. In Dangishta, the nitrate correlated positively to land use and associated N fertilizer (Table S2). Shallow groundwater wells in homestead and cultivated land-use areas had high nitrate concentrations (Table S3 in Supplementary Materials A). However, the amount of N fertilizer input was not correlated to nitrate concentration in Robit Bata, indicating a fast transport downwards in the interflow of nitrate from the application area.

Figure 12 displays the ratio of chloride and nitrate for each month. It aids in understanding the relative importance of transport and denitrification processes on the nitrogen concentrations in the aquifer. Almost the entire Robit Bata aquifer and the northwestern portion of the Dangishta aquifer have chloride and nitrate ratios of less than three all year long (dark blue color). In addition, in June, when fertilizers were applied and nitrate increased (Figure 8A), the whole watershed had a ratio below three. During the remainder of the rain phase, variable nitrate and chloride are added with the recharge, while denitrification occurs at the same time, while water moves downhill, making interpreting the pattern of the ratios difficult. After September, when the rain phase has ended, there is little recharge, and the ratio increase can be attributed to denitrification in the water that moves down to the river. Figure 12A shows that the ratio is the greatest in the eastern portion of Dangishta. Thus, denitrification occurs in the region. Many wells are located near the river where the groundwater is shallow (Figure 5A). As noted in Section 3.5, the concentration of $N\text{-}NO_3^-$ in the surface water is approximately half of that in the surface water, confirming the denitrification effect.

In Robit Bata, travel times are short; consequently, applied nitrate is transported rapidly out of the watershed. Thus, little groundwater is mixed with the recharge water. Most of the applied nitrogen fertilizer left the watershed within two months, as seen in Figure 10C of nitrate concentration in the stream. Nitrate concentration in the groundwater remained likely from irrigation water that leached the fertilizers into groundwater.

The reduced iron concentrations in groundwater confirm differences in denitrification as well. The negative correlation of Fe^{2+} with nitrate concentration in Tables 3 and 4 for Dangishta can be seen by comparing Figures 8 and 9, where Fe^{2+} and $N\text{-}NO_3^-$ juxtapose each other. When $N\text{-}NO_3^-$ concentration increased, Fe^{2+} concentration decreased. It is especially obvious in the dry phase.

In Robit Bata, reduced iron correlated insignificantly, and multiple linear regression indicates that it only contributed 1% in explaining the variation of nitrate. It implies less loss of nitrate due to denitrification. Rather, the nitrate is affected by its rapid transport out of the watershed. The short residence time of the groundwater in Robit Bata explains why chloride, a function of the rainfall-related transport, explains the nitrate concentration (Tables 3 and 4). It also explains why, in Figure 8B, the NO_3^- peak concentration increase in July after the June fertilizer application is hardly visible after a month. In Figure 12, almost the entire Robit Bata aquifer and the northwestern portion of the Dangishta aquifer have chloride and nitrate ratios of less than three during the whole year.

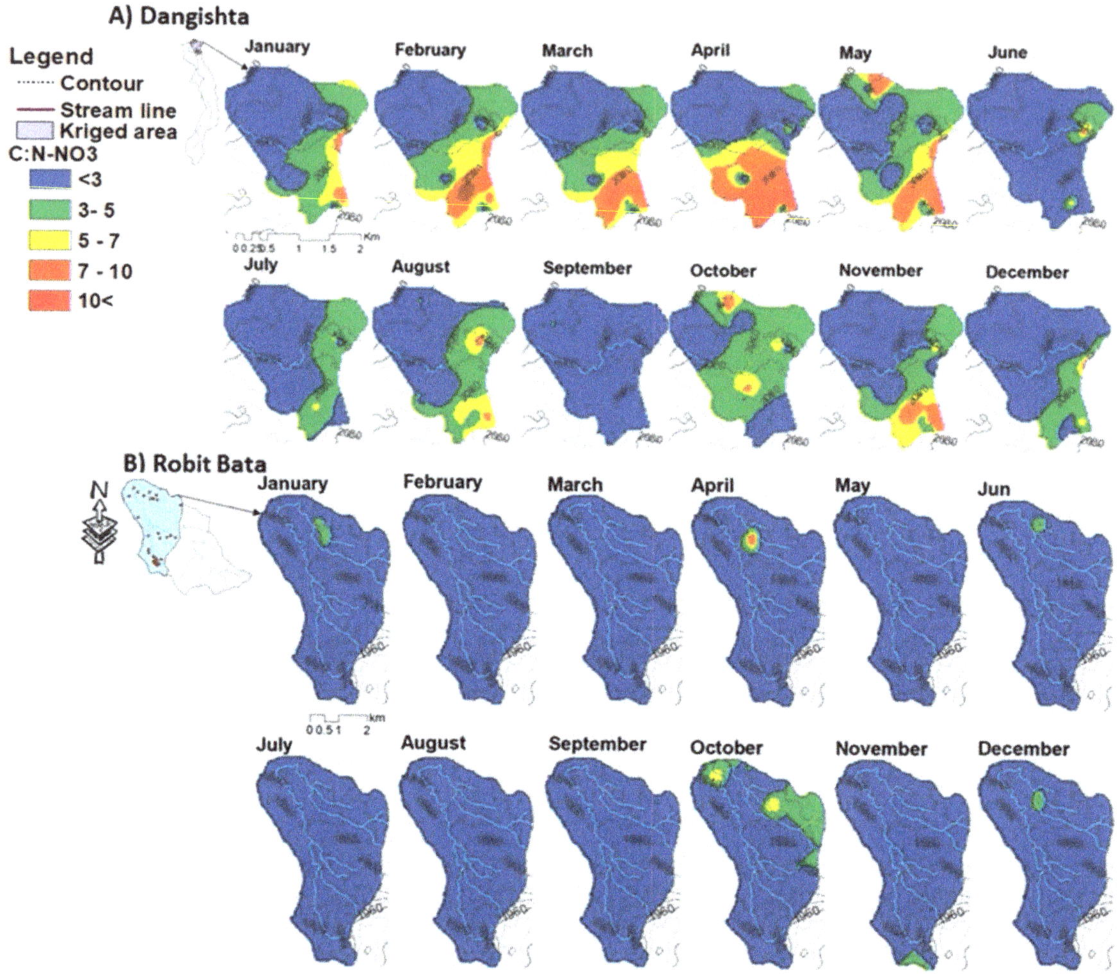

Figure 12. Monthly, two-year-averaged kriged ratio of chloride to nitrate concentration (2017 to 2018) in the aquifers of (**A**) Dangishta and (**B**) Robit Bata watersheds.

5. Conclusions

This study investigated nitrate concentration in groundwater in two contrasting volcanic watersheds (Dangishta and Robit Bata). Dangishta had lava intrusion dikes in the valley bottom that blocked subsurface flow, maintaining a permanent groundwater table. In Robit Bata, transmissivity was high, and volcanic dikes were absent. As a result, the groundwater residence time was short, the water table dropped rapidly, the river stopped flowing three to four months after the rain phase ended, and only near faults was groundwater available.

The average nitrate concentration in Dangishta was 4.6 mg N L^{-1} and not significantly different from Robit Bata, with average concentrations of 4.0 mg N L^{-1}. However, the monthly concentrations between the two watersheds varied greatly due to geohydrology. Interpretation of the results is extremely complex because, in the hillslopes aquifers, the water flows under gravity down the river. Consequently, wells do not sample the same water from one month to the next. Despite these complications, we concluded that in Robit Bata, due to the short residence time, nitrate concentration only increased the month after fertilizer application and was minimally affected by denitrification. In Dangishta, nitrate concentration remained high during the rain phase after fertilizer application because water moved slower than in Robit Bata. Therefore, in Dangishta, denitrification, besides transport out of the watershed, played a significant role in the decrease in nitrate concentrations.

Implications

Understanding the nitrate transport and controlling factors in groundwater is important to evaluate the fate of nitrogen derived from N fertilizer and to design best management practices. Since agricultural intensification has been advocated as one of the options to fulfill the food demand, N fertilizer application has increased by twofold in the last two decades to the current magnitude of 43 kg N ha^{-1} in Ethiopia [25]. Although this rate application is low compared with the application in temperate and sub-tropic regions [54], N-NO_3^- loss is high with the current application in the sub-humid tropical highland area because leaching is enhanced with monsoon rains. Above 10 mg N-NO_3^- L^{-1} concentrations were recorded for several months in wells used as drinking water. Akale et al. [50] also reported a similar order of nitrate concentrations in the groundwater of a sub-humid watershed. Nitrate concentrations in a rural catchment in Tigray [55] with a semi-arid climate and an annual rainfall of less than 1000 mm were much smaller than in this study. Therefore, best practices that improve fertilizer leaching loss such as applying at the right rate, time, and place should be implemented for sub-humid and humid highland regions.

Supplementary Materials: The following supporting information can be downloaded at: https://www.mdpi.com/article/10.3390/hydrology10040082/s1, Supplemental Materials A: Tables S1 and S2 contain a correlation coefficient between nitrate, land use, and cover around the well. Table S3 shows averaged nitrate concentration and N-fertilizer input in in buffer area around wells. Table S4 details the threshold concentrations for identifying redox processes in groundwater. Table S5 contains the monthly and annual precipitation for the two watershed. Figure S1 shows land-use land cover. Figure S2 details the assumption test for the MLR analysis model—Supplemental Materials B: an excel file with detailed information on nitrate and other groundwater constituents.

Author Contributions: F.K.S. contributed to the experimental design, data analysis, interpretation, and writing—original draft preparation; S.A.T. contributed to project administration, funding acquisition, and experimental design, and revised the manuscript for scientific content; P.S. contributed to project administration, acquisition of the financing, and experimental design; T.S.S. contributed to data analysis, interpretation, and writing—review and editing of the manuscript for the scientific content. All authors have read and agreed to the published version of the manuscript.

Funding: The study was made possible through funds provided by the U.S. Agency for International Development under the Feed the Future Evaluation of the Relationship between Sustainably Intensified Production Systems and Farm Family Nutrition (SIPS-IN) project (AID-OAA-L-14-00006) with additional support from the Feed the Future Innovation Lab for Small Scale Irrigation (contract no. AID-OAA-A-13-0005). Finally, the first author received the Queen Elizabeth Scholarship (QES) funding for a three-month stay at the University of Calgary to analyze his experimental data. Also received a grant from Cornell University for a three-month stay for the manuscript. The contents of this manuscript are the authors' responsibility and do not necessarily reflect the views of USAID, the United States government, and Bahir Dar and Cornell Universities.

Data Availability Statement: The raw data supporting this manuscript are provided in Supplementary Materials A and B.

Conflicts of Interest: The authors declare no conflict of interest.

References

1. Suthar, S. Contaminated drinking water and rural health perspectives in Rajasthan, India: An overview of recent case studies. *Environ. Monit. Assess.* **2011**, *173*, 837–849. [CrossRef] [PubMed]
2. Shukla, S.; Saxena, A. Global Status of Nitrate Contamination in Groundwater: Its Occurrence, Health Impacts, and Mitigation Measures. In *Handbook of Environmental Materials Management*; Springer: Cham, Switzerland, 2019; pp. 869–888.

3. Antiguedad, I.; Zabaleta, A.; Martinez-Santos, M.; Ruiz, E.; Uriarte, J.; Morales, T.; Comin, F.A.; Carranza, F.; Español, C.; Navarro, E.; et al. A simple multi-criteria approach to delimitate nitrate attenuation zones in alluvial floodplains. Four cases in south-western Europe. *Ecol. Eng.* **2017**, *103*, 315–331. [CrossRef]
4. Gheysari, M.; Mirlatifi, S.M.; Homaee, M.; Asadi, M.E.; Hoogenboom, G. Nitrate leaching in a silage maize field under different irrigation and nitrogen fertilizer rates. *Agric. Water Manag.* **2009**, *96*, 946–954. [CrossRef]
5. Jangam, C.; Pujari, P. Impact of on-site sanitation systems on groundwater sources in a coastal aquifer in Chennai, India. *Environ. Sci. Pollut. Res.* **2019**, *26*, 2079–2088. [CrossRef]
6. Burow, K.R.; Shelton, J.L.; Dubrovsky, N.M. *Occurrence of Nitrate and Pesticides in Ground water Beneath Three Agricultural Land-Use Settings in the Eastern San Joaquin Valley, California, 1993–1995*; US Department of the Interior, US Geological Survey: Reston, VA, USA, 1998; Volume 97.
7. Burow, K.R.; Nolan, B.T.; Rupert, M.G.; Dubrovsky, N.M. Nitrate in groundwater of the United States, 1991−2003. *Environ. Sci. Technol.* **2010**, *44*, 4988–4997. [CrossRef]
8. Banks, E.W.; Cook, P.G.; Owor, M.; Okullo, J.; Kebede, S.; Nedaw, D.; Mleta, P.; Fallas, H.; Gooddy, D.; John MacAllister, D.; et al. Environmental tracers to evaluate groundwater residence times and water quality risk in shallow unconfined aquifers in sub Saharan Africa. *J. Hydrol.* **2021**, *598*, 125753. [CrossRef]
9. MacAllister, D.J.; MacDonald, A.M.; Kebede, S.; Godfrey, S.; Calow, R. Comparative performance of rural water supplies during drought. *Nat. Commun.* **2020**, *11*, 1099. [CrossRef]
10. Plumpton, H.; Cornforth, R. Groundwater Country Profiles for Seven UpGro Consortium Project Countries in Africa. Available online: https://upgro.org/country-profiles/ethiopia/ (accessed on 23 October 2022).
11. Pavelic, P.; Giordano, M.; Keraita, B.N.; Ramesh, V.; Rao, T. *Groundwater Availability and Use in Sub-Saharan Africa: A review of 15 Countries*; International Water Management Institute: Colombo, Sri Lanka, 2012; p. 224.
12. Fan, A.M.; Steinberg, V.E. Health Implications of Nitrate and Nitrite in Drinking Water: An Update on Methemoglobinemia Occurrence and Reproductive and Developmental Toxicity. *Regul. Toxicol. Pharm.* **1996**, *23*, 35–43. [CrossRef]
13. WHO. *Guidelines for Drinking-Water Quality*, 3rd ed.; World Health Organization: Geneva, Switzerland, 2004; Volume 1, p. 494.
14. Fenta, M.C.; Anteneh, Z.L.; Szanyi, J.; Walker, D. Hydrogeological framework of the volcanic aquifers and groundwater quality in Dangila Town and the surrounding area, Northwest Ethiopia. *Groundw. Sustain. Dev.* **2020**, *11*, 100408. [CrossRef]
15. Berhanu, K.G.; Tegegn, A.M.; Aragaw, T.T.; Angualie, G.S.; Yismaw, A.B. The Secret of the Main Campus Water-Wells, Arba Minch University, Ethiopia. *J. Environ. Public Health* **2021**, *2021*, 4248505. [CrossRef]
16. Asfaw, D.; Mengistu, D. Modeling megech watershed aquifer vulnerability to pollution using modified DRASTIC model for sustainable groundwater management, Northwestern Ethiopia. *Groundw. Sustain. Dev.* **2020**, *11*, 100375. [CrossRef]
17. Tefera, A.K.; Wassie, A.B.; Sinshaw, B.G.; Defersha, D.T.; Takele, T.A.; Atanaw, S.B.; Tesfaye, A.T.; Getu, E.; Fenta, H.M.; Atinkut, H.B.; et al. Groundwater quality evaluation of the alluvial aquifers using GIS and water quality indices in the Upper Blue Nile Basin, Ethiopia. *Groundw. Sustain. Dev.* **2021**, *14*, 100636. [CrossRef]
18. Schepp, C.; Diekkrüger, B.; Becker, M. Hillslope Hydrology in a Deeply Weathered Saprolite and Associated Nitrate Transport to a Valley Bottom Wetland in Central Uganda. *Hydrology* **2022**, *9*, 229. [CrossRef]
19. Yimam, A.Y.; Bekele, A.M.; Nakawuka, P.; Schmitter, P.; Tilahun, S.A. Rainfall-Runoff Process and Groundwater Recharge in the Upper Blue Nile Basin: The Case of Dangishta Watershed. In Proceedings of the International Conference on Advances of Science and Technology, Bahir Dar, Ethiopia, 5–7 October 2018; pp. 536–549.
20. Tilahun, S.A.; Yilak, D.L.; Schmitter, P.; Zimale, F.A.; Langan, S.; Barron, J.; Parlange, J.Y.; Steenhuis, T.S. Establishing irrigation potential of a hillside aquifer in the African highlands. *Hydrol. Process.* **2020**, *34*, 1741–1753. [CrossRef]
21. Adem, A.A.; Addis, G.G.; Aynalem, D.W.; Tilahun, S.A.; Mekuria, W.; Azeze, M.; Steenhuis, T.S.J.W. Hydrogeology of Volcanic Highlands Affects Prioritization of Land Management Practices. *Water* **2020**, *12*, 2702. [CrossRef]
22. Azagegn, T.; Asrat, A.; Ayenew, T.; Kebede, S. Litho-structural control on interbasin groundwater transfer in central Ethiopia. *J. Afr. Earth Sci.* **2015**, *101*, 383–395. [CrossRef]
23. Dagnew, D.C.; Guzman, C.D.; Akale, A.T.; Tebebu, T.Y.; Zegeye, A.D.; Mekuria, W.; Tilahun, S.A.; Steenhuis, T.S. Effects of land use on catchment runoff and soil loss in the sub-humid Ethiopian highlands. *Ecohydrol. Hydrobiol.* **2017**, *17*, 274–282. [CrossRef]
24. Droppelmann, K.J.; Snapp, S.S.; Waddington, S.R. Sustainable intensification options for smallholder maize-based farming systems in sub-Saharan Africa. *Food Secur.* **2017**, *9*, 133–150. [CrossRef]
25. CSA. *Farm Management Practices*; The Federal Democratic Republic of Ethiopia Central Statistical Agency (CSA): Addis Ababa, Ethiopia, 2021.
26. Nigate, F.; Van Camp, M.; Kebede, S.; Walraevens, K. Hydrologic interconnection between the volcanic aquifer and springs, Lake Tana basin on the Upper Blue Nile. *J. Afr. Earth Sci.* **2016**, *121*, 154–167. [CrossRef]
27. BCEOM. *Abay River Basin Integrated Master Plan, Main Report*; Ministry of Water Resource (MOWR): Addis Ababa, Ethiopia, 1999.
28. Abidela, M.H.; Muche, H.; Schmitter, P.; Nakawuka, P.; Tilahun, S.A.; Langan, S.; Barron, J.; Steenhuis, T.S. Deep Tillage Improves Degraded Soils in the (Sub) Humid Ethiopian Highlands. *Land* **2019**, *8*, 159. [CrossRef]
29. Sishu, F.K.; Bekele, A.M.; Schmitter, P.; Tilahun, S.A.; Steenhuis, T.S. Phosphorus Export From Two Contrasting Rural Watersheds in the (Sub) Humid Ethiopian Highlands. *Front. Earth Sci.* **2021**, *9*, 1212. [CrossRef]

30. Yimam, A.Y.; Assefa, T.T.; Sishu, F.K.; Tilahun, S.A.; Reyes, M.R.; Prasad, P.V.V. Estimating Surface and Groundwater Irrigation Potential under Different Conservation Agricultural Practices and Irrigation Systems in the Ethiopian Highlands. *Water* **2021**, *13*, 1645. [CrossRef]
31. Getahun, A. Agricultural systems in Ethiopia. *Agric. Syst.* **1978**, *3*, 281–293. [CrossRef]
32. Amejo, A.G.; Gebere, Y.M.; Kassa, H.; Tana, T. Characterization of smallholder mixed crop–livestock systems in integration with spatial information: In case Ethiopia. *Cogent Food Agric.* **2019**, *5*, 1565299. [CrossRef]
33. Sishu, F.K.; Thegaye, E.K.; Schmitter, P.; Habtu, N.G.; Tilahun, S.A.; Steenhuis, T.S. Endosulfan Pesticide Dissipation and Residue Levels in Khat and Onion in a Sub-Humid Region of Ethiopia. In Proceedings of the International Conference on Advances of Science and Technology, Bahir Dar, Ethiopia, 2–4 August 2019; pp. 16–28.
34. Brown, E.; Skougstad, M.W.; Fishman, M. *Methods for Collection and Analysis of Water Samples for Dissolved Minerals and Gases*; U.S. Government Printing Office: Washington, DC, USA, 1970.
35. Baird, R.B.; Eaton, A.D.; Rice, E.W.; Bridgewater, L. *Standard Methods for the Examination of Water and Wastewater*, 23rd ed.; American Public Health Association, American Water Works Association, Water Environment Federation: Washington, DC, USA, 2017; Volume 23.
36. Palin, A.T. Palintest Water Analysis Technology. Available online: https://www.palintest.com/parameters/ (accessed on 27 November 2021).
37. ADSEW. Land Use Development and Management Planning. Available online: http://www.adswe.com.et/ (accessed on 19 March 2023).
38. McLay, C.D.A.; Dragten, R.; Sparling, G.; Selvarajah, N. Predicting groundwater nitrate concentrations in a region of mixed agricultural land use: A comparison of three approaches. *Environ. Pollut.* **2001**, *115*, 191–204. [CrossRef]
39. Koh, D.-C.; Kim, E.-Y.; Ryu, J.-S.; Ko, K.-S. Factors controlling groundwater chemistry in an agricultural area with complex topographic and land use patterns in mid-western South Korea. *Hydrol. Process.* **2009**, *23*, 2915–2928. [CrossRef]
40. Ki, M.-G.; Koh, D.-C.; Yoon, H.; Kim, H.-S. Temporal variability of nitrate concentration in groundwater affected by intensive agricultural activities in a rural area of Hongseong, South Korea. *Environ. Earth Sci.* **2015**, *74*, 6147–6161. [CrossRef]
41. McMahon, P.; Cowdery, T.; Chapelle, F.; Jurgens, B. *Redox Conditions in Selected Principal Aquifers of the United States*; 2327–6932; US Geological Survey: Reston, VA, USA, 2009.
42. Krige, D.G. A statistical approach to some basic mine valuation problems on the Witwatersrand. *J. South. Afr. Inst. Min. Metall.* **1951**, *52*, 119–139.
43. Matheron, G. Principles of geostatistics. *Econ. Geol.* **1963**, *58*, 1246–1266. [CrossRef]
44. Goovaerts, P. Geostatistics in soil science: State-of-the-art and perspectives. *Geoderma* **1999**, *89*, 1–45. [CrossRef]
45. Alemie, T.C.; Tilahun, S.A.; Ochoa-Tocachi, B.F.; Schmitter, P.; Buytaert, W.; Parlange, J.Y.; Steenhuis, T.S. Predicting shallow groundwater tables for sloping highland aquifers. *Water Resour. Res.* **2019**, *55*, 11088–11100. [CrossRef]
46. Woldemariyam, F.; Ayenew, T. Identification of hydrogeochemical processes in groundwater of Dawa River basin, southern Ethiopia. *Environ. Monit. Assess.* **2016**, *188*, 481. [CrossRef]
47. Enku, T.; Melesse, A.M.; Ayana, E.K.; Tilahun, S.A.; Abate, M.; Steenhuis, T.S. Groundwater Evaporation and Recharge for a Floodplain in a Sub-humid Monsoon Climate in Ethiopia. *Land Degrad. Dev.* **2017**, *28*, 1831–1841. [CrossRef]
48. Walker, D.; Parkin, G.; Gowing, J.; Haile, A.T. Development of a Hydrogeological Conceptual Model for Shallow Aquifers in the Data Scarce Upper Blue Nile Basin. *Hydrology* **2019**, *6*, 43. [CrossRef]
49. Katou, H. Determining competitive nitrate and chloride adsorption in an andisol by the unsaturated transient flow method. *Soil Sci. Plant Nutr.* **2004**, *50*, 119–127. [CrossRef]
50. Akale, A.T.; Moges, M.A.; Dagnew, D.C.; Tilahun, S.A.; Steenhuis, T.S. Assessment of Nitrate in Wells and Springs in the North Central Ethiopian Highlands. *Water* **2018**, *10*, 476. [CrossRef]
51. Lentz, R.D.; Lehrsch, G.A. Temporal changes in δ18O and δ15N of nitrate nitrogen and H_2O in shallow groundwater: Transit time and nitrate-source implications for an irrigated tract in southern Idaho. *Agric. Water Manag.* **2019**, *212*, 126–135. [CrossRef]
52. Castro-Barros, C.M.; Jia, M.; van Loosdrecht, M.C.M.; Volcke, E.I.P.; Winkler, M.K.H. Evaluating the potential for dissimilatory nitrate reduction by anammox bacteria for municipal wastewater treatment. *Bioresour. Technol.* **2017**, *233*, 363–372. [CrossRef]
53. Setargie, T.A.; Tilahun, S.A.; Schmitter, P.; Moges, M.A.; Gurmessa, S.K.; Tsunekawa, A.; Tsubo, M.; Berihun, M.L.; Fenta, A.A.; Haregeweyn, N. Characterizing shallow groundwater in hillslope aquifers using isotopic signatures: A case study in the Upper Blue Nile basin, Ethiopia. *J. Hydrol. Reg. Stud.* **2021**, *37*, 100901. [CrossRef]
54. Abebe, T.G.; Tamtam, M.R.; Abebe, A.A.; Abtemariam, K.A.; Shigut, T.G.; Dejen, Y.A.; Haile, E.G. Growing Use and Impacts of Chemical Fertilizers and Assessing Alternative Organic Fertilizer Sources in Ethiopia. *Appl. Environ. Soil Sci.* **2022**, *2022*, 4738416. [CrossRef]
55. Brhane, G.K. Characterization of hydro chemistry and groundwater quality evaluation for drinking purpose in Adigrat area, Tigray, northern Ethiopia. *Water Sci.* **2018**, *32*, 213–229. [CrossRef]

hydrology

MDPI

Article

Effects of a Gravel Pit Lake on Groundwater Hydrodynamic

Janja Vrzel [1,*], Hans Kupfersberger [1,*], Carlos Andres Rivera Villarreyes [2], Johann Fank [1,†] and Leander Wieser [1]

1 JR-AquaConSol GmbH, Steyrergasse 21, 8010 Graz, Austria; j.fank@aon.at (J.F.); leander.wieser@jr-aquaconsol.at (L.W.)
2 DHI WASY GmbH, Volmerstraße 8, 12489 Berlin, Germany; cvi@dhigroup.com
* Correspondence: janja.vrzel@jr-aquaconsol.at (J.V.); hans.kupfersberger@jr-aquaconsol.at (H.K.); Tel.: +43-664-80-45-45-26 (J.V.); +43-664-80-45-45-13 (H.K.)
† Johann Fank was retired from the affiliation.

Abstract: In Europe, 1132 Mt of sand and gravel were mined in 2019, which causes major changes to the hydrogeological cycle. Such changes may lead to significantly raised or lowered groundwater levels. Therefore, the aggregate sector has to ensure that impacts on existing environmental and water infrastructures are kept to a minimum in the post-mining phase. Such risk assessments are often made by empirical methods, which are based on assumptions that do not meet real aquifer conditions. To investigate this effect, predictions by empirical and numerical methods about hydraulic head changes caused by a pit lake were compared. Wrobel's equation, which is based on Sichardt's equation, was used as the empirical method, while a numerical groundwater flow model has been solved by means of the finite-element method in FEFLOW. The empirical method provides significantly smaller ranges of increased/decreased groundwater levels caused by the gravel pit lake as the numerical method. The underestimation of the empirical results was related to the finding that field measurements during pumping tests show a larger extent of groundwater drawdown than calculations with the Sichardt's equation. Simplifications of the 2D model approach have been evaluated against hydraulic head changes derived from a 3D groundwater model. Our results clearly show that the faster and cheaper empirical method—Wrobel's equation, which is often preferred over the more expensive and time-consuming numerical method, underestimates the drawdown area. This is especially critical when the assignment of mining permits is based on such computations. Therefore, we recommend using numerical models in the pre-mining phase to accurately compute the extent of a gravel/sand excavation's impacts on hydraulic head and hence more effective protection of groundwater and other related environmental systems.

Keywords: gravel pit lake; Wrobel's equation; empirical method; numerical method; drawdown; mounding; FEFLOW

1. Introduction

The aggregate industry, including extraction of gravel and sand from fluvial deposits by open-pit mechanical extraction, as well as from lakes, rivers, and sea by dredging, plays a key role in European's economy by providing materials for the construction sector [1–3]. European countries (EU-28) extracted even 1132 Mt of sand and gravel in 2019 [4].

A gravel pit lake (pit lake), which is the subject of this paper, remains as the final landform of the open-pit gravel mining. On one hand, pit lakes may present environmental risks, including a risk to drinking water supplies; on the other hand, they have the potential to become a beneficial end-use resource [5]. Groundwater flow through pit lakes, which can be called flow-through lakes [6], leads to significant impacts on the environment such as morphological changes, changes of (bio)chemical cycles in lake water and groundwater [7], changes in groundwater quality due to air- and water-born contaminants [8,9], harming existing terrestrial areas [10], creating valuable freshwater ecosystems [10], groundwater warming due to its exposure to air temperature [5], increased freshwater loss due to faster evaporation from the surface water, and slower evapotranspiration in a vegetated land [7], among others.

Changes in the hydrogeological cycle can cause raising or lowering of groundwater levels that lead to deteriorated conditions for farmers, water suppliers, groundwater-dependent ecosystems, or other groundwater-dependent sectors. A pit lake leads to a drawdown and mounding that occur up-gradient and down-gradient from a pit lake, respectively [11]. Thus, before the excavation (pre-mining phase), mining companies have to provide the expected impacts of the pit lake on local and regional hydrogeology. For such purposes, hydrogeologists use methods such as simple empirical equations, complex analytical solutions, or numerical methods [12].

Different empirical and numerical methods, with their specific advantages and uncertainties due to their limitations, deliver non-uniform results. Empirical solutions in hydrogeology often assume a homogeneous and isotropic hydraulic conductivity distribution, horizontal flow (and thus a horizontal aquifer bottom), infinite horizontal extent, and limited boundary conditions [13]. Such approximations of the real world make it difficult or even impossible to obtain reliable results based on empirical approaches to calculate groundwater drawdown and mounding by pit lakes. However, empirical approaches are still often used due to a lack of resources and/or measurement datasets in engineering practice.

As an alternative, numerical methods provide a non-unique solution for more complex systems, which is based on the understanding of a problem being solved [14,15] (p. 78). In studies of physical phenomena, partial-differential equations (e.g., groundwater flow equation) can be solved with different numerical approximations. In this way, a complex problem, such as transient groundwater flow under a heterogeneous or anisotropic hydraulic conductivity distribution, multiple boundary conditions and a diverse elevation of the model's base can be defined [13]. Several software packages are available for solving numerical methods in hydrogeology. In this paper, FEFLOW from MIKE Powered by DHI is applied, in which the Finite-Element Method (FEM) is implemented.

A concern about potential conflicts among mining companies, stakeholders, and society due to inconsistent assessments of a pit lake's effects on groundwater is already mentioned in Arnold et al. (2003) [13]. Arnold et al. (2003) [13] discuss the limitations of an analytical solution and numerical models for predicting the extent of a drawdown caused by mining aggregate below the groundwater table in a hypothetical sand and gravel aquifer. Specifically, they studied a 2D analytical method derived by Marinelli and Niccoli (2000) [16] and carried out numerical modeling using MODFLOW-2000. Kandelous and Šimůnek (2010) [17] also mention the discrepancy between analytical and numerical results when calculating dimensions of the wetting zone, due to many assumptions in analytical methods that may not fully represent the observed reality. Nevertheless, Marinelli and Niccoli (2000) [16] and Yihdego (2018) [18] claim that the groundwater inflow to a pit lake can be calculated by a simple analytical method during the initial stages of the mine development, while numerical modeling may be more appropriate for such studies at its advanced stages.

In addition to the aforementioned analytical equation derived by Marinelli and Niccoli (2000) [16] for assessing the drawdown extent caused by open pit lakes, there are other empirical equations for this purpose as well. For example, Yihdego (2018) [18] modified Sichardt's equation for this purpose, while Wrobel (1980) [19] derived it from Sichardt's and Lübbe's equations. The latter is commonly used in German-speaking countries (so-called D-A-CH region). The derivation of Wrobel's equation considers hydrogeological conditions near Munich in Germany and is based on a limited field dataset. To the author's knowledge, there is no other study that has proved its applicability to any other location or under different hydrogeological conditions, as it has been conducted originally by Wrobel in 1980.

In this paper, we, therefore, use the empirical method derived by Wrobel (1980) [19] and numerical modeling with FEFLOW for computing drawdown and mounding areas caused by gravel mining below the water table and provide a comparison of both results. The extensions of drawdown and mounding are the key criteria for the assignment of mining permits for an open pit. In the numerical simulations, the impacts of saturated thickness and slope of the hydraulic head on the alteration of the hydraulic head caused by a pit lake are examined. Additionally, 2D and 3D models in FEFLOW with identical hydrogeological parameters, geometry, and mesh discretization were compared, in order to understand how the chosen complexity of a model influences results and their deviations from the empirical method.

2. Methods

2.1. Empirical Method by Wrobel (1980)

Wrobel (1980) [19] has derived Equation (1) for estimating the area of drawdown caused by a pit lake in a saturated zone.

$$R = 1500 \times s \times \sqrt{K} \times logB, \quad (1)$$

where R is the distance between the edge of the pit lake and the point where the groundwater level reaches its undisturbed level again [L]; s is the difference between pre-mining and post-mining groundwater levels at the upstream edge of the pit lake [L]; K is hydraulic conductivity [LT^{-1}], B is the width of a pit lake [L], and 1500 is an empirically derived scaling parameter (for definition of parameters see also Figure 1). The length of the pit lake and the slope of hydraulic head is considered in Equation (1) via the parameter s (Figure 1).

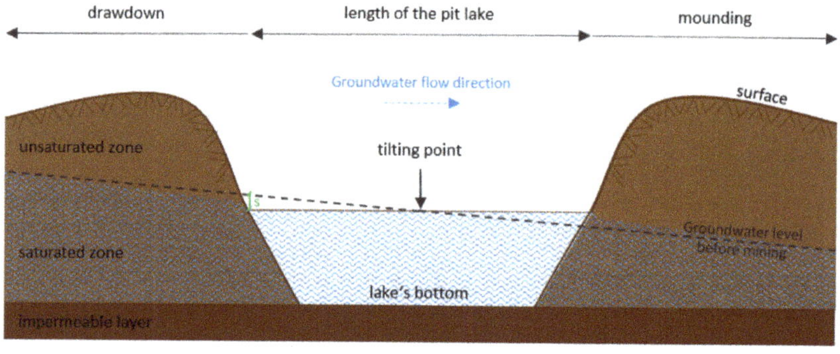

Figure 1. Graphical presentation of parameters in the Wrobel's equation (adopted after [20]).

Wrobel's Equation (1) is based on Lübbe's Equation (2) [11] that in turn was delineated from Sichardt's Equation (3) [21]. Equation (3) is used for estimating the radius of influence (R_{ZOI}) caused by pumping from vertical wells. Equations (1) and (2) differ in the consideration of the pit lake's width. In Wrobel's equation, it can be defined for a specific case via the parameter B, while this is not the case in Lübbe's equation, where it is considered via the scaling parameter of 10,000 [11,19].

$$R = 10,000 \times s \times \sqrt{K}, \quad (2)$$

$$R = 3000 \times s \times \sqrt{K}. \quad (3)$$

Wrobel (1980) [19] and Lübbe (1977) [11] claim that their Equations (1) and (2) are suitable for case studies, where the following assumptions are met: (1) horizontal groundwater flow in an unconfined aquifer that can be solved with a two-dimensional (2D) steady-state approach; (2) the width of the lake is perpendicular to the groundwater flow; (3) homogeneous properties (K) and constant thickness of the saturated zone are given. Lübbe (1977) [11] (p. 180) also writes that Equation (2) may be reliable only for a drawdown within a maximal distance of 100 m from a pit lake.

It is important to remember, that the empirical equations of Wrobel (1980) [19] and Lübbe (1977) [11] have been developed using a limited set of groundwater level measurements at two specific pit lake sites. The influence of the pit lake length, slope of the hydraulic head, and hydraulic conductivity on drawdown estimations have not been investigated into further detail.

2.2. Conceptual Model

The numerical groundwater flow model is based on a purely theoretical conceptual model with homogenous properties of the saturated zone and geometry presented in Figure 2.

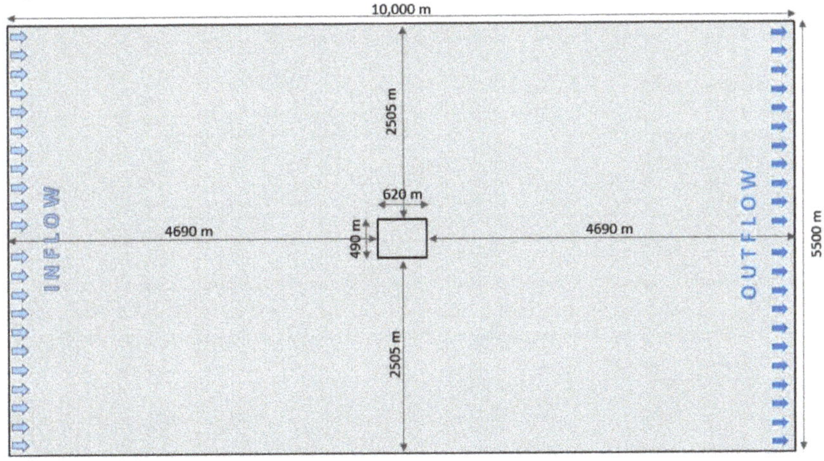

Figure 2. Geometry of the conceptual model.

In the rectangular model domain (10,000 m × 5500 m), a pit lake (620 m × 490 m) is located in its center. The long sides of the model domain (10,000 m) and the rectangular pit lake (620 m) are parallel to the groundwater flow direction. Groundwater inflow and outflow are defined by Dirichlet boundary conditions on the left and right sides of the model, respectively. The model is unconfined with hydraulic conductivity (K) of 5×10^{-3} ms^{-1} and effective porosity of 20%. The fully saturated void volume within the pit lake is mimicked by a hydraulic parametrization consisting of high hydraulic conductivity (100 ms^{-1}) and effective porosity of 100% (Figure 3).

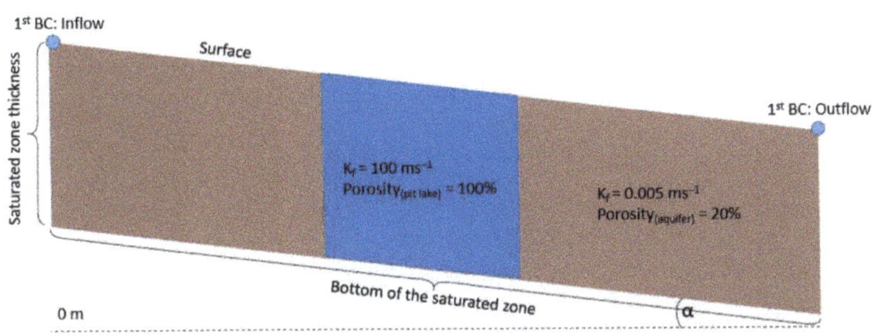

Figure 3. Longitudinal model profile.

Numerical modeling has been performed by FEFLOW (version 7.5) from MIKE Powered by DHI (www.mikepoweredbydhi.com (accessed on 25 May 2023)), which is the one of most widely used software for groundwater flow modeling. FEFLOW solves the groundwater flow equation with the finite-element numerical method [22]. The two-dimensional

(2D) model has 2,200,000 quadrilateral elements with a spatial resolution of 5 m × 5 m. In a 2D model, it is assumed that the pit lake extends in the vertical direction from the surface to the bottom of the saturated zone (Figure 3).

Unlike the empirical Equation (1), the slope of the hydraulic head and, therefore, aquifer thickness, can be defined in a numerical model. Several scenarios were considered in this study. They vary in the thickness of the saturated zone and the hydraulic head slope. In order to exclude effects of parameters, which are not considered in Equation (1), for instance, diverse saturated zone thicknesses and slope of the bottom, the hydraulic head and the bottom are parallel in the pre-mining phase in our study. The overview of scenarios is shown in Table 1. Furthermore, a three-dimensional (3D) FEFLOW model was developed following the details of Scenario 2 in order to evaluate the potential effects of assumptions that must be made in a 2D model (see Section 4.3).

Table 1. Scenarios for numerical models.

Slope of the Hydraulic Head α	Saturated Zone Thickness	
	20 m	**80 m**
2‰	Scenario 1	Scenario 3
4‰	Scenario 2	Scenario 4

Each scenario considered two calculations: (a) pre-mining state without the pond in the model (h_{PreMS}) and (b) post-mining phase with the fully developed pond in the model (h_{PostMS}). The difference (ΔH) of the resulting hydraulic heads is determined for each node of the FEFLOW's mesh as follows:

$$\Delta H = h_{PostMS} - h_{PreMS} \tag{4}$$

The ΔH is called drawdown when $h_{PostMS} < h_{PreMS}$ and mounding when $h_{PostMS} > h_{PreMS}$. For ease of comparison, we selected the 20 cm isoline as the single measure to evaluate the impacts of the scenarios on the hydraulic head.

2.3. Numerical Methods: Governing Groundwater Flow Equation in FEFLOW

The governing differential equation for 3D unconfined steady-state groundwater flow under anisotropic and heterogeneous conditions can be expressed as follows [15] (p. 77):

$$\frac{\partial}{\partial x}K_x\left(h\frac{\partial h}{\partial x}\right) + \frac{\partial}{\partial y}K_y\left(h\frac{\partial h}{\partial y}\right) + \frac{\partial}{\partial z}K_z\left(h\frac{\partial h}{\partial z}\right) + W = 0, \tag{5}$$

where Kx, Ky, and Kz are the principal components of the hydraulic conductivity tensor K [LT^{-1}], W is a volumetric flux per unit volume and includes sources and/or sinks [LT^{-1}], and $\frac{\partial h}{\partial x}$, $\frac{\partial h}{\partial y}$, and $\frac{\partial h}{\partial z}$ are the components of the vector grad h—the gradient of hydraulic head (h) [L], which implies the Darcy's law as follows:

$$q = -K \times grad\ h = -K \times \sin\alpha, \tag{6}$$

where q is a specific flux [LT^{-1}] and α slope of the phreatic surface [−] [15] (p. 77), [23] (p. 361).

The saturated zone in an unconfined porous medium with a phreatic surface forms a layer with a time-depended thickness $B(x_1, x_2, t)$. In this case, the bottom $f^B(x_1, x_2)$ is considered stationary, while the surface of the layer (which is equal to the hydraulic head in an unconfined porous medium, $h(x_1, x_2, t)$ can change in time [22] (p. 131):

$$B(x_1, x_2, t) = h(x_1, x_2, t) - f^B(x_1, x_2). \tag{7}$$

In order to reduce the 3D groundwater flow equation to a 2D essentially horizontal equation in an unconfined porous medium, where the gravity effects are ignored, the horizontal extent of a regional flow field must be much bigger than the thickness of the saturated zone [22] (p. 64). Such a relationship guarantees a small slope of the phreatic surface (α), which is needed for the Dupuit–Forchheimer simplification of the specific Darcy flux (Equation (6)). In addition to the dominantly horizontal flow, the Dupuit–Forchheimer simplification assumes a homogeneous and isotropic porous medium. Nevertheless, the vertical flow is not ignored even though vertical hydraulic head gradients are assumed to be negligible [15] (pp. 120–124).

Various variants of the 2D groundwater flow equation (e.g., Boussinesq equation, Laplace equation) have been later derived under the Dupuit–Forchheimer assumptions [23] (pp. 377–378). Nevertheless, Equation (8) is the basic vertically averaged (2D) equation for a steady-state groundwater flow in an unconfined porous medium under Dupuit–Forchheimer simplifications implemented in FEFLOW [22] (p. 407).

$$\overline{Q} = \nabla \times \overline{q}, \tag{8}$$

$$\overline{q} = -\boldsymbol{BK} \times \nabla \boldsymbol{h}, \tag{9}$$

where $\boldsymbol{h}$ is hydraulic head [L], $\overline{q} = \boldsymbol{B} \times \boldsymbol{q}$ is depth-integrated Darcy velocity [LT^{-1}], $\boldsymbol{B}$ is a thickness [L] of a saturated zone ($\boldsymbol{h} - f^{B}$), $\overline{Q}$ is the depth-integrated source/sink term [$-$], $\boldsymbol{K}$ is tensor of hydraulic conductivity [L^2T^{-1}], and f^{B} is the bottom bounding surface [L].

2.4. Pumping Test

Five pumping tests have been performed in the quaternary aquifer of the Lower Mur catchment in Austria. The selected wells named Gössendorf, Mureck A, Mureck B, Donnersdorf, and Bad Radkersburg are shown in Figure 4. Details of the data analysis and interpretation are discussed elsewhere [24,25].

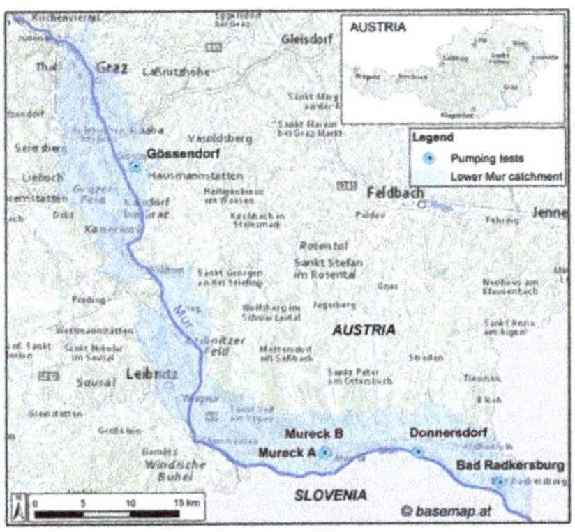

Figure 4. Locations of wells in the Lower Mur catchment in Austria, where pumping tests have been performed.

The purpose of the pumping tests in this study is to compare the radius of influence (R_{ZOI}) observed during pumping tests to the R_{ZOI} computed by Equation (3).

3. Results

3.1. Empirical Method after Wrobel

Equation (1) was solved for two scenarios that differ in slopes of the hydraulic head. The results are presented in Table 2. The calculations considered the following parameters: B = 490 m, K = 5 × 10^{-3} ms^{-1} and length of the pit lake of 620 m. Parameter R in Wrobel's equation returns drawdown extensions in only one direction, which is parallel to the given regional groundwater flow and is measured upstream from the left pit lake's edge.

Table 2. Scenarios for numerical model runs.

Slope of the Hydraulic Head [‰]	Vertical Drawdown *s* [m]	Horizontal Drawdown *R* [m]
2	0.6	171.2
4	1.2	342.4

In order to make empirical and numerical results comparable the vertical drawdown *s* is calculated from the hydraulic head slope (α) and the pit lake's length (620 m). As expected, due to the linear relationship between parameters *R* and *s* in Equation (1), the horizontal drawdown increases by a factor of two if the slope of the hydraulic head is doubled.

3.2. Numerical Simulations

Figure 5 shows the 20 cm isolines of drawdown and mounding (explained in Section 2.2) that result from the four scenarios listed in Table 1. The shape of the drawdown area expands widely transversal to the direction of groundwater flow, when the hydraulic head slope is 4‰, while it expands quite evenly in all directions (~950 m) as the hydraulic head gradient is only 2‰. The mounding area is smaller than the drawdown area for all four scenarios. When the slope of the hydraulic head slope is 4‰, the mounding shape is more circular compared to an ellipsoid shape of the drawdown area upstream of the pit lake. Moreover, the larger saturated zone thickness leads to a greater extent of the mounding area for the same hydraulic head slope, which is the opposite for the drawdown area. For the scenario with increased hydraulic head slope, the drawdown isolines show a clear anisotropic behavior which is more pronounced in the case of the smaller saturated zone thickness. Overall, the hydraulic head slope shows a more pronounced impact on the hydraulic head distortion by the pit lake than the saturated zone thickness. Increasing saturated zone thickness leads to a greater extent of the mounding area in the flow direction, but to a lesser extent transversal to the flow direction in the drawdown area.

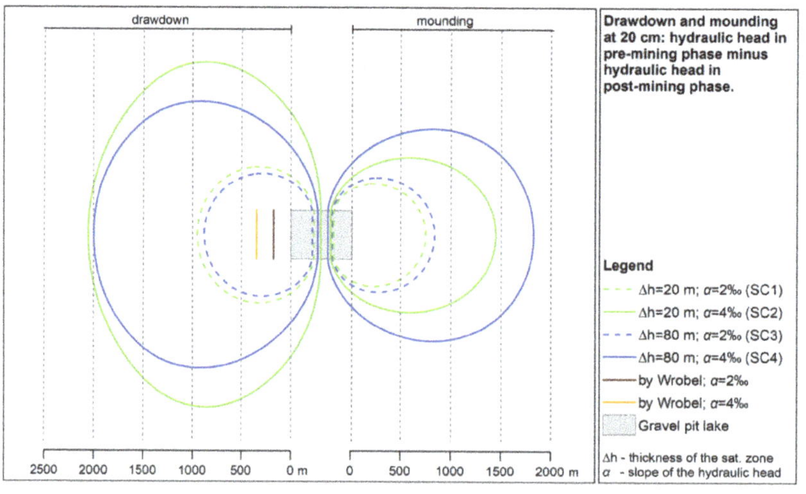

Figure 5. Drawdown calculated by Wrobel's equation, drawdown/mounding at 20 cm for Scenarios 1–4 (Table 1).

Empirical results presented in Section 3.1 were compared with numerically computed extensions of the drawdown. In general, numerical results are approximately six-times larger than empirical results, deviations between them increase with the increasing hydraulic head slope. For instance, drawdown extensions in scenarios with α = 4‰ are ~2050 m (numerical result) and only 342 m (empirical result). In scenarios with α = 2‰ extensions of the drawdown are ~950 m (numerical result) and 171 m (empirical result).

Figure 6 shows the differences of the hydraulic heads in all four scenarios (ΔH calculated by Equation (4)), following a longitudinal profile from the inflow to the outflow boundaries. The maximal drawdown and mounding are clearly on the edges of the pond. The lowest hydraulic head is approximately 0.6 m or 1.4 m below the initial hydraulic head for the hydraulic head gradients 2‰ and 4‰, respectively. The ΔH in the mounding area is slightly smaller, i.e., the hydraulic head is 0.6 m or 1.2 m above the initial hydraulic head for the hydraulic head gradients of 2‰ and 4‰, respectively.

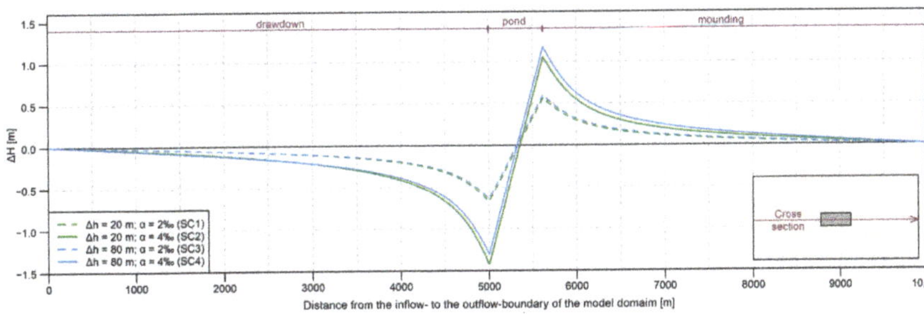

Figure 6. Drawdown and mounding in a vertical cross section for Scenarios 1–4 (Table 1), computed with FEFLOW (Δh thickness of the saturated zone, α-slope of the hydraulic head).

4. Discussion

4.1. Comparison of Groundwater Drawdowns Calculated with Empirical and Numerical Methods

To evaluate the results of the drawdown due to a pit lake by applying the empirical equation after Wrobel (1980) [19], a 2D groundwater flow model was built in FEFLOW. The numerical model approximates as closely as possible the conditions on which Equation (1) was derived such as homogeneous hydraulic conductivity value, steady-state conditions, and longer face of the rectangular gravel pit lake being perpendicular to the groundwater flow direction. In addition, the model has a parallel bottom and surface with a slight inclination. Results as described in Section 3.2, of such a simple numerical model significantly differ from those calculated by Equation (1). In fact, the difference between the hydraulic heads at the pre-mining and the post-mining stages (i.e., ΔH in Equation (4)) is significantly underestimated by Equation (1) in comparison to FEFLOW results (see Section 3.2).

It was observed that the hydraulic head changes (i.e., ΔH in Equation (4)) are more sensible on the gradient of the hydraulic head than on the thicknesses of the saturated zone. Figure 5 shows a smaller horizontal extension of the drawdown area and a larger horizontal extension of the mounding area when the saturated zone is thicker, and a larger ΔH when the hydraulic head is steeper. The ΔH at the edge of the pit lake is up to approximately 0.8 m larger when the hydraulic head slope is 4‰ in comparison to the hydraulic head slope of 2‰ (Figure 6).

Although a laminar flow is assumed in the 2D model, Wang et al. (2013) [26] demonstrated that the Dupuit–Forchheimer assumptions give a good approximation of hydraulic head for an unconfined aquifer, whose gradient is less than 5.7° (100‰). Therefore, it can be concluded that the error associated to the sloping base of the saturated zone in our study is negligible. Furthermore, Equation (1) considers very roughly the saturated thickness and the hydraulic head gradient via the scaling parameter. Such a simplification can obviously lead to inaccurate and potentially misleading results. Nevertheless, the numerical method has a great advantage over the empirical method in respect to the ability of incorporating spatially distributed dataset into the analysis.

In order to see an effect of one of the spatially distributed datasets on the drawdown and mounding areas, additional scenarios with a horizontal bottom slope, which often appears in the real world, were calculated. Due to the horizontal bottom (0 m), these scenarios differ from those in Table 1 in the uneven thickness of the saturated zone, which decreases from the inflow- to the outflow-borders of the model domain. The results show larger mounding area in comparison to the drawdown area. Their difference increases with the larger initial slope of the hydraulic head. Also, drawdown area is smaller and mounding area is larger in comparison to the results where the bottom slope is equal to the hydraulic head slope. This leads to a slightly smaller discrepancy of the drawdown area for scenarios with the horizontal bottom from the drawdown area computed by Equation (1).

The tilting line denotes the line at which the lake water level in the post-mining phase and the groundwater level in the pre-mining phase have the same elevation. According to Lübbe (1977) [11], the tilting line should be located in the center of a pit lake (perpendicular to the water flow direction). In the longitudinal cross section in Figure 7, the hydraulic heads for the pre- and post-mining phases do not exactly follow Lübbe's concept [11]. Instead, the tilting line derived by the FEFLOW's computations is shifted about 45 m from the center of the pit lake in the direction of the model's outflow border in Scenario 2. The reason is that the 2D flow Equation (5) in FEFLOW under the Dupuit–Forchheimer assumptions is non-linear, and it follows a second-order derivative. The Dupuit–Forchheimer assumptions are described in Section 2.3. Shifted tilting lines with a parabolic surface of the hydraulic head are also shown in Jost et al. (2023) [27].

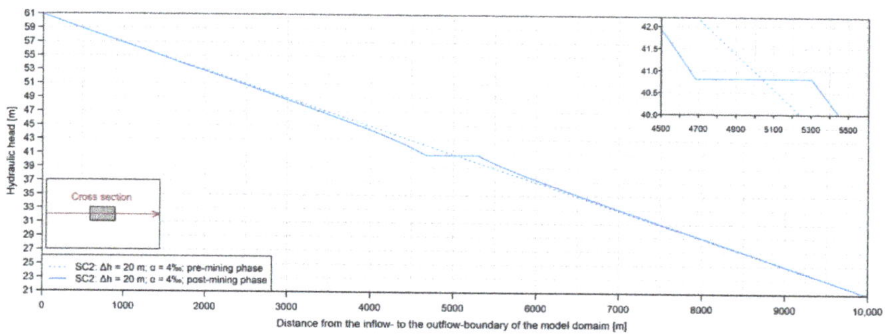

Figure 7. Hydraulic heads in the longitudinal cross section in pre- and post-mining phases for Scenario 2 (Δh thickness of the saturated zone, α-slope of the hydraulic head).

In addition to the thickness of the saturated zone and the hydraulic head slope, the extensions of drawdown and mounding areas also depend on the local hydraulic conductivity distribution, the shape of the pit lake, its depth and hydraulic connection to any other surface water bodies [7,27,28]. Although the evaluation of each potential effect of a gravel pit lake on groundwater is not the subject of this paper, they can be accurately implemented in a numerical model. Jost et al. (2023) [27] studied the influences of geometrical, hydrodynamical and meteorological factors on the groundwater level and water balances. The potential errors by ignoring these parameters have to be considered during the result analysis of an individual project.

A numerical model requires additional information for its parametrization, which may be considered impractical due to the frequent lack of required data, high costs, and time-consuming work. Therefore, it is considered a useful tool only in the advanced stages of gravel mining, when more hydrogeological datasets are available for a study area [12,13,18]. Nevertheless, our study shows that in the context of pit lakes, even extremely simple 2D numerical models can provide more accurate results than the empirical approach. Therefore, we propose to use simple numerical models already in the initial stages of gravel mining to properly evaluate potential environmental impacts associated with the pit lake. As soon as new data is available, such models can be calibrated (or re-calibrated) on demand.

4.2. Radius of Influence Calculated by Sichardt's Equation Compared to Field Data from Pumping Tests

As described in Section 2.1, Wrobel's equation (Equation (1)) is based on Sichardt's equation (Equation (3)). The latter is used to calculate the extent to which the hydraulic head is influenced by pumping from a vertical well (the radius of influence R_{ZOI}). However, larger radii of influence in pumping tests than those calculated by Equation (3) have been reported by Yihdego (2018) [18] and Fuleccia (2015) [12]. Desens and Houben (2022) [29] proposed correction factors for Equation (3), which are intended for case studies with different porosities and saturated zone thicknesses.

For the reasons mentioned above, five pumping tests, as shown in Figure 4, were performed in the quaternary aquifer of the Lower Mur catchment in Austria. The saturated thicknesses measured at the five wells range between 4 m and 13 m, the observed vertical drawdowns vary from 0.8 m to 1.4 m and the resulting hydraulic conductivities fluctuate between 2×10^{-3} ms^{-1} and 9×10^{-3} ms^{-1}. Thus, the hydrogeological conditions observed at these five wells are similar to the ones considered in the derivation of the empirical Equation (1).

The radii of influence (R_{ZOI}) at the five wells caused by the groundwater withdrawal have been delineated from groundwater level readings obtained at the observation wells. It turned out that these R_{ZOI} were between 1.25 and 6 times larger than the values calculated by Equation (3). In order to fit R_{ZOI} calculated by Equation (3) with field data from the five pumping tests, an individual scaling factor for Equation (3) was adjusted accordingly to each pumping well. The new scaling factors fluctuate between 3750 and 18,000, which clearly deviates from its original value of 3000.

The extensions of the horizontal drawdown calculated by Equation (1) are also smaller compared to the numerical results driven by FEFLOW (Figure 5). The drawdown of 20 cm is approximately up to six times more distant from the pit lake in the 2D model compared to Equation (1). Additionally, it has to be considered that in the evaluation of the groundwater flow model results the drawdown isoline at 20 cm was investigated. This means that the actual drawdown area extends even wider. Hence, it can be assumed that the empirical Equations (1) and (3) fundamentally underestimate the extension of the drawdown area related to a pit lake or groundwater pumping.

These results underline our hypothesis that a unique scaling factor in Equation (1) cannot adequately account for the various subsurface characteristics, which are relevant for calculating the drawdown area. For example, the length of the pit lake and the slope of the hydraulic head are only superficially integrated in Equation (1) through the parameter s (see Section 2.1), although they can affect the range of drawdown. Furthermore, the shape of the pit lake, its position with respect to the groundwater flow direction and thickness of the saturated zone, which affect the drawdown as well, cannot be incorporated in Equation (1). All these facts make Equation (1) inadequate for calculating the drawdown area caused by a specific pit lake in complex hydrogeological conditions.

4.3. 2D vs. 3D Numerical Model in FEFLOW

Groundwater flow is in principle a three-dimensional (3D) problem, thus a simplified 2D model under Dupuit–Forchheimer assumptions has certain limitations (see Section 2.3). Nevertheless, the 2D model often provides sufficient results, especially when it comes to large regional models. A 2D model requires significantly less time and costs for its implementation. Also, shorter running times give better chances to have a good calibration and facilitate uncertainty analysis [15] (p. 122). The use of a 2D model can be often questionable, especially when the saturated zone has a non-horizontal surface and bottom and heterogeneous distributions of hydrogeological parameters. For instance, Bresciani et al. (2014) [30] have questioned whether the Dupuit–Forchheimer assumption can be suitable for predicting the groundwater seepage area in hillslopes.

A further analysis was carried out with a 3D numerical model to evaluate the simplifications considered in the 2D model. This analysis is based only on Scenario 2 (see Table 1) and the computed areas of drawdown and mounding between the two models have been compared as shown in Figure 8. In contrast to the 2D model, a 3D model does not neglect the vertical hydraulic gradient, thus it can offer full flexibility regarding the heterogeneity of the study domain and implementation of a pit lake's depth. This is particularly important when the bottom of the pit lake does not coincide with the aquifer bottom.

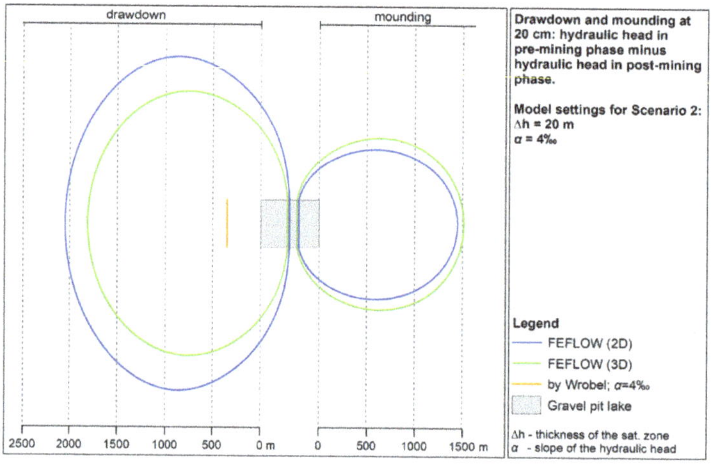

Figure 8. Hydraulic head changes for Scenario 2 in the 2D and four-layer 3D groundwater models.

For the comparison, a four-layer 3D model was built in FEFLOW. Here, the pit lake in the post-mining phase is extended from the surface to the bottom of the saturated zone analogue to the 2D model. The horizontal discretization in all the layers is the same as in the 2D FEFLOW model, i.e., quadrilateral elements of 5 m.

Figure 8 shows small deviations between the sizes of the drawdown and mounding areas at 20 cm in the 2D and 3D models. On one hand, the extension of the drawdown area is up to 180 m in the direction of groundwater flow and 300 m perpendicular to the groundwater flow direction larger in the 2D model than in the 3D model. The shapes of the drawdown isolines follow an elliptical form in both modeling cases. On the other hand, the extension of the mounding areas is similar for both models and exhibits a circular form. In comparison to the 2D model, the 3D modeling approach slightly reduces the underestimation of the drawdown area calculated by the empirical Wrobel's equation (see yellow line in Figure 8). The extensions of drawdown areas in the 2D and 3D models deviate for about 210 m.

The tilting line in the 3D model is also not located in the center of the pit lake, which indicates the non-linearity of the 3D model.

5. Conclusions

The assignment of the mining rights in pit lakes often depends on the extension of hydraulic head changes. These estimations are typically based on simplified empirical equations. However, such an empirical approach cannot account for real hydrogeological complexities and thus can lead to underestimation of the affected areas [12]. An underestimation of the drawdown zones may deteriorate groundwater conditions, for instance through contamination and groundwater warming [31]. For this reason, this paper compared groundwater drawdowns caused by an open gravel pit lake that is derived by the empirical Wrobel's equation and numerical methods. Results show significantly underestimated drawdown area, which is calculated by the empirical Wrobel's equation compared to the results obtained from a two-dimensional (2D) groundwater flow model. Numerical computations demonstrate that the slope has a larger effect on the extent of the drawdown and mounding areas as the thickness of the saturated zone. Additionally, the potential error of using 2D model for solving a three-dimensional (3D) groundwater flow problem was evaluated with a 3D FEFLOW model. The modeling exercises showed that the 2D numerical approach leads to larger drawdown areas, whereas the mounding area was marginally wider in the 3D model.

Based on this study can be said that Wrobel's equation is not appropriate for calculating drawdowns caused by pit lakes under heterogeneous conditions. Wrobel's equation considers the slope only indirectly by calculating the drawdown and it does not include the thickness of the saturated zone. It has been even shown that Sichardt's equation, from which Wrobel's equation has been derived, provides misleading results too. Thus, it is worth investing in numerical models for the purpose of risk assessment, because they do provide more comprehensive and spatially distributed results. In fact, they can be used as a base for selecting more effective measures for groundwater protection. Many study sites meet requirements for using a much less demanding 2D model instead of a complicated 3D model. The 3D model provides more precise results than the 2D model, while the 2D model, despite its simplifications, often proves to be a good compromise, especially when it comes to regional models.

The findings of this paper have to be investigated further, e.g., to evaluate the impact of transient conditions or subsurface heterogeneity. Nevertheless, we recommend the stakeholders avoid using empirical equations for risk assessment analysis of pit lakes.

Author Contributions: Conceptualization, H.K. and J.V.; methodology, H.K. and J.V.; formal analysis, J.V., H.K. and C.A.R.V.; calculations—FEFLOW, J.V.; calculation—Pumping Tests, J.F. and L.W.; writing—original draft preparation, J.V.; writing—review and editing, J.V., H.K., C.A.R.V. and J.F.; visualization, J.V. All authors have read and agreed to the published version of the manuscript.

Funding: This research received no external funding.

Data Availability Statement: No new data were created or analyzed in this study. Data sharing is not applicable to this article.

Acknowledgments: We acknowledge the DHI WASY GmbH support group for fruitful discussions about our early results.

Conflicts of Interest: The authors declare no conflict of interest.

References

1. Eynard, U.; Georgitzikis, K.; Wittmer, D.; El Latunussa, C.; Torres de Matos, C.; Mancini, L.; Unguru, M.; Blagoeva, D.; Bobba, D.; Pavel, C.; et al. *Study on the EU's List of Critical Raw Materials (2020): Non-Critical Raw Materials Factsheets (Final)*; European Commission: Brussels, Belgium, 2020. [CrossRef]
2. Highley, D.; Harrison, D.; Cameron, D.; Lusty, P.; Cowley, J.; Rayner, D. *Mineral Planning Factsheets: Construction Aggregates*; Brown, T., Ed.; BGS: London, UK, 2019.
3. EEA (European Environmental Agency). *Effectiveness of Environmental Taxes and Charges for Managing Sand, Gravel and Rock Extraction in Selected EU Countries*; EEA Report, No. 2/2008; EEA: Copenhagen, Denmark, 2008. [CrossRef]
4. UEPG (European Aggregates Association). Annual Review 2020–2021. Available online: https://www.aggregates-europe.eu/wp-content/uploads/2023/03/Final_-_UEPG-AR2020_2021-V05_spreads72dpiLowQReduced.pdf (accessed on 3 May 2023).
5. Schultze, M.; Vandenberg, J.; McCullough, C.D.; Castendyk, D. The Future Direction of Pit Lakes: Part 1, Research Needs. *Mine Water Environ.* **2022**, *41*, 533–543. [CrossRef]
6. Kidmose, J.; Engesgaard, P.; Nilsson, B.; Laier, T.; Looms, M.C. Spatial Distribution of Seepage at a Flow-Through Lake: Lake Hampen, Western Denmark. *Vadose Zone J.* **2011**, *10*, 110–124. [CrossRef]
7. Mollema, P.; Antonellini, M. Water and (Bio)Chemical Cycling in Gravel Pit Lakes: A Review and Outlook. *Earth Sci. Rev.* **2016**, *159*, 247–270. [CrossRef]
8. Muellegger, C.; Weilhartner, A.; Battin, T.J.; Hofmann, T. Positive and Negative Impacts of Five Austrian Gravel Pit Lakes on Groundwater Quality. *Sci. Total Environ.* **2013**, *443*, 14–23. [CrossRef]
9. Castendyk, D.N.; Eary, L.E.; Balistrieri, L.S. Modeling and Management of Pit Lake Water Chemistry 1: Theory. *Appl. Geochem.* **2015**, *57*, 267–288. [CrossRef]
10. Søndergaard, M.; Lauridsen, T.L.; Johansson, L.S.; Jeppesen, E. Gravel Pit Lakes in Denmark: Chemical and Biological State. *Sci. Total Environ.* **2018**, *612*, 9–17. [CrossRef] [PubMed]
11. Lübbe, E. *Schriftenreiche des Kuratoriums für Wasser und Kulturbauwesen, Heft 29, Baggerseen, Bestandsaufnahme, Hydrologie und planerische Konsequenzen*; Paul Parey: Hamburg/Berlin, Germany, 1977; ISBN 3-490-02997-6.

12. Fileccia, A. Some Simple Procedures for the Calculation of the Influence Radius and Well Head Protection Areas (Theoretical Approach and a Field Case for a Water Table Aquifer in an Alluvial Plain). *AS/IT JGW* **2015**, *4*, 7–23. [CrossRef]
13. Arnold, L.R.; Langer, W.H.; Paschke, S.S. *Analytical and Numerical Simulation of the Steady-State Hydrologic Effects of Mining Aggregate in Hypothetical Sand-and-Gravel and Fractured Crystalline-Rock Aquifers*; Water-Resource Investigations Report 02-4267; USGS: Denver, CO, USA, 2003. Available online: https://pubs.usgs.gov/wri/2002/4267/report.pdf (accessed on 23 May 2023).
14. Brownlee, J. Machine Learning Mastery. Analytical vs Numerical Solutions in Machine Learning. Available online: https://machinelearningmastery.com/analytical-vs-numerical-solutions-in-machine-learning/ (accessed on 26 April 2023).
15. Anderson, M.P.; Woessner, W.W.; Hunt, R.J. *Applied Groundwater Modelling: Simulation of Flow and Advective Transport*, 2nd ed.; Elsevier Inc.: Sand Diego, CA, USA, 2015. [CrossRef]
16. Marinelli, F.; Niccoli, W.L. Simple Analytical Equations for Estimating Ground Water Inflow to a Mine Pit. *Ground Water* **2000**, *38*, 311–314. [CrossRef]
17. Kandelous, M.K.; Šimůnek, J. Comparison of Numerical, Analytical, and Empirical Models to Estimate Wetting Patterns for Surface and Subsurface Drip Irrigation. *Irrig. Sci.* **2010**, *28*, 435–444. [CrossRef]
18. Yihdego, Y. Engineering and Enviro-Management Value of Radius of Influence Estimate from Mining Excavation. *J. Appl. Water Eng. Res.* **2018**, *6*, 329–337. [CrossRef]
19. Wrobel, J.P. Wechselbeziehungen zwischen Baggerseen und Grundwasser in gut durchlässigen Schottern. *GWF Wasser/Abwasser* **1980**, *121*, 165–173.
20. Niemeyer, R. *Hydrologische Untersuchungen an Baggerseen und Alternativen der Folgenutzung*; Lehrstuhl für Landwirtschaftl. Wasserbau u. Kulturtechnik, Inst. für Städtebau, Bodenordnung u. Kulturtechnik d. Univ.: Bonn, Germany, 1978.
21. Kyrieleis, W.; Sichardt, W. *Grundwasserabsenkung bei Fundierungsarbeiten*; Julius Springer: Berlin, Germany, 1930. (In German)
22. Diersch, H.-J.G. *FEFLOW—Finite Element Modeling of Flow, Mass and Heat Transport in Porous and Fractured Media*; Springer: Berlin/Heidelberg, Germany, 2014. [CrossRef]
23. Bear, J. *Dynamics of Fluids in Porous Media*; Dover Publications, Inc.: New York, NY, USA, 1972.
24. Fank, J.; Wieser, L. *Mureck VFB2 Neuerrichtung, Hydrogeologisches Gutachten*; JR-AquaConsol GmbH: Graz, Austria, 2021. (In German)
25. Fank, J.; Mach, J. *SI-MUR-AT, Optimierung Brunnenfeld Mureck des Wasserverbandes "Wasserversorgung Grenzland Südost"*; Grundwasserhydrologisches Gutachten; JR-AquaConsol GmbH: Graz, Austria, 2018. (In German)
26. Wang, Q.; Zhan, H.; Tang, Z. A New Package in MODFLOW to Simulate Unconfined Groundwater Flow in Sloping Aquifers. *Groundwater* **2014**, *52*, 924–935. [CrossRef] [PubMed]
27. Jost, A.; Wang, S.; Verbeke, T.; Colleoni, F.; Flipo, N. Hydrodynamic Relationships Between Gravel Pit Lakes and Aquifers: Brief Review and Insights from Numerical Investigations. *CR GEOSCI* **2023**, *355*, 1–25. [CrossRef]
28. Hofman, T.; Müllegger, C. *Einfluss von Nassbaggerungen auf die Oberflächen- und Grundwasserqualität*; Abschlussbericht; Vienna University: Vienna, Austria, 2011. (In German)
29. Desens, A.; Houben, G.J. Jenseits von Sichardt—Empirische Formeln zur Bestimmung der Absenkreichweite eines Brunnens und ein Verbesserungsvorschlag. *Grundwasser* **2022**, *27*, 131–141. [CrossRef]
30. Bresciani, E.; Davy, P.; de Dreuzy, J.R. Is the Dupuit assumption suitable for predicting the groundwater seepage area in hillslopes? *Water Resour. Res.* **2014**, *50*, 2394–2406. [CrossRef]
31. Ahern, J.A. Ground-Water Capture-Zone Delineation: Method Comparison in Synthetic Case Studies and a Field Example on Front Wainwright, Alaska. Master's Thesis, University of Alaska Fairbanks, Fairbanks, Alaska, 2005. Available online: https://scholarworks.alaska.edu/handle/11122/6942 (accessed on 23 May 2023).

hydrology MDPI

Article

ANN-Based Predictors of ASR Well Recovery Effectiveness in Unconfined Aquifers

Saeid Masoudiashtiani [1,*] and Richard C. Peralta [2]

1 Water Resources Professional III, Hydrology Bureau, Office of the State Engineer (OSE), Santa Fe, NM 87504-5102, USA
2 Civil and Environmental Engineering Department, Utah State University, 4110 Old Main Hill, Logan, UT 84322-4110, USA; richard.peralta@usu.edu
* Correspondence: saeid.masoudiashtiani@usu.edu

Abstract: In this study, we present artificial neural networks (ANNs) to aid in a reconnaissance evaluation of an aquifer storage and recovery (ASR) well. Recovery effectiveness (REN) is the proportion of ASR-injected water recovered during subsequent extraction from the same well. ANN-based predictors allow rapid REN prediction without requiring preparation for and execution of solute transport simulations. REN helps estimate blended water quality resulting from a conservative solute in an aquifer, extraction for environmental protection, and other uses, respectively. Assume that into an isotropic homogenous portion of an unconfined, one-layer aquifer, extra surface water is injected at a steady rate during two wet months (61 days) through a fully penetrating ASR well. And then, water is extracted from the well at the same steady rate during three dry months (91-day period of high demand). The presented dimensionless input parameters were designed to be calibrated within the ANNs to match REN values. The values result from groundwater flow and solute transport simulations for ranges of impact factors of unconfined aquifers. The ANNs calibrated the weighting coefficients associated with the input parameters to predict the achievable REN of an ASR well. The ASR steadily injects extra surface water during periods of water availability and, subsequently, steadily extracts groundwater for use. The total extraction volume equaled the total injection volume at the end of extraction day 61. Subsequently, continuing extraction presumes a pre-existing groundwater right.

Keywords: aquifer storage and recovery; unconfined aquifer; MODFLOW; MT3DMS; artificial neural network

1. Introduction

1.1. Recovery Effectiveness (REN) of Aquifer Storage and Recovery (ASR) Well

The application of water resources management to increase water sustainability involves controlling water movement to (a) increase beneficial use and (b) avoid damage to life and nature. The managed aquifer recharge (MAR) method helps achieve long-term water sustainability by using available stormwater runoff, treated wastewater, or other surface water [1]. As a relatively inexpensive way to improve water supply, one of the MAR tools that injects water into an aquifer is aquifer storage and recovery (ASR) wells. The wells are especially useful where there is limited space, high population density, and increasing demand for groundwater [2,3]. ASR is a tool for integrating surface water and groundwater. It can help provide water storage in regions that have periods of both drought and intense precipitation [4]. ASR involves storing available excess surface water within an aquifer and subsequently recovering the water when advantageous. Note, ASR systems' fundamental concept and structure design are based on one vertical well for injection and extraction. Another MAR tool, ASTR (Aquifer Storage Transfer and Recovery) system, uses one vertical well for injection and another one in a different location for extraction.

If available, extra surface water is injected into an aquifer via an ASR well. After the availability of surface water for injection ceases, the same ASR well immediately begins extracting groundwater for irrigation use. REN represents the proportion of injected water molecules contained within ASR extracted water. REN cannot exceed a value of one.

REN differs in recovery efficiency (RE) used to quantify the performance of ASR wells. The wells inject water into brackish, saline, or coastal aquifers and extract blended water that does not need treatment before its intended use [5,6]. RE can exceed a value of one as it is the ratio of the extracted volume that does not need treatment to the injectate volume [7–13].

REN is important in situations requiring knowledge of how much of the injected water is subsequently extracted or how much remains in the aquifer. Sometimes, modest mixing of (un)treated surface water or stormwater with groundwater in an aquifer is legally and environmentally acceptable. Predicting REN helps estimate the degree of blending that will occur. Forghani and Peralta [14] used a counting molecule method to predict the effectiveness (REN) of ASR well recovery instead of assessing the escape of injectate. The relation between REN and RE is:

$$\mathrm{RE} = \mathrm{REN}(\mathrm{V_E}) + \frac{\mathrm{V(E, amb)}}{\mathrm{V_{inj}}} \tag{1}$$

where $\mathrm{V(E, amb)}$ is the volume of ambient groundwater in the extracted water, $\mathrm{V_E}$ is the total volume of extracted water, and $\mathrm{V_{inj}}$ is the total volume of injectate.

Situations in which injected water and native groundwater have similar quality are particularly challenging for predicting REN in the field. Field tracer tests have been performed to predict injectate recovery [15–17] but involve much uncertainty [18].

Thus, a simplified procedure or tool is needed to predict REN for ASR operations in unconfined aquifers, which are nonlinear systems. In recent years, a type of artificial intelligence (AI), artificial neural network (ANN), has been broadly used to predict groundwater quality and quantity [14,19–21]. Here, the application of this AI type helps quantify injectate recovery (REN) for ASR systems in unconfined aquifers.

1.2. Application of Artificial Neural Network (ANN)

An artificial neural network (ANN) simplifies biological neural networks to simulate human cognitive abilities via a mathematical structure [22]. ANN can develop theoretical or empirical relationships between input(s) and output(s) data to model nonlinear water resource systems [23,24]. This tool is used in hydrology, hydraulics, and water resources management for various tasks, such as forecasting water and groundwater levels and quality, modeling sediment, estimating rainfall–runoff, and managing floods [20]. The drawbacks of ANN [25] involve defining the optimal network structure and training its parameters. Thus, optimization algorithms are necessary to discover global and optimal solutions within complex parameter spaces. The algorithms help mitigate overparameterization and ensure error minimization. But it is important to note that achieving a low training error does not always guarantee good performance [25].

As published studies have shown, the calibration of different ANN models can be used to simulate and forecast water table fluctuations. Such models improve water supply planning in areas where aquifer information is not available [26–28]. In Taiwan, backpropagation of ANN forecasted the variation in groundwater quality resulting from seawater salinization and arsenic pollutant factors [29]. For predicting iodine levels of groundwater in China, an ANN was developed to aid environmental management because traditional analytical techniques are time consuming, difficult, and expensive [30]. Banerjee et al. [31] calibrated ANN to present a simpler and more accurate alternative to SUTRA (Saturated–Unsaturated Transport) numerical model techniques for groundwater salinity prediction. In general, understanding ANN structure and operation and nonlinear hydrologic processes is crucial for modeling rainfall–runoff, streamflow, groundwater management, water quality simulation, and precipitation [32].

For ASR operation within confined freshwater aquifers, ANN development predicts REN, and is an alternative to solute transport numerical simulation [14]. The present study tackles injectate-recovery (REN) prediction of ASR for specific times within unconfined aquifers and uses the ANN tool for the prediction in such nonlinear systems.

1.3. ANN Structure

An ANN structure to predict responses of nonlinear systems includes components such as: (a) input(s); (b) neuron(s) involving different numbers and arrangements; (c) output(s); and (d) a transfer (activation) function to interconnect the input parameters and neuron(s) to output(s). Figure 1 shows three layers (minimum number of layers) illustrating the positions of the ANN components: (a) the input layer; (b) one hidden layer; and (c) the output layer. In the layers, they are n inputs (I1 to In), n neurons (N1 to Nn), and an output (O1), respectively (Figure 1).

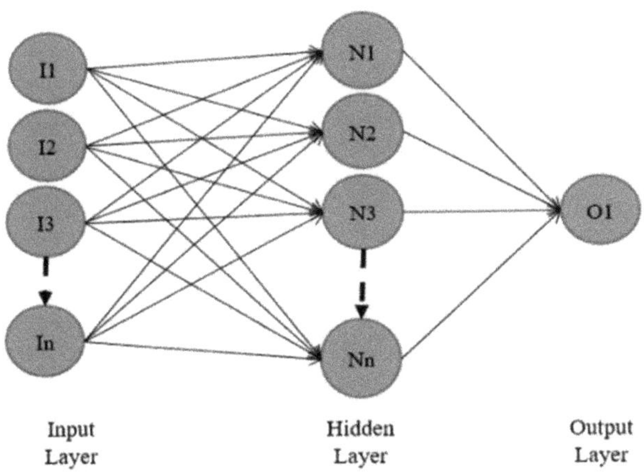

Figure 1. Schematic diagram of ANN structure including the input layer, one hidden layer, and the output layer.

Defining inputs within the input layer is based on their impacts on the output(s) for prediction of the output(s). For instance, Sahoo et al. [24] applied known and important input parameters relatively to predict pesticide occurrence in wells. Forghani and Peralta [14] utilized input parameters that impact REN prediction for confined freshwater aquifers.

In published studies, the optimum number of neurons in an ANN has been defined as less than the number of input parameters [33]. Notably, increasing the number of neurons in a hidden layer and the number of hidden layer(s) can result in (a) a significant nondecreasing in root mean square error (RMSE) [29] and (b) losing the generalization ability of the ANN prediction for a new data set [24,34,35]. Thus, finding the optimum number of neurons and hidden layers in an ANN structure involves multiple "trials and errors" [24].

In the position of the hidden layer, ANN uses a linear, non-linear, or logic transfer (activation) function that provides an output corresponding to a weighted summation of the input parameters [31,36]. Among various transfer functions, sigmoid function ($f(u) = \frac{1}{1+e^{-u}}$) is a typical function having values of responses from 0 to 1 for a range of input parameters [19,36].

In this study, the presented suite of ANN using values of dimensionless analytical parameters (terms) is a tool to predict REN for specific times. The ANN-based predictors assist in estimating the retrieval of injectate, which is not simply the inverse process of the injectate's movement in the down gradient. Employed ranges of impact factor values exceed the ranges of values reported for a representative 756 km^2 (288 mi^2) shallow unconfined aquifer in an intermountain valley in the U.S. Great Basin. Herein, the number of neurons and hidden layers for ANN are one and one, respectively. The numbers were discovered by using a trial-and-error method in reference to statistical analyses. In short, REN evaluation involves estimating solute concentrations within ASR-extracted water and solute remaining within the aquifer. Water users and environmental protection scientists are interested in both.

2. Materials and Methods

2.1. Parameters and Procedures for REN Simulation and Prediction

2.1.1. Overview

Section 2.1 provides an overview of assumptions, parameters, and procedures employed to derive REN values first via simulation and then prediction. It is assumedly desirable to increase groundwater availability during a dry period (June–August) by injecting available excess surface water into an unconfined aquifer during a preceding wet period (April–May). This section defines the factors that impact REN and identifies the ranges of impact factor values that exist in Salt Lake Valley, Utah.

After the selection of approximately half-month durations for flow and solute transport simulations, standard methods are used in estimating (a) the advective plume length after two months (61 days) of injection, (b) the longitudinal dispersivity, (c) the Courant number, (d) the maximum time step size, (e) the total number of simulation time steps, and (f) the time steps per each stress period. These estimates enable preparing a groundwater aquifer model domain sufficiently large that assumed injection and extraction rates will not appreciably affect any employed boundary conditions.

To cover the ranges of REN-affecting factors, 48,000 unique input files were prepared for flow and solute transport simulations that were then processed in parallel. From the simulation results, the procedure required computing a simulated REN response, identifying impact factor combinations, and developing an artificial neural network (ANN) for the REN prediction. The ANN uses values of dimensionless analytical terms containing the impact factors to obtain the ANN-associated weighting coefficient values for the prediction. Finally, a statistical comparison of simulated versus ANN-based predicted values of REN determined the accuracy of the ANN-based predictors.

2.1.2. Selection of REN Impact Factors and Their Value Ranges

Based upon work by Fetter [37], Bedient et al. [38], Pavelic et al. [11], Ward et al. [12,13], Bakker [5], Brown et al. [7], Smith et al. [4], and Forghani and Peralta [14], eight impact factors that affect REN are considered: (1) initial aquifer (background) hydraulic gradient; (2) horizontal hydraulic conductivity; (3) initial (original) aquifer saturated thickness; (4) porosity; (5) specific yield; (6) steady rates of injection and extraction; (7) durations of injection and extraction; and (8) well diameter. A total of 48,000 unique sets of impact factor data were obtained by applying the Table 1 value ranges for groundwater flow and solute transport simulations.

The ranges of background hydraulic gradient, horizontal hydraulic conductivity, and initial aquifer saturated thickness used here include the values employed for Layer 1 of a Salt Lake Valley groundwater model [39]. To set the 0.1–0.6 range of porosity values, this study also relies upon values from Gelhar et al. [40] and Heath [41]. To determine the specific yield range, the porosity range and a 0.375–0.95 range of ((specific yield)/porosity) quotients (Table 1) are used. The 5.451 to 327.06 m^3/d (1 to 60 gpm) steady injection rate range approximates the range of one percent of the average local Red Butte Creek flow from 2014 through 2016 during April and May, which are the months of greatest streamflow [42]. The range of steady extraction rates is equal but opposite in sign to the injection rate range. The extraction duration is three months (Table 1).

Table 1. Impact factor value ranges.

Impact Factor	Range (SI)	Range (English)
Background hydraulic gradient	0.00001–0.015	0.00001–0.015
Horizontal hydraulic conductivity	4–20 (m/d)	13.124–65.61 (ft/d)
Initial aquifer saturated thickness	8–46 (m)	26.25–150.91 (ft)
Porosity	0.1–0.6	0.1–0.6
(Specific yield)/(porosity)	0.375–0.95	0.375–0.95
Specific yield	0.0375–0.57	0.0375–0.57
Daily constant injection rate *	5.451–327.06 (m^3/d)	0.0022–0.132 (cfs) or 1–60 (gpm)
Well diameter	15.24 (cm)	6 (inch)

* Note, injection and extraction durations are two months (April and May) and three months (June, July, and August), respectively. Extraction begins when injection ceases.

2.1.3. Modeled System and Simulators

In essence, to allow computing REN with reasonable accuracy, the MODFLOW [43] finite difference flow model and its Multi-Node Well (MNW2) package and the MT3DMS solute transport model [44] are used. To distinguish the injectate from native groundwater and to provide a solute for transport simulation, a hypothetical 100 ppm concentration is assigned to the injectate. A fully penetrating ASR well in a homogenous, isotropic, freshwater, one-layer, unconfined aquifer is modeled, and extraction rates that are equal but opposite in sign to the injection rates are used. Specified constant-head boundaries on the eastern and western edges of a square model area, no-flow boundaries on the northern and southern edges, and the ASR well at the center are employed (Figures 2 and 3).

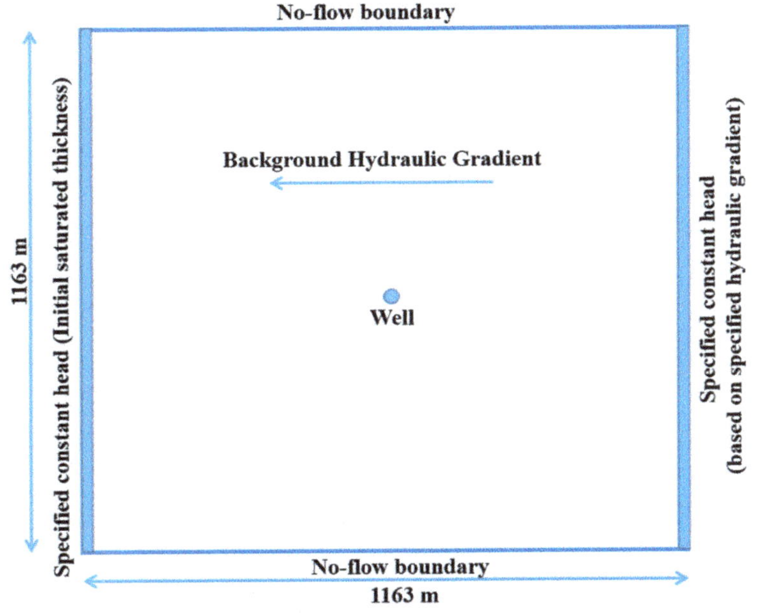

Figure 2. MODFLOW2005-MT3DMS model study area for ASR well, top view (not to scale).

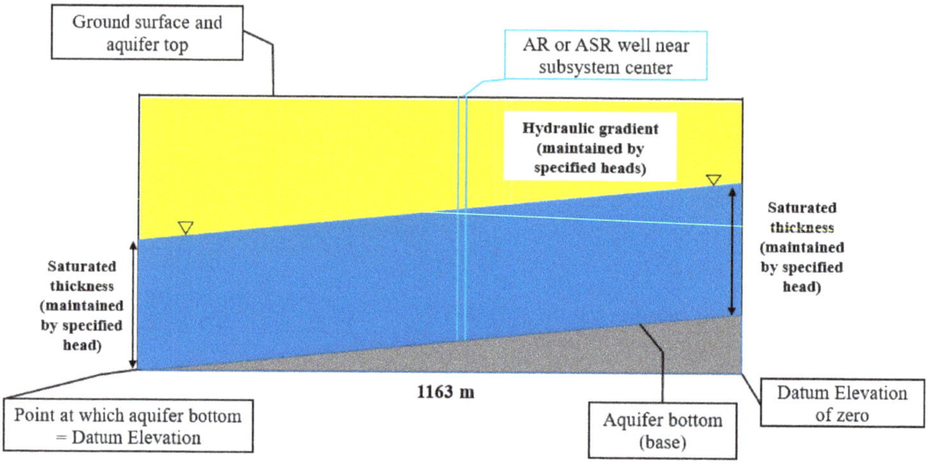

Figure 3. MODFLOW2005-MT3DMS model study area for ASR well, side view (not to scale).

Toward sizing the model and preparing model inputs, the greatest advective plume length that would occur by the end of the injection period for any set of Table 1 impact factor values is estimated. Ignoring gradient changes induced by an injection or groundwater mound, the greatest pore velocity would be 3 m/d (9.84 ft/d). After 61 days, this would yield a 183 m (600.4 ft) advective plume length. The assumed ground surface elevation of the model study area is a value of 100 m (328.08 ft). This results in (a) zero evapotranspiration loss of water from groundwater because the capillary fringe does not reach the overlying root zone. Note that the height of the capillary fringe in various soils does not exceed one meter [41] and (b) zero deep percolation into the same aquifer for this study. Because the unconfined aquifer includes a very thick initial saturated thickness, deep percolation might cause only a small relative change in transmissivity.

Longitudinal dispersion will lengthen the plume further. For an advective plume longer than one meter, the longitudinal dispersivity can be estimated as [40,45–47]:

$$\alpha_L = 0.83\left[\log_{10} L_p\right]^{2.414} \tag{2}$$

where α_L = longitudinal dispersivity (m) and L_p = advective plume length (m). If $L_p \leq 1$ m, $\alpha_L = 0.1L_p$.

MT3DMS uses the user-input Courant number (C) to control the advective process by decreasing oscillations, improving accuracy, and decreasing numerical dispersion. $C = (v \times \Delta t)/\Delta x$, where v = linear pore velocity, Δx = the grid cell dimension at the well location (0.5 m or 1.64 ft), and Δt is the maximum desirable time step size [48].

To determine the simulation time step size in days, suitable for the preferred spatial discretization, the grid Peclet number, P, as equaling 2C [48] is estimated. By assuming that P also equals $(\Delta x/\alpha_L)$, one can compute the maximum time step size desirable for use during injection.

The first estimation of the total number of time steps needed for the injection era was the integer result of dividing the total injection duration by the time step size. Those steps are partitioned equally into each injection stress period and then two more steps per period are added to increase the likelihood of successful simulation. All preliminary simulations for all data combinations used the same number of time steps for flow and solute transport simulations.

Preliminary simulations helped determine the horizontal domain size required to avoid appreciable boundary condition impact from groundwater pumping. Simulations employed the broadest extents of values of injection rate, extraction rate, horizontal hydraulic conductivity, initial (original) aquifer saturated thickness, and initial aquifer (background) hydraulic gradient. The resulting selected 1163 m by 1163 m (3815.62 ft. by 3815.62 ft.) model domain has 129 rows and 129 columns that transition smoothly from cell sizes of 10 m by 10 m (32.80 ft. by 32.80 ft.) to 0.5 m by 0.5 m (1.64 ft. by 1.64 ft.). In the center, the central smallest cell is a 7.62 cm (3 inch) radius vertical well. To have a uniform background hydraulic gradient and saturated thickness from the east to the west, the layer bottom elevation paralleled the desired initial water table.

MODFLOW2005 and MT3DMS simulations use 11 half-month stress periods to easily utilize available rainfall, streamflow, and plant water needs information. Period 1 simulates steady-state background heads. Transient periods 2–5 employ injection and periods 6–11 simulate extraction. Both models also use identical numbers of time steps per period, and that number can differ within a period. The MODFLOW2005 PCG solver uses a 0.01 m head change criterion and a 0.01 m residual convergence criterion. To simulate advection and dispersion, MT3DMS employs the total variation diminishing (TVD) package and the generalized conjugate gradient (GCG) solver. The injection stress period has varied time step durations. To avoid excessive processing time, the extraction period has a single time step (using even 300 time steps for extraction increases REN by less than 0.005).

From the impact factor value ranges in Table 1, five background hydraulic gradients, five hydraulic conductivities, eight injection or extraction rates, six porosity values, ten initial aquifer saturated thicknesses, and four (specific yield/porosity) ratios are used. For each of the 48,000 possible combinations, a different set of impact factor data for groundwater flow and solute transport simulations is prepared. As mentioned above, the 100 ppm of imaginary non-reactive solute (e.g., chloride or dissolved nitrate, etc.) is assigned to the injectate for distinguishing it from native groundwater. As commonly carried out in irrigation water management, chloride can be assumed to be a conservative contaminant. Freshwater, including rainwater, can dissolve halite in soil while recharging or passing through an aquifer. Also, under oxidizing conditions, dissolved nitrate is considered to be nonreactive and conservative [49]. Thus, in MTDMS flow and transport simulations in this study, the only simulated MT3DMS flow and transport simulations are advection and dispersion [40]. MT3DMS simulations provide both the mass of solute injected into the ASR well and the mass of solute recovered from the well. REN equals the extracted solute mass divided by the injected solute mass.

Because the processing time of a single MODFLOW2005-MT3DMS simulation might exceed an hour, parallel processing was used to drastically reduce the computing time [14,50,51]. The Message Passing Interface (MPI) [52–54] of the C++ programming language was utilized to implement parallel processing on multi-cored personal computers and on node clusters of the Center for High Performance Computing (CHPC) at the University of Utah, USA. About 47,000 of the 48,000 attempted MODFLOW2005-MT3DMS simulations were completed successfully. Each successful simulation provided the total mass of solute injected through the ASR well and the mass of solute recovered from the well. REN values were computed after 15, 30, 45, 61, 76, and 91 days of simulated extraction (REN15, REN30, REN45, REN61, REN76, and REN91, respectively).

2.2. Development and Evaluation of Dimensionless Analytical Parameters (Terms) for ANN-Based Predictors

2.2.1. Overview

This section describes the development of dimensionless analytical parameters (terms) as inputs for artificial neural networks (ANNs) that can rapidly predict REN values for use in lieu of numerical simulation models. Data employed for this section include the impact factors in Table 1 applied for the terms, resulting simulation outputs, and computed RENs. Section activities include (1) developing dimensionless analytical expressions that equal (a) a ratio of (extraction/plume) volumes, (b) a ratio of (capture zone/plume) widths, and (c) a ratio of (capture zone/plume) lengths; (2) developing artificial neural networks (ANNs) that apply the dimensionless analytical terms to predict REN values; and (3) defining statistical indices to evaluate ANN-based predictors. The first analytical expression (dimensionless volume) addresses a dimensionless volumetric ratio of the pumped volume that contains the plume volume at a specific time. Some initial trial and error tests for REN prediction were started using the dimensionless volume in regression processes. Regarding the results, it was concluded that the volume presents significant responses for REN prediction but is not sufficient. Also, logically hydraulic down gradient results in moving the plume from the ASR well location. Besides the dimensionless volume, the down-gradient width and length ratios were applied in ANN to modify how much of the plume volume was within the down-gradient capture zone for REN prediction. In the width and length ratios, a natural logarithm was applied to aid their non-linear magnitudes to be valued linearly for the prediction.

2.2.2. Development of Dimensionless Analytical Parameters (Terms)

The impact factors in Table 1 affect REN both in the field and as simulated using MODFLOW2005-MT3DMS. Some initial attempts for REN prediction were started using the impact factors as independent inputs within different artificial neural network (ANN) structures. However, these attempts did not meet the convergence criteria to predict REN. Then, dimensionless analytical terms were designed to be used by ANNs toward predicting dimensionless REN from simulation results. The terms consist of (a) a ratio of the volume of extracted water divided by the injectate plume volume at the time that extraction begins; (b) a natural logarithm of the ratio of the steady-state down-gradient capture zone width divided by the plume width at the end of the injection; and (c) a natural logarithm of the ratio of the steady-state down-gradient capture zone length divided by the advective plume length down-gradient at the end of the injection.

The volume of extracted water at a moment in time equals the extraction duration to that moment times the extraction rate. The volume of the injectate plume equals its 2D area times a representative vertical thickness. The area, A_p, of an elliptical injection plume having a normal or Gaussian concentration distribution, a length of $3\sigma_x$, and a width of $3\sigma_y$, is:

$$A_p = \pi 3\sigma_x 3\sigma_y \tag{3}$$

where σ_x is the standard deviation of concentration in the x direction, $(L) = \sqrt{2D_L t}$; σ_y is similarly defined; D_L and D_T are longitudinal and horizontal transverse dispersions (L^2/T) that respectively equal $\alpha_L \times v_x$ and $\alpha_T \times v_y$; α_L and α_T are horizontal longitudinal dispersivity and transverse dispersivity (L), respectively; v_x and v_y are the linear pore velocities in the longitudinal x and transverse y directions (L/T), respectively; and t is the injection duration (T) [37,38,55].

In the field, horizontal transverse dispersivity is typically an order of magnitude smaller than longitudinal dispersivity [40,44]. Freeze and Cherry [56] indicated that the above equations can be used for preliminary estimation of solute migration arising from small contaminant spills in simple hydrogeologic settings. Assuming the horizontal transverse dispersivity is one-tenth of the longitudinal dispersivity, by substitution, the plume area is:

$$A_p = \pi 18\sqrt{0.1}\alpha_L vt \tag{4}$$

where v = the linear pore velocity.

An estimate of the 61-day plume volume is the sum of the volumes of a cylinder and a cone (the cylinder and cone represent the initial or original aquifer saturated thickness and injection or groundwater mound height, respectively). The cylinder volume equals the product of the plume area and the initial (original) aquifer saturated thickness (b_{ist}). The cone (injection or groundwater mound) volume equals the plume area times $b_{im}/3$. Thus, the plume volume is:

$$V_p = A_p\left(b_{ist} + \frac{1}{3}b_{im}\right) = \left(\pi 18\sqrt{0.1}\alpha_L vt\right) \times \left(b_{ist} + \frac{1}{3}b_{im}\right) \tag{5}$$

where V_p is the plume volume, b_{ist} is the initial (original) aquifer saturated thickness, and b_{im} is the MODFLOW2005 injection or groundwater mound height that is water above the initial water table. For about 48,000 combinations of the impact factor values in Table 1, this study determined injection or groundwater mound heights by MODFLOW2005 simulation of the unconfined aquifer.

Herein, employing a two-step equation-based process with a regression equation in Appendix A to estimate the b_{im} avoids the need to run MODFLOW2005 for new combinations of aquifer conditions.

Equation (6) shows the definition of a dimensionless volume (DLV) as the ratio of the volume of the extracted water (V_{ext}) and the plume volume (V_p):

$$\mathrm{DLV} = \frac{V_{ext}}{V_p} \tag{6}$$

where $V_{ext} = Q \times t$; Q = pumping extraction rate; and t = extraction duration.

Because after any duration of extraction REN is a) a cumulative and nonlinear relative mass and b) between 0 and 1, concerning these REN properties, term 1 includes the sigmoid function (an S-shaped nonlinear function with a response between 0 and 1) of the DLV:

$$\text{Term 1} = \frac{1}{1+e^{-\mathrm{DLV}}} \tag{7}$$

By substituting Equations (5) and (6):

$$\text{Term 1} = \left[\frac{1}{1+e^{-\frac{V_{ext}}{Ap\left(b_{ist}+\frac{1}{3}b_{im}\right)}}}\right] \tag{8}$$

Term 2 is a natural logarithm of the ratio of the (steady-state down-gradient capture zone width)/(plume width at end of injection that is $3\sigma_y$) in Equation (9). The steady-state down-gradient capture zone width equals $\{Q/(4Kb_{ist}i)\}$ [57]. The plume width ($3\sigma_y$) is defined for Equation (3):

$$\text{Term 2} = \mathrm{LN}\left(\frac{\frac{Q}{4Kb_{ist}i}}{3 \times \sqrt{2 \times 0.1 \times \alpha_L \times v \times t_{inj}}}\right) \tag{9}$$

where LN is the natural logarithm; Q is the pumping extraction rate (L^3/T); K is the horizontal hydraulic conductivity (L/T); b_{ist} is the initial (original) aquifer saturated thickness (L); i is the initial aquifer (background) hydraulic gradient (-); α_L is the horizontal longitudinal dispersivity (L); v is the linear pore velocity (L/T); and t_{inj} is the injection duration, 61 days (T).

Term 3 is a natural logarithm of the ratio of the (steady-state down-gradient capture zone length)/(length of advective plume down-gradient at end of injection):

$$\text{Term 3} = \mathrm{LN}\left(\frac{\frac{Q}{2\pi Kb_{ist}i}}{\frac{Kit_{inj}}{n}}\right) \tag{10}$$

where LN is the natural logarithm; the steady-state down-gradient capture zone length is $\{Q/(2\pi \times K \times b_{ist} \times i)\}$ [57]; Q is the pumping extraction rate (L^3/T); π is 3.1416; K is the horizontal hydraulic conductivity (L/T); b_{ist} is the initial (original) aquifer saturated thickness (L) and the advective plume length down-gradient of the well at end of injection (L) is $\{(Kit_{inj})/n\}$; i is initial aquifer (background) hydraulic gradient; t_{inj} is the injection duration, 61 days (T); and n is porosity (-).

Therefore, by using values of such terms as inputs, artificial neural networks (ANNs) can predict injectate recovery (REN) for the specific times.

2.2.3. Development of ANN to Predict REN

As mentioned above, a presented artificial neural network (ANN) represents a mathematical relationship between the REN output and the dimensionless analytical terms computed for Table 1 ranges of impact factor values. In essence, each ANN uses the dimensionless ratios of volume, width, and length (respectively, terms 1–3) to predict REN for a specific extraction duration. Herein, ANN development employed the neuralnet package of the R software [36].

In advance, to find an optimal number of neurons, hidden layers, and outputs, the trial-and-error method was applied to execute preliminary runs of the ANN using (a) the impact factors as independent inputs within various ANN structures with different numbers of hidden layers, neurons, and one and multi-REN outputs and (b) the terms as inputs (note: the first term was defined via regression processes) within one ANN structure for predicting all six REN values. The properties of these runs were various numbers of (i) hidden layers and (ii) neurons of the hidden layers. In both sets of runs, the ANN could not meet the convergence criteria or predict REN accurately.

Thus, the presented ANN structure and properties for six REN values concerning accuracy and simplicity were defined (Figure 4) as follows: (a) three input values of terms 1, 2, and 3; (b) one output REN value for a specific time; (c) a network type of resilient backpropagation with weight backtracking; (d) a network threshold of 0.01 for training; (e) one network repetition for training; (f) one hidden layer; (g) one neuron in the hidden layer; (h) a logistic transfer (activation) function; (i) biases of 1 in input and hidden layers; and (j) a differentiable error function of "sum of squared errors". And the rest of the properties used were from the neuralnet default.

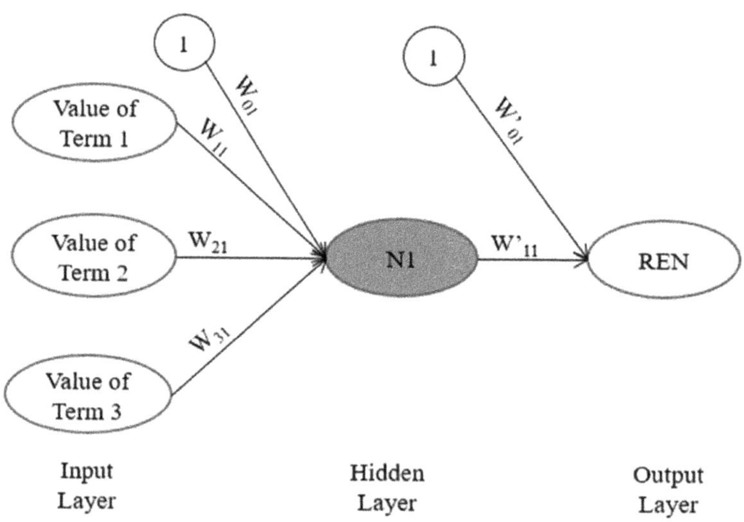

Figure 4. Schematic diagram of ANN structure including three-term inputs, one neuron, REN output for a specific time, and associated weighting coefficients.

The ANN structure in Figure 4 is named "feedforward" [58–61]; this structure transfers information from a previous layer to the next one. Here, all term 1–3 inputs and bias 1 connect to the neuron in the hidden layer that is linked to the output along with the hidden layer bias. The associated weighting coefficients represent the connection importance in the network. The value of the neuron in the hidden layer is calculated by using the sigmoid activation function on the sum of the products of the values of the bias, the terms (inputs), and their associated weights (Equation (11)).

$$N_1 = \frac{1}{1 + EXP[-(1 \times W_{01} + Term\ 1 \times W_{11} + Term\ 2 \times W_{21} + Term\ 3 \times W_{31})]} \quad (11)$$

Then, the REN value is calculated by the sum of the bias value multiplied by its associated weighting value and products of the neuron value and the value of its associated weight (Equation (12)).

$$REN = (1 \times W'_{01}) + (N_1 \times W'_{11}) \quad (12)$$

Herein, six ANN algorithms sharing the same structure predict REN values for the six specific times (15, 30 … , and 91 days after extraction) separately. For ANN calibration (training), the inputs (values of the terms 1 to 3) and the output (REN value for a specific time that is a result of MODFLOW2005-MT3DMS simulations) of the dataset were entered into the ANN model. The ANN model optimized the associated weighting values (W_{01}, W_{11}, W_{21}, W_{31}, W'_{01}, and W'_{11}) by minimizing the sum of squared errors that are differences between the simulated REN and the ANN-based predicted output (REN). For training the ANN, the whole dataset (the 47,000 MODFLOW2005-MT3DMS results or REN values) was split into 10 parts (named the 10-fold cross-validation method) randomly without replacement by applying a written code in R software (Figure 5). The 10-fold cross-validation method involves nine parts of the data set for training and one for testing or validation [58,60].

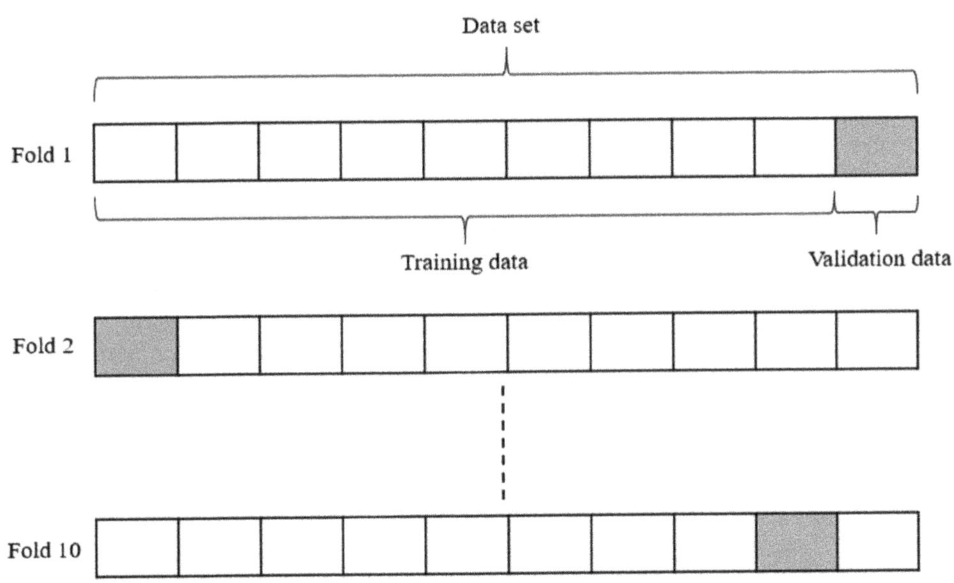

Figure 5. Scheme of 10-fold cross validation method for the data set of ANN.

Based upon the cross-validation method and values of the following statistical indices (Figure A1), the best fold presented the six optimal values of the associated weighting coefficients (W_{01}, W_{11}, W_{21}, W_{31}, W'_{01}, and W'_{11}) for the ANN to predict REN for a specific time. In total, six ANN algorithms sharing the same structure provided 36 values of the associated weighting coefficients for six REN predictions (REN15, REN30 … , and REN91). The required evaluation of the folds to select the optimal values of the weighting resulted in using some statistical indices that are presented in the following section.

2.2.4. Evaluation of Developed ANN-Based Predictors

The accuracy of the six ANN algorithms to predict six REN values was evaluated by applying statistical indices in Table A1 of Appendix A. Such indices helped select the optimal values of the associated weighting coefficients from the 10 folds of each ANN algorithm. The optimal values in each of the six ANN algorithms defined the best ANN for REN prediction. The values of the indices show the accuracy of the predicted REN by comparing it with the simulated REN (i.e., results of successful MODFLOW2005-MT3DMS simulations) for specific times. Although the predicted REN was not evaluated for a real case, the simulated REN was assumed to present the result of the case for the evaluation. After the reconnaissance evaluation, if a more detailed evaluation is needed for a site having heterogeneous data, one would not use the ANN. Instead, one would calibrate and use a heterogeneous flow and transport model.

3. Results and Discussion

Predicting recovery effectiveness (REN) for an ASR system in an unconfined aquifer requires simulating groundwater flow and solute transport, followed by evaluating the results. Herein, the developed artificial neural network (ANN) requiring values of the dimensionless terms predicts REN accurately for ASR implementers. The present study also statistically demonstrated the accuracy of the ANN-based predictors within the ranges of the impact factors and modeled ASR assumptions.

It is assumed that an ASR well is installed within an unconfined, homogenous, isotropic, freshwater, one-layer aquifer. Also, when extra surface water is available, 61 days (two wet months) of steady injection into the ASR well occurs, followed by 91 days (three dry months) of steady extraction from the same well. For six specified distinct times after extraction begins, the ANN-based predictors were developed to predict REN values in lieu of the numerical simulations. Each of the six ANN algorithms involves six associated weighting coefficients that were valued optimally. The neuralnet that used the 10-fold cross-validation method helped obtain optimal values of the coefficients. Here, Table 2 presents values of the weighting coefficients for the six ANN algorithms to predict the REN values after 15, 30, 45, 61, 76, and 91 days (REN15, REN30 . . . , and REN91).

Table 2. Optimal values of associated weighting coefficients for the six ANN algorithms to predict six REN values.

Weighting Coefficient/ Extraction Days	W_{01}	W_{11}	W_{21}	W_{31}	W'_{01}	W'_{11}
15	0.88776	1.36690	0.05449	1.26304	0.01797	0.22883
30	0.42093	0.04244	0.14767	0.99647	0.01670	0.47337
45	0.05082	0.02403	0.05824	0.94451	0.00925	0.69328
61	−0.22883	0.15617	0.02508	0.91678	0.00437	0.85361
76	−0.35194	0.21135	0.03392	0.92816	0.00580	0.92971
91	0.34696	−0.16184	−0.06269	−0.97335	0.96797	−0.95680

For the evaluation, Table 3 statistically shows the accuracy of the ANN-based predictors. For all extraction durations, R^2 values exceed 0.9987. The root mean squared error (RMSE) of the prediction that ranged from 0.0029 to 0.0127 (gm/gm) always increases with extraction duration.

To illustrate a comparison of a REN simulation with the ANN-based prediction, Figures 6, 7, A1 and A2 in this section and Appendix A compare simulated versus ANN-based predicted REN values for validation (testing) data and for training and validation data, respectively. In each figure, the x-axis represents the simulated and assumedly accurate REN values. The y-axis represents the ANN-based predicted REN. Perfect prediction is represented by circles lying on the diagonal line (i.e., the orange dotted line) as shown in Figures 6 and 7 below.

Table 3. Statistical comparisons of ANN-based estimated versus simulated REN for validation (testing) data.

Extraction Days/Parameter	15	30	45	61	76	91	Interpretation Ranges
ME (gm/gm)	−0.0000	0.0000	0.0000	0.0000	0.0002	0.0002	$-\infty < \text{ME} < +\infty$; Perfect: 0
RMSE (gm/gm)	0.0029	0.0040	0.0081	0.0111	0.0116	0.0116	$0 \leq \text{RMSE} < +\infty$; Perfect: 0
PWRMSE (gm/gm)	0.0025	0.0036	0.0085	0.0120	0.0124	0.0127	$0 \leq \text{PWRMSE} < +\infty$; Perfect: 0
r (-)	0.9994	0.9997	0.9995	0.9994	0.9995	0.9995	$-1 \leq r \leq 1$; Perfect: 1 or −1
R^2 (-)	0.9987	0.9995	0.9991	0.9988	0.9990	0.9990	$0 \leq R^2 \leq 1$; Perfect: 1
E_{NS} (-)	0.9987	0.9995	0.9991	0.9988	0.9990	0.9990	$-\infty < E_{NS} \leq 1$; Perfect: 1
∣PBIAS∣ (%)	0.03	0.03	0.01	0.00	0.04	0.04	$\lvert\text{PBIAS}\rvert \leq 25\%$ very good
SI (%)	1.47	1.11	1.64	1.92	1.80	1.74	Perfect: SI < 20%; Operational: SI < 60%

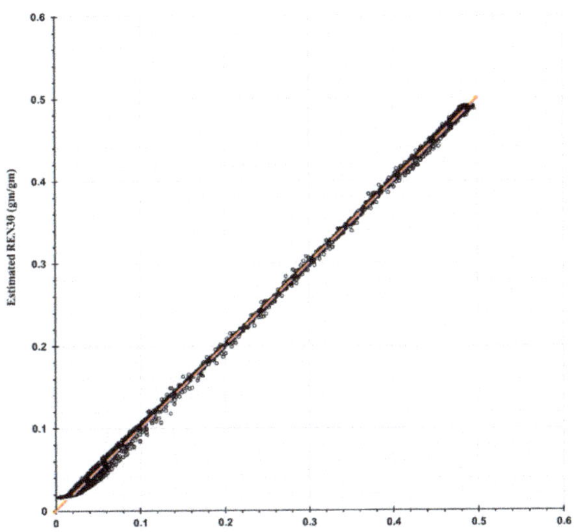

Figure 6. Comparison of ANN-based predicted versus simulated REN30 for validation (testing) data.

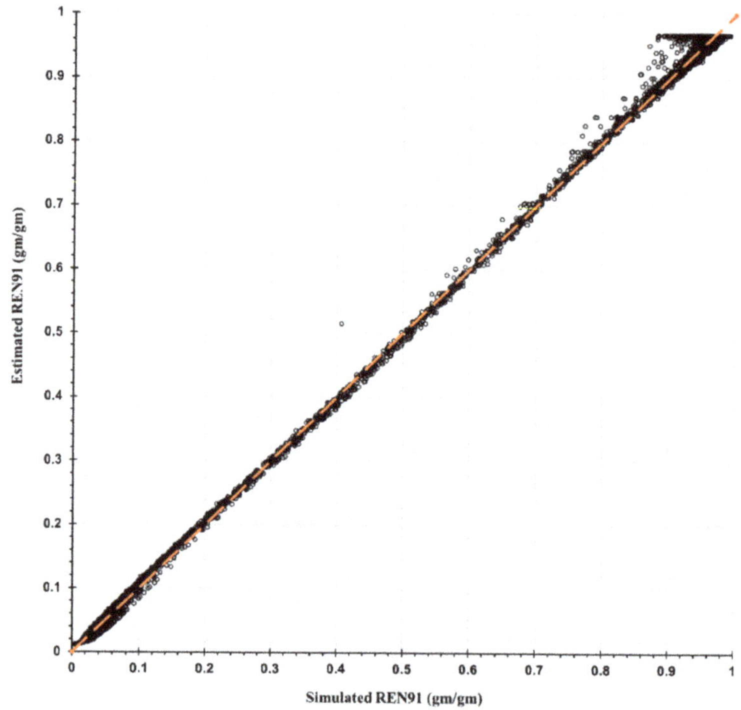

Figure 7. Comparison of ANN-based predicted versus simulated REN91 for validation (testing) data.

4. Conclusions

This study provides a rapid method to predict aquifer storage and recovery (ASR) and well recovery effectiveness (REN) using artificial neural network (ANN)-based predictors. Here, REN represents the proportion of injected water molecules contained in ASR-extracted water from the same well. This is useful for balancing goals of increasing utilizable water supply while protecting aquifer environmental quality. Also, this method improves the prediction of the mass of a conservative contaminant remaining in the aquifer after a season of injection and extraction. The predicted REN values enable predicting the concentration of the extracted water that is a blend of injectate and natural background groundwater. Using the presented ANNs that apply dimensionless terms enables predicting REN for one ASR well. The well fully penetrates a homogenous, isotropic, unconfined, one-layer aquifer. The REN is predicted after every half a month of a three-month extraction period. The ANNs can be applied using a wide range of impact factor values. These ranges are aquifer hydraulic conductivity (4–20 m/d or 13.124–65.61 ft/d), porosity (0.1–0.6), specific yield (0.0375–0.57), initial (original) aquifer saturated thickness (8–46 m or 26.25–150.91 ft), initial aquifer (background) hydraulic gradient (0.00001–0.015), and steady pumping rate of injection and extraction (5.451–327.06 m^3/d or 1–60 gpm). Statistics describing the accuracy of the predicted REN values after 15 to 91 days of extraction range from 0.9987 to 0.9995 for R^2 and 1.11 to 1.92% for scatter index (SI). The predictive REN root mean square error (RMSE) ranges from 0.0029 to 0.0116 (0.29% to 1.16%). Note, a REN could vary from near 0.00 to 1.00 (0.00 to 100%) in value.

The six ANNs share the same structure but differ in the values of the neurally associated weights. As shown statistically, the six ANN algorithms allow a user to predict REN rapidly without having to prepare for and execute numerical flow or solute transport simulations. The generated expressions make it simple and practical to evaluate ASR potential for a wide range of unconfined aquifer conditions. Assuming the presence of a pre-established groundwater right, the extraction can be continued accordingly. The simplicity of the ANNs can help water managers to rapidly perform a reconnaissance level evaluation for a candidate ASR well site. At a reconnaissance level, there might not be more than one observed hydraulic conductivity or transmissivity value in a candidate injection site.

Author Contributions: Conceptualization, S.M. and R.C.P.; methodology, S.M. and R.C.P.; validation, S.M. and R.C.P.; formal analysis, S.M.; investigation, S.M. and R.C.P.; resources, S.M.; data curation, S.M.; writing—original draft preparation, S.M. and R.C.P.; writing—review and editing, R.C.P.; visualization, S.M.; supervision, R.C.P.; project administration, R.C.P.; funding acquisition, R.C.P. All authors have read and agreed to the published version of the manuscript.

Funding: This research was funded by U.S. EPA-STAR project having grant number [83582401], the Civil and Environmental Engineering (CEE) Department, and Utah Water Research Laboratory (UWRL) at Utah State University. It was supported and approved as journal paper number 9588 by the Utah Agricultural Experiment Station at Utah State University.

Data Availability Statement: All data generated and used during the study appear in the published article. Related information is available from the corresponding author by request.

Acknowledgments: The authors gratefully acknowledge support from the Center for High Performance Computing (CHPC) at the University of Utah. The authors appreciate R. Ryan Dupont for his leadership of the EPA-STAR grant project, Marv Halling and David G. Tarboton, heads of the Civil and Environmental Department, and the Utah Water Research Laboratory at Utah State University, respectively.

Conflicts of Interest: The authors declare no conflict of interest.

Appendix A

As mentioned above, a two-step analytical equation-based process with a regression equation estimates MODFLOW2005-simulated injection or groundwater mound height (b_{im}) and helps avoid the need to run MODFLOW2005 for the b_{im} value. The two-step process employed (a) the Cooper and Jacob [62] straight-line method to compute the head change (s') in an equivalent confined aquifer and (b) the Jacob correction [63] to convert the computed confined aquifer head change into an unconfined aquifer head change. The Cooper and Jacob [62] straight-line method computes the head change in an equivalent confined aquifer by $s'=\left\{\frac{Q}{4\pi T}\mathrm{LN}\left(\frac{2.25Tt}{r^2S_y}\right)\right\}$, where Q is a constant positive pumping (here, constant negative injection flow), T is horizontal transmissivity, S_y is the unconfined aquifer specific yield, and t is the elapsed time after steady pumping began. The Jacob correction [63] converts the confined aquifer head change into an unconfined aquifer head change as $s= b\left(1-\sqrt{1-\left(\frac{2s'}{b}\right)}\right)$, where b is the initial (original) aquifer saturated thickness. This is appropriate because values of the late-time function (u_B) are much less than 0.01 for the impact factor values in Table 1 [62–67].

The regression equation with use of the two-step process estimates b_{im} values with a mean error (ME) of 0.000 (m); root mean square error (RMSE) of 0.005 (m); peak weighted root mean square error (PWRMSE) of 0.006 (m); R^2 of 0.9999; and percent of bias (PBIAS) of 0.000 (%). Thus, the regression equation was defined as:

b_{im} (MODFLOW injection or groundwater mound height estimated using a regression equation) = | {(1.026623 × mound height from the two-step analytical process) + 0.002061} |

Table A1. Statistical indices.

Parameter	Formula *	Range	Applied by
Mean Error (ME)	$\frac{1}{n}\sum_{i=1}^{n}(O_i - S_i)$	$-\infty < ME < +\infty$; Perfect: 0	Javan et al. [68]
Root Mean Square Error (RMSE)	$\sqrt{\frac{1}{n}\sum_{i=1}^{n}(O_i - S_i)^2}$	$0 \leq RMSE < +\infty$; Perfect: 0	Mentaschi et al. [69]; Javan et al. [68]; Jimeno-Sáez et al. [70]
Peak Weighted Root Mean Square Error (PWRMSE)	$\sqrt{\frac{1}{n}\sum_{i=1}^{n}(O_i - S_i)^2 \times \left(\frac{O_i+\overline{O}}{2\overline{O}}\right)}$	$0 \leq PWRMSE < +\infty$; Perfect: 0	Javan et al. [68]
Pearson's Correlation Coefficient (r)	$\frac{\sum_{i=1}^{n}(O_i-\overline{O})\times(S_i-\overline{S})}{\sqrt{\sum_{i=1}^{n}(O_i-\overline{O})^2\times\sum_{i=1}^{n}(S_i-\overline{S})^2}}$	$-1 \leq r \leq 1$; Perfect: 1 or -1	Moriasi et al. [71]; Javan et al. [68]
Coefficient of Determination (R^2)	$\frac{\left[\sum_{i=1}^{n}(O_i-\overline{O})\times(S_i-\overline{S})\right]^2}{\sum_{i=1}^{n}(O_i-\overline{O})^2\times\sum_{i=1}^{n}(S_i-\overline{S})^2}$	$0 \leq R^2 \leq 1$; Perfect: 1	Moriasi et al. [71]; Jimeno-Sáez et al. [70]
Nash–Sutcliffe Efficiency (NSE or E_{NS})	$1-\frac{\sum_{i=1}^{n}(O_i-S_i)^2}{\sum_{i=1}^{n}(O_i-\overline{O})^2}$	$-\infty < E_{NS} \leq 1$; Perfect: 1	Nash and Sutcliffe [72]; Moriasi et al. [71]; Javan et al. [68]; Jimeno-Sáez et al. [70]
Percent Bias (PBIAS)	$\frac{\sum_{i=1}^{n}(O_i-S_i)}{\sum_{i=1}^{n}O_i}\times 100$	$\lvert PBIAS \rvert \leq 25\%$ very good	Moriasi et al. [71]; Jimeno-Sáez et al. [70]
Scatter Index (SI)	$\sqrt{\frac{\sum_{i=1}^{n}\left[(S_i-\overline{S})-(O_i-\overline{O})\right]^2}{\sum_{i=1}^{n}O_i^2}}$	Perfect: SI < 20%; Operational: SI < 60%	Janssen and Komen [73]; Moriasi et al. [71]

* Note, n is the number of data pairs; O_i represents the observed MODFLOW2005-MT3DMS results and S_i represents the estimated values; $\overline{O}$ is the mean observed value; and $\overline{S}$ is the mean estimated value.

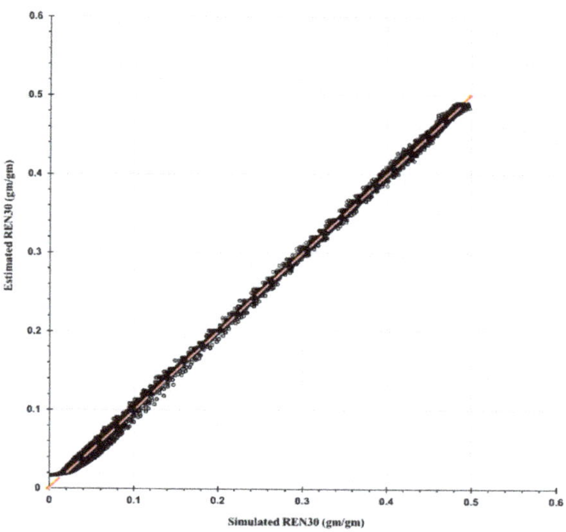

Figure A1. Comparison of ANN-based predicted versus simulated REN30 for training and validation (testing) data.

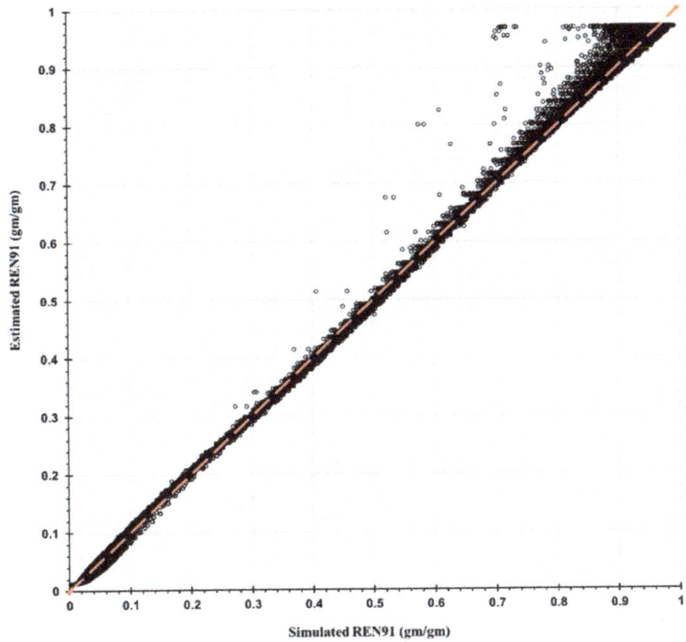

Figure A2. Comparison of ANN-based predicted versus simulated REN91 for training and validation (testing) data.

References

1. Alam, S.; Borthakur, A.; Ravi, S.; Gebremichael, M.; Mohanty, S.K. Managed aquifer recharge implementation criteria to achieve water sustainability. *Sci. Total Environ.* **2021**, *768*, 144992. [CrossRef] [PubMed]
2. Daus, A.; GSI Environmental Inc. Aquifer Storage and Recovery. Improving Water Supply Security in the Caribbean Opportunities and Challenges; Discussion paper No. IDB-DP-00712; Inter-American Development Bank (IDB) Publication, Water and Sanitation Division: 2019. Available online: https://publications.iadb.org/en/aquifer-storage-and-recovery-improving-water-supply-security-caribbean-opportunities-and-challenges (accessed on 26 October 2022).
3. U.S. EPA. Underground Injection Control, Aquifer Recharge, and Aquifer Storage and Recovery. 2021. Available online: https://www.epa.gov/uic/aquifer-recharge-and-aquifer-storage-and-recovery (accessed on 26 October 2022).
4. Smith, W.B.; Miller, G.R.; Sheng, Z. Assessing aquifer storage and recovery feasibility in the Gulf Coastal Plains of Texas. *J. Hydrol.* **2017**, *14*, 92–108. [CrossRef]
5. Bakker, M. Radial Dupuit interface flow to assess the aquifer storage and recovery potential of saltwater aquifers. *Hydrogeol. J.* **2010**, *18*, 107–115. [CrossRef]
6. Pyne, R.D.G. *Groundwater Recharge and Wells: A Guide to Aquifer Storage Recovery*; CRC Press: Boca Raton, FL, USA, 1995.
7. Brown, C.J.; Ward, J.; Mirecki, J. A Revised Brackish Water Aquifer Storage and Recovery (ASR) Site Selection Index for Water Resources Management. *Water Resour. Manag.* **2016**, *30*, 2465–2481. [CrossRef]
8. Kimbler, O.K.; Kazmann, R.G.; Whitehead, W.R. Cyclic storage of freshwater in saline aquifers. In *Louisiana Water Resources Research Institute Bulletin # 10*; Louisiana Water Resources Research Institute: Baton Rouge, LA, USA, 1975; pp. 75–78.
9. Lowry, C.S.; Anderson, M.P. An Assessment of Aquifer Storage Recovery Using Ground Water Flow Models. *Ground Water* **2006**, *44*, 661–667. [CrossRef]
10. Lu, C.; Du, P.; Chen, Y.; Luo, J. Recovery efficiency of aquifer storage and recovery (ASR) with mass transfer limitation. *Water Resour. Res.* **2011**, *47*, W08529. [CrossRef]
11. Pavelic, P.; Dillon, P.J.; Simmons, C.T. Multiscale Characterization of a Heterogeneous Aquifer Using an ASR Operation. *Groundwater* **2005**, *44*, 155–164. [CrossRef]
12. Ward, J.D.; Simmons, C.T.; Dillon, P.J. Variable-density modelling of multiple-cycle aquifer storage and recovery (ASR): Importance of anisotropy and layered heterogeneity in brackish aquifers. *J. Hydrol.* **2008**, *356*, 93–105. [CrossRef]
13. Ward, J.D.; Simmons, C.T.; Dillon, P.J.; Pavelic, P. Integrated assessment of lateral flow, density effects and dispersion in aquifer storage and recovery. *J. Hydrol.* **2009**, *370*, 83–99. [CrossRef]
14. Forghani, A.; Peralta, R.C. Intelligent performance evaluation of aquifer storage and recovery systems in freshwater aquifers. *J. Hydrol.* **2018**, *563*, 599–608. [CrossRef]

15. Bockelmann, A.; Zamfirescu, D.; Ptak, T.; Grathwohl, P.; Teutsch, G. Quantification of mass fluxes and natural attenuation rates at an industrial site with a limited monitoring network: A case study. *J. Contam. Hydrol.* **2003**, *60*, 97–121. [CrossRef] [PubMed]
16. Ptak, T.; Piepenbrink, M.; Martac, E. Tracer tests for the investigation of heterogeneous porous media and stochastic modelling of flow and transport—A review of some recent developments. *J. Hydrol.* **2004**, *294*, 122–163. [CrossRef]
17. Visser, A.; Singleton, M.J.; Esser, B.K. *Xenon Tracer Test at Woodland Aquifer Storage and Recovery Well: A Report to West Yost Associates LLNL-TR-652313*; Lawrence Livermore National Laboratory: Livermore, CA, USA, 2014.
18. Fitts, C.R. Uncertainty in deterministic groundwater transport models due to the assumption of macrodispersive mixing: Evidence from the Cape Cod (Massachusetts, U.S.A.) and Borden (Ontario, Canada) tracer tests. *Contam. Hydrol.* **1996**, *23*, 69–84. [CrossRef]
19. Khaki, M.; Yusoff, I.; Islami, N. Application of the artificial neural network and neuro-fuzzy system for assessment of groundwater quality. *Clean Soil Air Water* **2015**, *43*, 551–560. [CrossRef]
20. Nordin, N.F.C.; Mohd, N.S.; Koting, S.; Ismail, Z.; Sherif, M.; EL-Shafie, A. Groundwater quality forecasting modelling using artificial intelligence: A review. *Groundw. Sustain. Dev.* **2021**, *14*, 100643. [CrossRef]
21. Sakizadeh, M. Artificial intelligence for the prediction of water quality index in groundwater systems. *Model. Earth Syst. Environ.* **2016**, *2*, 8. [CrossRef]
22. Haykin, S. *Neural Networks: A Comprehensive Foundation*, 2nd ed.; Prentice Hall: Upper Saddle River, NJ, USA, 1994.
23. Sahoo, G.B.; Ray, C.; Mehnert, E.; Keefer, D.A. Application of artificial neural networks to assess pesticide contamination in shallow groundwater. *Sci. Total Environ.* **2006**, *367*, 234–251. [CrossRef] [PubMed]
24. Sahoo, G.B.; Ray, C.; Wade, H.F. Pesticide prediction in ground water in North Carolina domestic wells using artificial neural networks. *Ecol. Model.* **2005**, *183*, 29–46. [CrossRef]
25. De Vos, N.J.; Rientjes, T.H.M. Constraints of artificial neural networks for rainfall-runoff modelling: Trade-offs in hydrological state representation and model evaluation. *Hydrol. Earth Syst. Sci. Discuss.* **2005**, *2*, 365–415. [CrossRef]
26. Coulibaly, P.; Anctil, F.; Aravena, R.; Bobee, B. Artificial neural network modeling of water table depth fluctuations. *Water Resour. Res.* **2001**, *37*, 885–896. [CrossRef]
27. Daliakopoulos, I.N.; Coulibaly, P.; Tsanis, I.K. Groundwater level forecasting using artificial neural networks. *J. Hydrol.* **2005**, *309*, 229–240. [CrossRef]
28. Malik, A.; Bhagwat, A. Modelling groundwater level fluctuations in urban areas using artificial neural network. *Groundw. Sustain. Dev.* **2021**, *12*, 100484. [CrossRef]
29. Kuo, Y.; Liu, C.; Lin, K. Evaluation of the ability of an artificial neural network model to assess the variation of groundwater quality in an area of blackfoot disease in Taiwan. *Water Res.* **2004**, *38*, 148–158. [CrossRef] [PubMed]
30. Liu, H.; Li, J.; Cao, H.; Xie, X.; Wang, Y. Prediction modeling of geogenic iodine contaminated groundwater throughout China. *J. Environ. Manag.* **2022**, *303*, 114249. [CrossRef]
31. Banerjee, P.; Singh, V.S.; Chatttopadhyay, K.; Chandra, P.C.; Singh, B. Artificial neural network model as a potential alternative for groundwater salinity forecasting. *J. Hydrol.* **2011**, *398*, 212–220. [CrossRef]
32. Govindaraju, R.S. Artificial Neural Networks in Hydrology. II: Hydrologic Applications, By the ASCE Task Committee on Application of Artificial Neural Networks in Hydrology. *J. Hydrol. Eng.* **2000**, *5*, 124–137. [CrossRef]
33. Maren, A.; Harston, C.; Pap, R. *Handbook of Neural Computing Applications*; Academic Press: San Diego, CA, USA, 1990.
34. Hagan, M.T.; Demuth, H.B.; Beale, M.H. *Neural Network Design*; PWS Publishing Co.: Boston, MA, USA, 1996.
35. Mezard, M.; Nadal, J.P. Learning in feedforward layered networks: The tiling algorithm. *J. Phys. A Math. Gen.* **1989**, *22*, 2191–2203. [CrossRef]
36. Gunther, F.; Fritsch, S. Neuralnet: Training of Neural Networks. *R J.* **2010**, *2*, 30–38. [CrossRef]
37. Fetter, C.W. *Contaminant Hydrogeology*, 2nd ed.; Prentice-Hall Inc.: Upper Saddle River, NJ, USA, 1999; pp. 73–74.
38. Bedient, P.B.; Rifai, H.S.; Newell, C.J. *Ground Water Contamination, Transport and Remediation*, 2nd ed.; Prentice-Hall Inc.: Upper Saddle River, NJ, USA, 1999; pp. 179–180.
39. Lambert, P.M. Numerical Simulation of Ground-Water Flow in Basin-Fill Material in Salt Lake Valley, Utah. United States Geological Survey, Technical Publication No. 110-B 1995. Available online: https://pubs.er.usgs.gov/publication/70179464 (accessed on 24 August 2007).
40. Gelhar, L.W.; Welty, C.; Rehfeldt, K.R. A Critical Review of Data on Field-Scale Dispersion in Aquifers. *Water Resour. Res.* **1992**, *28*, 1955–1974. [CrossRef]
41. Heath, R.C. Basic Ground-Water Hydrology. United States Geological Survey. *Water-Supply Pap.* **1983**, *2200*, 84.
42. iUTAH, Innovate Urban Transitions and Arid Region Hydro-Sustainability. 2012. Available online: https://iutahepscor.org/ (accessed on 13 July 2012).
43. McDonald, M.G.; Harbaugh, A.W. *A Modular Three-Dimensional Finite-Difference Ground-Water Flow Model*; United States Geological Survey Report, Techniques of Water-Resources Investigations 06-A1; US Geological Survey: Liston, VA, USA, 1988. [CrossRef]
44. Zheng, C.; Wang, P.P. MT3DMS: A Modular Three-Dimensional Multispecies Transport Model for Simulation of Advection, Dispersion, and Chemical Reactions of Contaminants in Groundwater Systems: Documentation and User's Guide. In *U.S. Army Engineer Research and Development Center Cataloging-in-Publication Data, Final Report, Contract Report SERDP-99-1*; US Army Engineer Research and Development Center: Vicksburg, MS, USA, 1999.
45. U.S. EPA. Online Tools for Site Assessment Calculation. U.S. Environmental Protection Agency. 2019. Available online: https://www3.epa.gov/ceampubl/learn2model/part-two/onsite/longdisp.html (accessed on 31 August 2021).

46. Wilson, J.L.; Conrad, S.H.; Mason, W.R.; Peplinski, W.; Hagan, E. *Laboratory Investigation of Residual Liquid Organics*; 600/6-90/004; United States Environmental Protection Agency: Ada, OK, USA, 1990.
47. Xu, M.; Eckstein, Y. Use of Weighted Least-Squares Method in Evaluation of the Relationship between Dispersivity and Field Scale. *Ground Water* **1995**, *33*, 905–908. [CrossRef]
48. Daus, A.D.; Frind, E.O.; Sudicky, E.A. Comparative error analysis in finite element formulations of the advection-dispersion equation. *Adv. Water Resour.* **1985**, *8*, 86–95. [CrossRef]
49. Macpherson, G.L.; Townsend, M.A. Perspectives on Sustainable Development of Water Resources in Kansas, Chapter 5: Water Chemistry and Sustainable Yield. Kansas Geological Survey Bulletin 239. 1998. Available online: https://www.kgs.ku.edu/Publications/Bulletins/239/Macpherson/index.html (accessed on 27 May 2013).
50. Gropp, W.; Lusk, E.; Skjellum, A. *Using MPI: Portable Parallel Programming with the Message-Passing Interface*, 3rd ed.; MIT Press: Cambridge, MA, USA, 2014.
51. Ketabchi, H.; Ataie-Ashtiani, B. Assessment of a parallel evolutionary optimization approach for efficient management of coastal aquifers. *Environ. Model. Softw.* **2015**, *74*, 21–38. [CrossRef]
52. Neal, J.C.; Fewtrell, T.J.; Bates, P.D.; Wright, N.G. A comparison of three parallelization methods for 2D flood inundation models. *Environ. Model. Softw.* **2010**, *25*, 398–411. [CrossRef]
53. Sloan, J.D. *High Performance Linux Clusters: With OSCAR, Rocks, OpenMosix, and MPI: A Comprehensive Getting-Started Guide*; O'Reilly Media, Inc.: Newton, MA, USA, 2009.
54. Snir, M.; Otto, S.; Huss-Lederman, S.; Walker, D.; Dongarra, J. *MPI: The Complete Reference*; The MIT Press: Cambridge, MA, USA, 1996.
55. Bear, J. Some Experiments in Dispersion. *J. Geophys. Res.* **1961**, *66*, 2455–2467. [CrossRef]
56. Freeze, R.A.; Cherry, J.A. *Groundwater*; Prentice Hall Inc.: Upper Saddle River, NJ, USA, 1979; p. 604.
57. U.S. EPA. *A Systematic Approach for Evaluation of Capture Zones at Pump and Treat Systems*; EPA/600/R-08/003; U.S. Environmental Protection Agency: Washington, DC, USA, 2008.
58. Dreyfus, G. *Neural Networks: Methodology and Applications*; Springer: Berlin/Heidelberg, Germany, 2005; pp. 135–137.
59. Hornik, K.; Stinchcombe, M.; White, H. Multilayer feedforward networks are universal approximators. *Neural Netw.* **1989**, *2*, 359–366. [CrossRef]
60. Priddy, K.L.; Keller, P.E. *Artificial Neural Networks: An Introduction*; SPIE PRESS, The International Society for Optical Engineering: Bellingham, WA, USA, 2005.
61. Schmidhuber, J. Deep learning in neural networks: An overview. *Neural Netw.* **2015**, *61*, 85–117. [CrossRef]
62. Cooper, H.H., Jr.; Jacob, C.E. A Generalized Graphical Method for Evaluating Formation Constants and Summarizing Well-Field History. *Trans. Am. Geophys. Union* **1946**, *27*, 526–534. [CrossRef]
63. Jacob, C.E. Notes on determining permeability by pumping tests under water table conditions. United States Geological. Survey Open-file Report, Effective radius of drawdown test to determine artesian well. *Am. Soc. Civil Eng. Proc.* **1944**, *72*, 629–646.
64. Fitts, C.R. *Groundwater Science; Academic Press*; Elsevier Science: Cambridge, MA, USA, 2002; pp. 221–235.
65. Huisman, L. *Groundwater Recovery*; Winchester Press and the Macmillan Press: New York, NY, USA, 1972; pp. 206–211.
66. Neuman, S.P. Analysis of Pumping Test Data from Anisotropic Unconfined Aquifers Considering Delayed Gravity Response. *Water Resour. Res.* **1975**, *11*, 329–342. [CrossRef]
67. Schwartz, F.W.; Zhang, H. *Fundamentals of Groundwater*; John Wiley and Sons Inc.: Hoboken, NJ, USA, 2003; pp. 258–270.
68. Javan, K.; Fallah Haghgoo Lialestani, M.R.; Nejadhossein, M. A comparison of ANN and HSPF models for runoff simulation in Gharehsoo River watershed, Iran. *Model. Earth Syst. Environ.* **2015**, *1*, 41. [CrossRef]
69. Mentaschi, L.; Besio, G.; Cassola, F.; Mazzino, A. Problems in RMSE-based wave model validations. *Ocean Model.* **2013**, *72*, 53–58. [CrossRef]
70. Jimeno-Sáez, P.; Senent-Aparicio, J.; Pérez-Sánchez, J.; Pulido-Velazquez, D. A Comparison of SWAT and ANN Models for Daily Runoff Simulation in Different Climatic Zones of Peninsular Spain. *Water* **2018**, *10*, 192. [CrossRef]
71. Moriasi, D.N.; Arnold, J.G.; Van Liew, M.W.; Bingner, R.L.; Harmel, R.D.; Veith, T.L. Model Evaluation Guidelines for Systematic Quantification of Accuracy in Watershed Simulations. *Soil Water Div. Am. Soc. Agric. Biol. Eng.* **2007**, *50*, 885–900. [CrossRef]
72. Nash, J.E.; Sutcliffe, J.V. River flow forecasting through conceptual models: Part 1, A discussion of principles. *J. Hydrol.* **1970**, *10*, 282–290. [CrossRef]
73. Janssen, P.A.E.M.; Komen, G.J. An Operational Coupled Hybrid Wave Prediction Model. *Geophys. Res.* **1984**, *89*, 3635–3654. [CrossRef]

hydrology

MDPI

Article

Groundwater Bodies Subdivision in Corsica: A Critical Approach Based on Multivariate Water Quality Criteria Using Large Database

Hajar Lazar [1], Meryem Ayach [1], Abdoul-Azize Barry [2], Ismail Mohsine [1], Abdessamad Touiouine [3], Frédéric Huneau [4], Christophe Mori [4], Émilie Garel [4], Ilias Kacimi [1], Vincent Valles [5,6] and Laurent Barbiero [7,*]

1 Geosciences, Water and Environment Laboratory, Faculty of Sciences Rabat, Mohammed V University, Avenue Ibn Batouta, Rabat 10100, Morocco; hajar_lazar2@um5.ac.ma (H.L.); meryem_ayach@um5.ac.ma (M.A.); ismail.mohsine@um5r.ac.ma (I.M.); i.kacimi@um5r.ac.ma (I.K.)
2 Geoscience and Environment Laboratory (LaGE), Department of Earth Sciences, Joseph KI-ZERBO University, Ouagadougou 7021, Burkina Faso; abdoul-azize.barry@alumni.univ-avignon.fr
3 Laboratory of Geosciences, Faculty of Sciences, Ibn Tofaïl University, BP 133, Kenitra 14000, Morocco; tiouiouine.a-samad@taalim.ma
4 Département d'Hydrogéologie, Université de Corse Pascal Paoli, Campus Grimaldi, BP52, 20250 Corte, France; huneau@univ-corse.fr (F.H.); mori@univ-corse.fr (C.M.); garel_e@univ-corse.fr (É.G.)
5 Mixed Research Unit EMMAH (Environnement Méditerranéen et Modélisation des Agro-Hydrosystèmes), Hydrogeology Laboratory, Avignon University, 84916 Avignon, France
6 Faculté des Sciences et Techniques (FSTBM), BP 523, Beni Mellal 23000, Morocco
7 Institut de Recherche Pour le Développement, Géoscience Environnement Toulouse, CNRS, University of Toulouse, Observatoire Midi-Pyrénées, UMR 5563, 14 Avenue Edouard Belin, 31400 Toulouse, France
* Correspondence: laurent.barbiero@get.omp.eu

Abstract: The cross-referencing of two databases, namely the compartmentalization into groundwater bodies (GWB) and the quality monitoring (2830 observations including 15 physico-chemical and bacteriological parameters, on 662 collection points and over a period of 27 years) is applied to better understand the diversity of the waters of the island of Corsica (France) and to facilitate the surveillance and quality monitoring of the groundwater resource. Data conditioning (log-transformation), dimensional reduction (PCA), classification (AHC) and then quantification of the information lost during grouping (ANOVA), highlight the need to sub-divide the groundwater bodies in the crystalline part of the island in order to take better account of lithological diversity and other environmental factors (slope, altitude, soil thickness, etc.). The compartmentalization into 15 units, mainly based on structural geology, provides less information than the grouping into 12 units after subdivision of the crystalline region. The diversity of the waters in terms of chemical and bacteriological composition is discussed, and the results encourage a review of the compartmentalization of the island's GWBs, with a view to more targeted monitoring based on this diversity.

Keywords: groundwater bodies; water quality monitoring; database; Corsica; France

1. Introduction

Groundwater, which is less sensitive to pollution than surface water, is used primarily to supply networks for human consumption [1,2], and a proper management of this resource requires an inventory of its quantity and quality. Previous studies have shown that the identification and understanding of spatial and temporal patterns of groundwater quality are greatly improved if reactive processes are taken into account at all stages of monitoring, including data design and analysis [3–5]. Therefore, research has focused both on the assessment of existing monitoring networks and on modelling for the siting of new wells, using sometimes computationally intensive procedures [6–8]. Against this backdrop, the European Community's Framework Directive of 12 December 2006 (2006/118/EC) encouraged member countries to map not only surface water but also groundwater bodies (GWBs) with a view to monitoring and protecting them [9]. The mapping of these GWBs has stimulated considerable research efforts in the EU Member States [10–17]. This inventory has been referenced according to the major European watersheds (Rhône, Rhine, Danube, Loire, Seine, etc.). In order to cover all territories, small coastal rivers and island areas have been attached to the nearest major basin. In France, for example, we refer to the Rhône-Méditerranée-Corse basin, which includes the Rhône basin, the basins of small rivers flowing into the Mediterranean and the island of Corsica. Independently of this inventory, the Regional Health Agencies have been carrying out regular checks on the quality of water intended for human consumption, both surface water and groundwater, throughout France for over 30 years. All this geo-referenced data conveys information acquired during the water's journey from the surface to the aquifer, and has gradually been incorporated into a database called Sise-Eaux. The result is a large mass of information on water quality, particularly groundwater, which can be cross-referenced with GWB mapping [18]. Our research team has tackled this task with the aim of gaining a better understanding of the processes responsible for the variability (spatial and temporal) of water characteristics, in order also to facilitate the management and protection of the resource by regional health agencies [19–22]. From a practical point of view, it is also necessary to find a spatial unit for optimized monitoring and surveillance [18]. The method proposed for cross-referencing these two sources of information consists of a dimensional reduction of the data using principal component analysis, a characterization of the average groundwater quality for each GWB and on each factorial axis, and then a grouping of GWBs by degree of similarity of water quality. This method has been applied over a wide area, including the French administrative regions of Provence-Alpes-Côte d'Azur and Occitanie. The island of Corsica has a marked geological contrast between a crystalline western part and an eastern part made up of several metamorphic formations and a few small sedimentary zones. In this context, despite a certain degree of lithological heterogeneity, which may have an impact on the characteristics of the groundwater, only 3 GWBs cover the whole of the crystalline zone, with a much larger surface area than the GWB in the other sectors of the island.

The aim of this work is to assess and quantify, by analysis of variance based on physico-chemical and bacteriological parameters, whether such a division of GWBs in Corsica is relevant, or whether an alternative division more suited to monitoring and surveillance of groundwater resources can be developed.

2. Materials and Methods

2.1. Corsica Island

The island of Corsica in the Mediterranean Sea (Figure 1a), often referred to as the mountain in the sea, is a mountainous massif 180 km long along the north–south axis and 82 km at its widest, with a total surface area of 8722 km^2. The average altitude is 568 m, rising to 2706 (Monte Cinto), with many peaks over 2000 m. The steep gradients and very pronounced relief mean that the area is highly differentiated according to catchment area. From a geological point of view (Figure 1b), the island is clearly structured, with the "Hercynian Corsica" to the west, an essentially crystalline complex composed mainly of granodiorites and monzogranites, alkaline granites and volcanic formations to the north-west, but also a basic tholeiitic complex [23]. The east of the island, generally referred to as "Alpine Corsica", is occupied by metamorphic formations, particularly schists. Carbonate-quartz sedimentary formations are found in the extreme south of the island. In the French reference system for groundwater bodies (https://services.sandre.eaufrance.fr/geo/sandre, accessed on 7 February 2022), groundwater bodies are referenced by a unique code such as FREGxxx, where FR refers to France, E denotes the Corsica Island and G refers to groundwater resource. In this frame, crystalline Corsica has been divided into three large GWBs, namely FREG619, FREG620 and FREG621, while the smaller Alpine Corsica region is covered by 11 GWBs. The sedimentary part of the extreme south comprises a single GWB (FREG131) (Figure 1c). Given the disproportion in size between the GWBs delineated within the crystalline and alpine regions of Corsica, the three GWBs of the crystalline region were subdivided into sub-GWBs on the basis of hydrographic basins using the island's DTM (Figure 1d). The aim was to take better account of lithological variability within the crystalline region. The GWBs of the FREG619, FREG620 and FREG621 crystalline regions were therefore subdivided into 15, 8 and 5 sub-GWBs, respectively.

2.2. Sise-Eaux Database Extraction

The Sise-Eaux database (https://data.eaufrance.fr/concept/sise-eaux accessed on 15 March 2021) is supplied with the results of health inspections of water points supplying local communities [24,25]. It covers both untreated water, taken from catchment areas or boreholes, and treated water, i.e., disinfected by chlorination or another process and possibly decanted or filtered. In this work, only untreated (bulk) water has been considered. For each water sample taken, several analysis reports could have been realized, namely, complete with several hundreds of parameters, standard with about 30 parameters, or a routine follow-up, with only about ten parameters. Thus, data extraction from the whole island over 27 years (from April 1993 to September 2020) results in a matrix with a total of 5395 water analyses involving 23 parameters, but with some empty cells that need to be conditioned before processing (Figure 1a). After the manual correction of errors during data entry in the database and the elimination of empty cells, a full matrix consisting of 2830 observations and 15 parameters was used for this study. The 15 parameters selected are major ions (Ca, Mg, Na, SO_4, Cl, HCO_3), electrical conductivity (EC), bacteriological parameters (total Coliforms (Col.), revivable aerobic bacteria at 22 °C and 37 °C (Aer.22, Aer.37), also including the fecal contamination parameters *Enterococci* and *Escherichia coli* (*Ent.*, *E. coli*), nitrate ions (NO_3), and two trace metals (Fe and Mn). For some sampling points, the geographical coordinates were not entered in the database. These coordinates were retrieved from the Infoterre database of the French Geological Survey (https://infoterre.brgm.fr/ accessed on 15 March 2021). Each water sample was then assigned to a GWB, or sub-GWB for the crystalline region, based on its geographic coordinates and depth. Ultimately, these 2830 water samples came from 662 sampling points, i.e., an average of 4.3 samples for each sampling point, and spread over 40 (sub-)GWBs (Figure 1d).

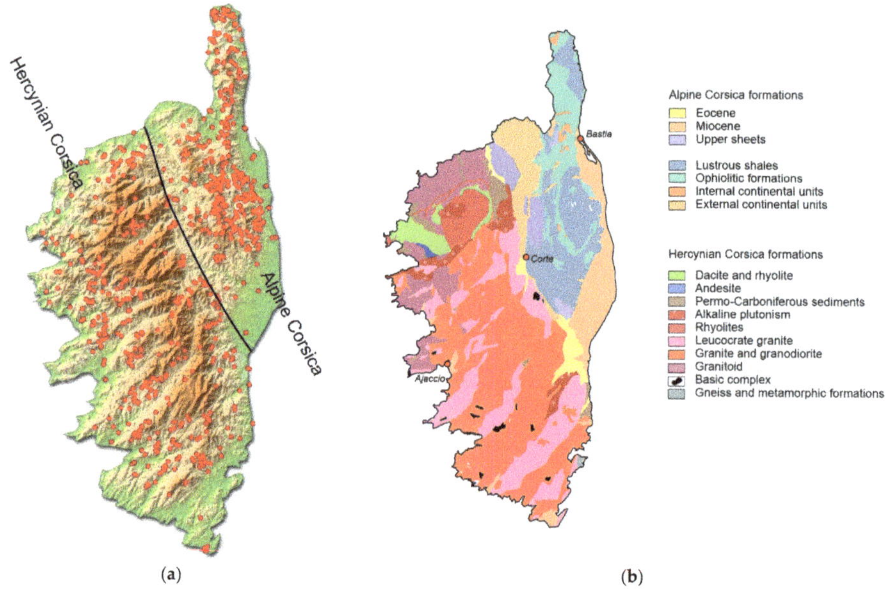

(a) (b)

Figure 1. *Cont.*

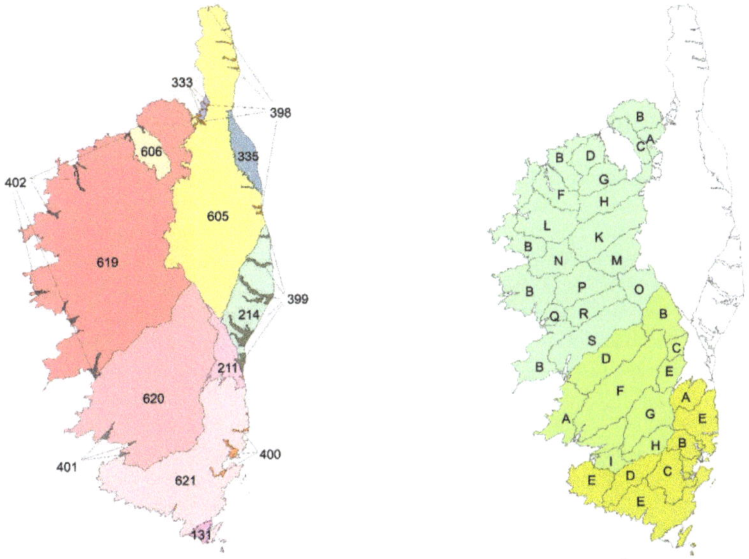

(c) (d)

Figure 1. (**a**) Distribution of groundwater sample collection points; (**b**) main geological formations; (**c**) compartmentalization of FREG groundwater bodies according to the French Geological Survey; (**d**) subdivision of the three main groundwater bodies (FREG619, FREG620 and FREG621) of the crystalline region according to river basins.

2.3. Data Treatments

The Kolmogorov–Smirnov normality test, which is adapted to high-dimensional statistical distributions [26], was applied to compare the empirical cumulative distribution function to a specified normal distribution, in this case one with the same mean and standard deviation. Following this test, the next step was to perform a logarithmic conditioning of the data as suggested by Jabrane et al. [20], according to the formula: $y = \log(x + DL)$, x being the value of the physico-chemical or bacteriological parameter X, and DL the detection limit of this same parameter. The aim of this transformation was to bring the distributions of each parameter closer to a normal distribution and to limit the weight of extreme values likely to obscure certain processes responsible for the variability within the dataset [22,27]. The normality tests revealed the presence of two observations (out of 2830) which clearly showed a data entry error. These two observations were therefore removed from the study.

A principal component analysis (PCA) was then performed on the log-transformed data to reduce the dimensionality of the data hyperspace [19,28] and to identify and rank the sources of variability in the dataset [29]. The PCA was based on the correlation matrix and thus considered the reduced centered variables, which allowed the integration of parameters of very diverse natures and units. Furthermore, it was carried out by diagonalization of the correlation matrix. Under this condition, the factorial axes were perpendicular to each other, and were therefore theoretically associated with independent processes responsible for the variability of water quality. The factorial axes, totaling about 95% of the information, were retained. The last factorial axes, explaining a small percentage of the variance, and generally considered as statistical noise [30], were eliminated. The reduction in dimensionality was estimated using the Bartlett sphericity test [31].

The mean values of the coordinates of each GWB (and sub-GWB) on each factorial axis were then calculated, thus characterizing each of these hydrogeological units. An unsupervised ascending hierarchical clustering (UHC) was then used to classify the hydrogeological units by degree of similarity, all parameters considered. The result is a group of GWBs containing GWBs with similar groundwater quality. By way of comparison, this process was carried out by considering only the GWBs delineated by the French Geologic Survey, i.e., with only three GWBs covering the entire crystalline region of the island, and then subdividing this region into sub-GWBs. At this stage, its sampling point, the GWB or sub-GWB, and the group of GWBs and sub-GWBs to which it belongs characterize each water sample. The choice of the number of groups was guided by the variation of the ratio between the inter-group and intra-group variability in order to maximize the intra-group homogeneity and the inter-group heterogeneity. The results are iteratively collated to produce a dendrogram. The choice of where to cut the phenom line was determined by a compromise between, on the one hand, the break in the slope in the relationship between the number of clusters and dissimilarity, and on the other hand, the requirements in terms of managing the number of spatial units generated. Indeed, too many space units would be unmanageable in terms of monitoring and protection by the health agency.

Finally, an analysis of variance (ANOVA) was carried out to measure the variance at each grouping scale using the R^2 coefficient, which is the ratio between the inter-class variance, i.e., the variance explained, and the total variance [32,33]. This analysis measures the effectiveness of the hydrogeological unit grouping method chosen by measuring the loss of information during grouping from the sampling point scale to the (sub)GWB groups [22]. The variance explained at the sampling point scale indicates the spatial variability at this scale, and the deviation of the R^2 from unity indicates the variance linked to temporal variability (since there are on average 4.3 samples per sampling point), added to a variability linked to analytical imprecision deemed negligible. The loss of information accompanying clustering was measured by the R^2 value indicating the spatial variability at each nested scale subGWB, GWB, subGWB group and GWB group.

3. Results

3.1. Principal Component Analysis

The distribution of the inertia of the different factorial axes is shown in Table 1. PCA allowed a significant reduction in dimensionality. The first principal component clearly stood out from the others, accounting for 37.1% of the total variance. About half of the information contained in the database (49.3%) was concentrated in the first factorial plan, which is substantial compared with the 15 parameters in the dataset. The first five factorial axes had eigenvalues greater than unity, meaning that they concentrated more information than a single parameter [29]. The first eight axes, which accounted for 94.7% of the information, were retained for further hierarchical clustering calculations. The remaining 5.3% was considered as statistical noise and information generated by minor sources of variability.

Table 1. Inertia of the factorial axes of the PCA carried out on 2828 observations and 15 parameters.

	PC1	PC2	PC3	PC4	PC5	PC6	PC7
Explained variance %	37.1	12.3	12.1	9.5	8	6.5	6
Cumulative explained variance %	37.1	49.4	61.5	70.9	79	85.5	91.5
Eigenvalue	5.56	1.84	1.82	1.42	1.2	0.98	0.91

The contribution of the parameters to the first four PCs is shown in Figure 2. The first factorial axis explained 37.1% of the variance and contrasted mineralized water, positively scored with CE and major ions (Ca, Mg, Na, Cl, SO_4, HCO_3), with poorly mineralized water contaminated by *E. coli* and *Enterococcus*. The second factorial axis (12.3% of the variance) contrasted water marked by revivable bacteria with water with a chloride-sodium chemical profile marked by the presence of metals and fecal contamination. The third PC

(12.1% of the variance) was also affected by the presence of revivable bacteria but also fecal contamination (*E. coli* and *Ent.*) and the presence of metals, again in a predominantly sodium chloride chemical context, while the fourth PC had positive coordinates for fecal contamination but negative for metals. Bartlett's sphericity test showed a calculated X^2 of 40143, much stronger than the critical X^2 (=130, significance level of 0.05 and p-value < 0.0001), which highlighted the fact that the switch from 15 initial parameters to eight retained PCs, i.e., an eight-dimensional hyperspace, was a significant dimensional contraction with a loss of only 5.3% of the information.

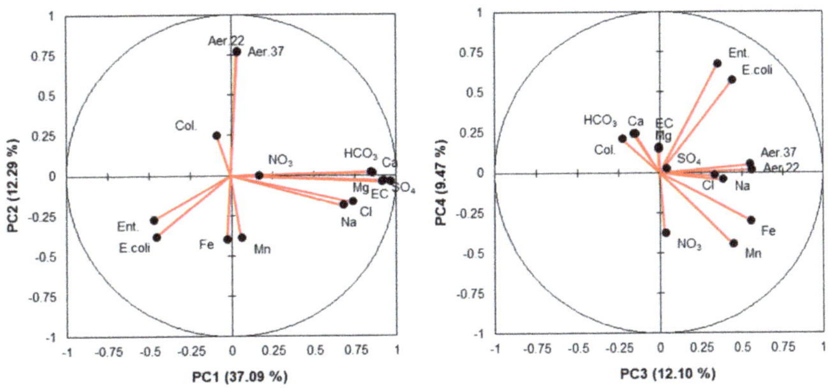

Figure 2. Distribution of the parameters on the four first principal components.

3.2. Unsupervised Agglomerative Hierarchical Clustering

The clustering diagrams obtained from the first eight factorial axes for the 15 GWBs and the 40 GWBs + sub-GWBs are shown in Figures 3 and 4, respectively. The GWB coded FREG333, for which there were few water analyses, was withdrawn from the study at this clustering stage. Clustering of the GWBs reveals six groups of GWBs. The distribution of these groups highlights an opposition between crystalline Corsica, which constitutes a large undifferentiated group covering approximately two thirds of the island, and the rest of the territory (Figure 5a).

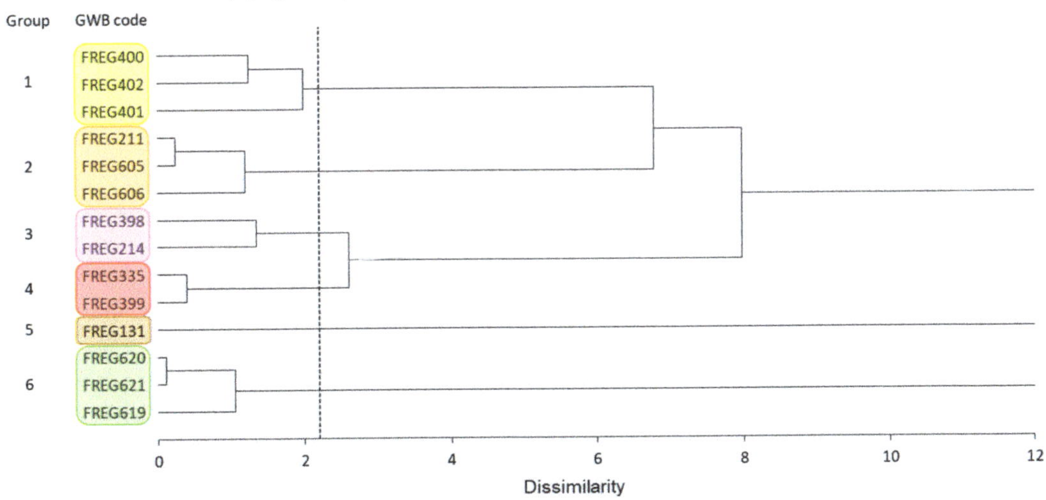

Figure 3. Dendrogram resulting from the agglomerating hierarchical clustering based on 14 GWB, individualizing six groups of groundwater bodies.

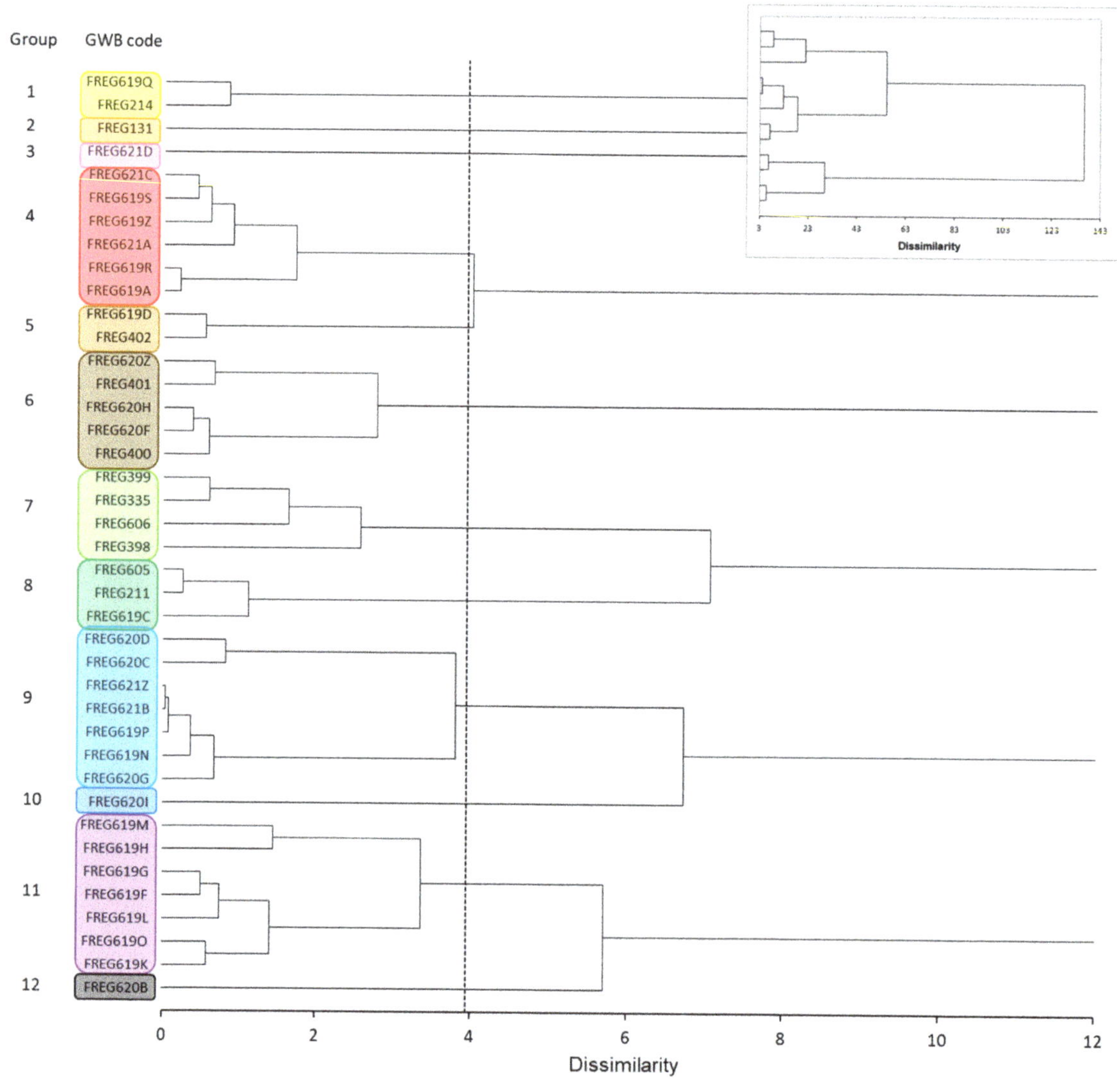

Figure 4. Dendrogram resulting from the agglomerating hierarchical clustering, based on 40 GWB and sub-GWB, individualizing 12 groups of groundwater bodies.

By subdividing the GWBs of the crystalline region, we obtained 40 units over the whole of the island, and the hierarchical classification resulted in 12 distinct groups, the distribution of which is presented in Figure 5b. This classification has lessened the diversity of waters within Alpine Corsica, with several previously distinct GWBs now grouped together in group 8. On the other hand, a diversity of waters appeared within the crystalline region of the Hercynian Corsica, with sub-GWBs grouped together in small patches (groups 5, 6 and 11) or dispersed (groups 3, 4, 7, 9). Some sub-GWBs stood out from the others, such as the GWB in the sedimentary material at the southern extremity of the island (group 2) and two sub-GWBs from FREG620, which formed groups 10 and 12.

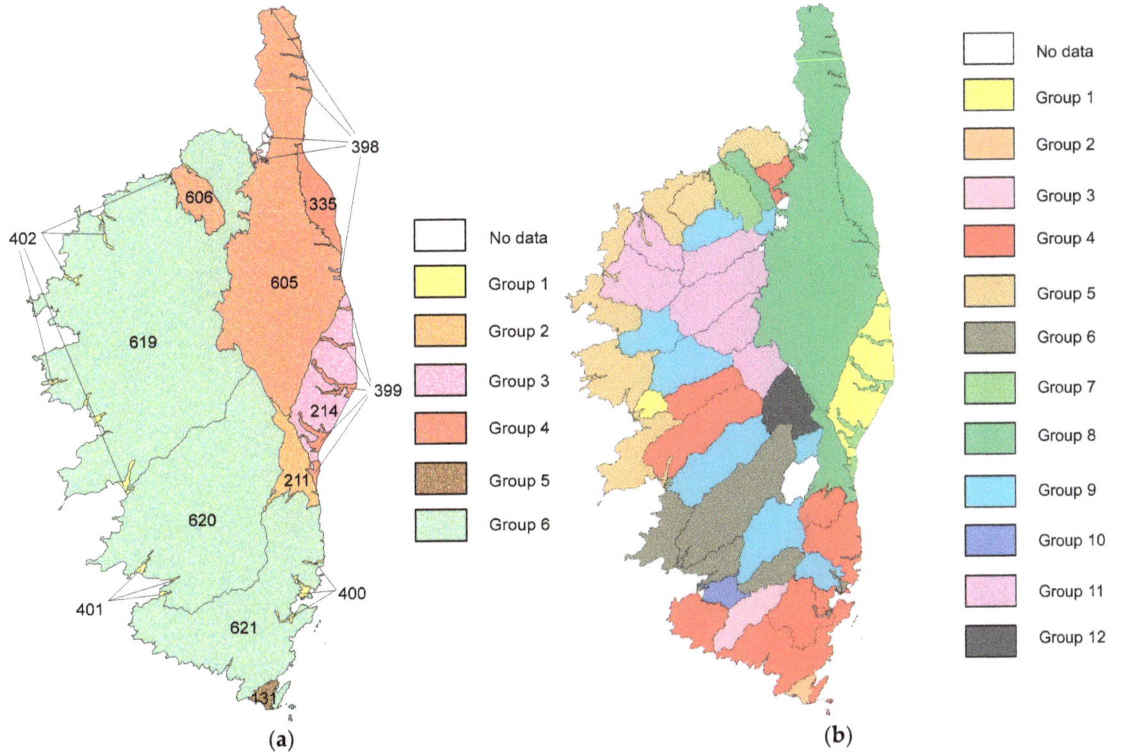

Figure 5. Distribution of (**a**) the 6 GWB groups from clustering based on 14 groundwater bodies and (**b**) the 12 subGWB groups from clustering based on 40 sub-groundwater bodies.

The distribution of the groups obtained in the PC1-PC2 and PC3-PC4 score plots is shown in Figure 6, and the mean values of the parameters for each group are summarized in Table 2.

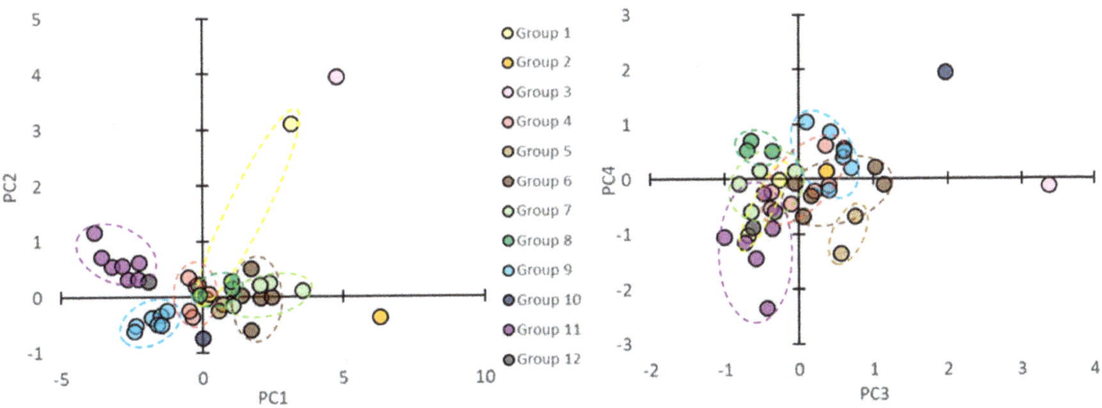

Figure 6. Distribution of the 12 sub-GWB groups in the PC1-PC2 and PC3-PC4 score plots.

Table 2. Mean values of the parameters for each group of sub-GWBs.

	Groups											
	1	2	3	4	5	6	7	8	9	10	11	12
Ent.	0.15	0.00	0.00	0.45	0.17	0.32	0.12	0.49	1.12	1.56	0.56	0.08
E. coli	0.05	0.08	0.00	0.45	0.11	0.39	0.05	0.35	1.20	2.06	0.36	0.11
Col.	0.00	0.00	0.00	0.00	0.02	0.00	0.00	0.09	0.00	0.00	0.12	0.00
Aer.22	0.00	0.00	0.55	0.01	0.02	0.02	0.00	0.01	0.00	0.06	0.01	0.00
Aer.37	0.00	0.00	0.27	0.00	0.02	0.02	0.00	0.01	0.00	0.05	0.01	0.00
EC	2.52	3.03	2.75	2.16	2.23	2.40	2.52	2.39	2.00	2.28	1.77	1.93
Ca	1.31	2.11	1.55	0.82	0.92	1.09	1.53	1.44	0.70	1.05	0.46	0.82
Mg	1.04	1.01	1.26	0.55	0.64	0.79	0.99	0.82	0.37	0.72	0.11	0.25
Cl	1.55	2.15	1.90	1.32	1.39	1.57	1.24	1.09	1.12	1.41	0.84	0.83
SO_4	1.28	1.58	1.34	0.82	0.92	1.03	1.12	0.95	0.62	0.86	0.55	0.46
Na	1.44	1.88	1.69	1.17	1.21	1.39	1.06	0.89	0.99	1.23	0.74	0.82
HCO_3	1.91	2.57	2.15	1.49	1.56	1.74	2.17	2.03	1.36	1.70	1.08	1.51
NO_3	1.23	0.87	0.67	0.25	0.29	0.26	0.56	0.15	0.17	0.17	0.35	−0.24
Fe	1.04	0.98	1.05	1.13	1.47	1.28	1.18	1.18	1.29	1.46	1.20	1.43
Mn	1.04	0.98	1.04	1.05	1.35	1.07	1.08	1.10	1.08	1.06	1.11	1.24

Groups 9 and 10, located on crystalline rocks in Hercynian Corsica, showed the highest levels of fecal contamination. Group 10 comprises a single sub-GWB, reflecting a local specificity. It had the same characteristics as group 9, but differed in that it contained an abundance of revivable bacteria and even higher levels of *E. coli* contamination than group 9. Both groups 9 and 10 are located in the same geographical environment, i.e., mountainous areas, and therefore sloping, on granite, leucocratic granite and granodiorite. All these factors (altitude, slope, erosion, shallow soils) are even more marked in groups 11 and 12, which correspond to high mountain areas (Monte Cinto, Monte Corona) and ridge lines (Marmano and Pietra Piana forests) with skeletal or even absent soils. Group 11 is located on a more acidic lithology of rhyolites and alkaline plutonism. In contrast, group 2, the only small area in Corsica on limestone (FREG131), is characterized by very low fecal contamination and higher mineral load. Group 3, also made up of a single sub-GWB (FREG621D) on granite, corresponds to a wide, low-altitude, non-wooded valley.

3.3. Explained Variance Measured from R^2

The temporal variability for each parameter is shown in Figure 7. This was low for major ions and electrical conductivity, intermediate for fecal contamination parameters (*E. coli* and Ent.) and nitrates, and high for coliform parameters, revivable bacteria and metals. As a result, the proportion of spatial variability for the latter was reduced. The spatial variance explained for each parameter according to the level of grouping scale is displayed in Table 3, together with the corresponding numbers. All the results were related to the spatial variability at the scale of the collection points.

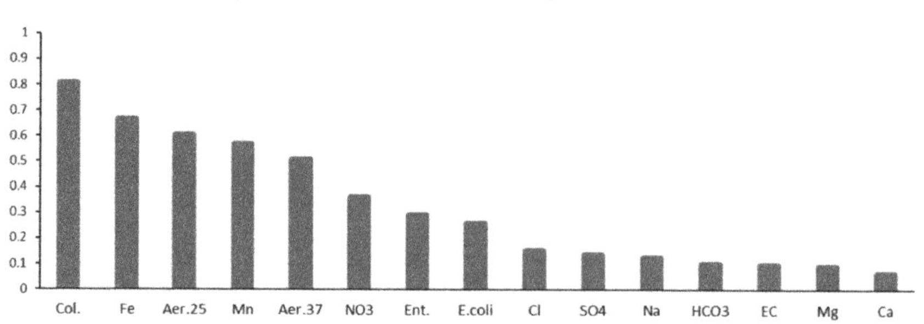

Figure 7. Temporal variance of each water quality parameter within the dataset (2828 water samples collected over 27 years).

Table 3. Spatial variance explained for each parameter according to the level of grouping scale.

	Explained Variance			
	Group Sub-GWB	**GWB**	**Sub-GWB**	**Sampling Points**
Number of units	12	15	40	662
Colif.	0.194	0.171	0.292	1
Fe	0.134	0.125	0.283	1
Aer. 25 °C	0.064	0.007	0.045	1
Mn	0.176	0.139	0.243	1
Aer. 37 °C	0.100	0.004	0.122	1
NO_3	0.145	0.266	0.392	1
Enter.	0.328	0.327	0.529	1
E. coli	0.394	0.359	0.593	1
Cl	0.327	0.278	0.552	1
SO_4	0.445	0.375	0.540	1
Na	0.304	0.330	0.581	1
HCO_3	0.508	0.581	0.706	1
EC	0.528	0.439	0.630	1
Mg	0.442	0.405	0.573	1
Ca	0.516	0.595	0.706	1

Grouping the GWBs into 40 units, i.e., considering the GWBs combined with the sub-GWBs of the crystalline region, resulted in a fairly significant reduction in the variance explained, mainly for bacteriological parameters and metals, less marked for nitrates, and even less for fecal contamination parameters and major ions. Grouping into the 15GWBs delineated by the French Geological Survey, or into 12 groups of (sub-)GWBs, was accompanied by a slight additional reduction in the explained variance. However, for most parameters, grouping into 12 units resulted in a higher proportion of explained variance than grouping into 15 units.

4. Discussion

4.1. A multifactorial Information Set

The distribution of the PC inertia, and in particular the fact that the first five eigenvalues are greater than unity, underlines that the information contained in the dataset cannot be summed up in one or two parameters, but on the contrary that this information is clearly multifactorial. The contrast between mineralized water and water with low mineral content but subject to fecal contamination (*E. coli* and *Enterococci*) has been classically observed in the processing of databases in Mediterranean climates [18–22]. During rainy periods, runoff water, which is poorly mineralized but turbid and loaded with bacteria, contaminates vulnerable or poorly protected water catchment structures. In contrast, during low-water periods, the catchment structures only collect groundwater, which is more mineralized, filtered by the soil and less affected by surface bacterial contamination. This interpretation is consistent with the high temporal variability of bacteriological parameters (Figure 7). However, the high contribution of fecal contamination parameters on the first four principal components suggests a more complex determinism of the spatial and temporal variations in these parameters. As the factorial axes are orthogonal in the hyperspace of the data, they cannot reflect correlated mechanisms, and each PC represents a process that is independent of the others. The pathways of fecal contamination cannot therefore be reduced solely to runoff induced by heavy rainfall, but are a priori the result of a diversity of pathways, as has already been suggested for other regions [19]. The factors are multiple and interdependent, with environmental factors influencing the presence or absence of extensive livestock farming (particularly pigs), which should contribute to this result.

4.2. A Sub-GWB Division Better Suited to Monitoring the Resource

The large undifferentiated group of GWBs covering the crystalline region (Figure 5a) contrasts with the medium to very small groups in the other regions. This highlights the significant disparity in terms of the area covered by the different GWB groups. The effectiveness of a mapping can be measured by the variance explained for a given number of spatial units. For the same number of units, the most relevant map containing the most spatial information is the one with the highest explained variance. In our case, the 15 GWBs (Figure 1c) convey less information than 12 groups of sub-GWBs (Figure 5b) without leading to an increase in the number of spatial units. The division of the territory into GWBs proposed by the French Geological Survey on the basis of structural geology is therefore less effective than the map of sub-GWB groups in terms of groundwater quality. This result is difficult to reconcile with the development of a strategy for monitoring and protecting water resources intended for human consumption. The size of the crystalline region and its division into three GWBs makes it impossible to distinguish the lithological variability within this zone, and to visualize the resulting variability in water quality. Such a division will necessarily lead to monitoring of the crystalline region that is ill suited to the diversity of the waters in this region. The GWB's classification into six groups, based largely on the island's structural geology, is disappointing: it is relevant, but trivial and too schematic for the zonation of Corsican groundwater. Crystalline Corsica has highly variable rock types, particularly in terms of alkali and earth alkali content. This means that the chemical composition of the water, the nature of the soil, the physico-chemical properties of the surface formations, and the susceptibility to erosion and the transport of bacteria are all very different depending on the lithology.

This result quantifies and demonstrates (1) the need to divide large GWBs into sub-GWBs and (2) the effectiveness of grouping sub-GWBs, which ultimately corresponds to a gain in information compared with dividing into 15 units, for most parameters, particularly for fecal contamination criteria, revivable bacteria, most major ions and electrical conductivity. Only the parameters Ca, HCO_3 and NO_3 do not follow this rule, but the low proportion of spatial variability partly explains this result for nitrates. This result, which is valid for almost all the parameters, demonstrates the relevance and interest of working on the scale of groups of sub-GWBs and justifies, a posteriori, the approach of dividing the three GWBs of the crystalline region. This better score for dividing the area into groups of sub-GWBs is probably not due to the fact of having chosen a topographical criterion for dividing crystalline Corsica, i.e., the river basins, but rather to the fact of indirectly integrating the lithological heterogeneity internal to crystalline Corsica, even if this choice can be partially justified by the impact on likely flow paths. In hindsight, it turns out that this division into basins reflects the lithology of crystalline Corsica quite well (compare, for example, Figure 1b,d, or Figures 1b and 5b). If we had chosen to focus directly on lithology, our results would probably have been just as good. This choice is valid for Corsica, but is not necessarily applicable to other sites.

4.3. A Result Closer to the Distribution of Lithology

The clustering result from the GWBs (Figure 5a) is in line with the structural geology of Corsica, since it is on this basis that the GWBs were delimited, whereas the clustering of the sub-GWBs (Figure 5b) gives a result close to the petrographic distribution, and thus seems to correspond to the reality of another geological information. This result makes sense. Within crystalline Corsica, the granitoid zones near the north-west coast are clearly differentiated in terms of water quality (group 5). Similarly, the areas of alkaline plutonism and rhyolites (group 11) stand out clearly from the granite and granodiorite areas (groups 4, 6 and 9).

4.4. Other Environmental Factors

Steep slopes and siliceous rock with a low alkaline and alkali-earth content result in very coarse-textured soils with a relatively low calcium level, which reduces the flocculation of colloids and facilitates the transfer of suspended matter, the main vector for the transport of bacteria. This explains the high level of fecal contamination in groups 9 and 10. The altitude and steep slopes also favor short-lived water-rock interaction and cooler temperatures, resulting in water with a low mineral content, which is very marked in groups 9 and 10, and even more so in groups 11 and 12. In contrast, group 2, which is on limestone and therefore rich in calcium, is more gently sloping and is characterized by very low fecal contamination, probably due to the flocculating role of calcium ions, the reduction in run-off (two major factors in solid and therefore bacterial transport), and a slow flow in the aquifer that encourages the self-purification process and the elimination of pollution. The limestone lithology, a relatively soluble rock, explains the higher mineral content. For group 3, with its fairly flat topography on granite, run-off is reduced and the soils are thicker, which limits fecal contamination, which is lower than in the surrounding units. At the same time, the shallower topography favors slower run-off and longer water–rock interaction times, resulting in a higher mineral content than in other areas of crystalline Corsica. In conclusion, lithological specificity not only affects the composition of water in terms of major ions. The relative abundance of alkaline earths or alkalis in the less siliceous lithologies probably results in greater structural stability in the soils derived from them, and consequently less vulnerability to bacterial transport, which may explain this differentiation in multi-parameter water quality between the different formations of crystalline Corsica.

To sum up, the division of the GWBs of the crystalline region into sub-GWBs has led to a better understanding of the diversity of waters, by revealing the role of:

- The nature of the soils, an important factor in the vulnerability of water catchment structures;
- The altitude, which has an impact on temperature and therefore on human land use, the type of forest, the intensity of pedogenesis, etc.;
- The geomorphology;
- The lithological diversity within the crystalline region, with an important aspect regarding the presence of flocculent cations.

While remaining on this purely lithological criterion, it appears that it would be relevant to distinguish zones with alkaline plutonism from granitic zones. The results of the study show that it would be possible to review the division of GWBs, particularly in crystalline Corsica.

4.5. Methodological Contribution

The present work highlights a series of methodological advances that we feel are relevant to bring together here:

- The study confirms that the method implemented, and progressively improved, is suitable for highlighting the information contained in Sise-Eaux type databases. Of course, the existence of such a database is a prerequisite for its application;
- Quantifying the loss of information that accompanies the grouping of GWBs makes it possible to measure the effectiveness of the method.
- This analysis method, developed for vast regions of the order of 60,000 to 80,000 km^2, can be deployed for regions around 10 times smaller, even if the European Commission recommends an even larger scale, that of the major European watersheds (Danube, Rhône, Rhine, Seine, Loire, Po, etc.);
- While it now appears that the information analysis method can be extended to other administrative regions, we do not know whether it can be extended to new parameters not taken into account here (pesticides, land use, fractured medium, water–rock contact time, etc.). The use of databases of much larger dimensions is currently being tested in the Provence-Alpes-Côte d'Azur administrative region (PACA, France);
- However, the proposed method has certain limitations. In particular, at this stage, the analysis is carried out parameter by parameter, which does not take into account the interrelations between several parameters. Access to a greater number of parameters

over a larger number of observations should make it possible to establish a typology of parameter behaviour, while separating their variability in space (range of each variable) and in time (common variability of parameters during rainy events, more or less marked seasons, or even multi-annual dry or wet periods). This typology of parameters would be a further step forward in the type of information extracted from large databases.

5. Conclusions

A method for mapping groundwater quality in France has been developed by our research group over the last few years, with the aim of optimizing the surveillance and monitoring of groundwater quality by regional health agencies. As the method has been applied to areas of varying size (Provence-Alpes-Côte d'Azur and Occitanie regions), it has become increasingly effective. Here, the method has been tested in the case of a much smaller region, the island of Corsica, and its application on the basis of GWBs delineated by the French Geologic Survey gives relevant but trivial results, which do not allow discerning of the variability of water quality within Hercynian crystalline Corsica. An adaptation is proposed. It consists of dividing the large GWBs of crystalline Corsica into sub-GWBs on the basis of hydrographic basins. Our study shows that it is possible, using analysis of variance, to quantify the evolution of the loss of spatial information when changes of scale are made, from the sampling point to the various groupings of GWBs. The analysis of variance shows a gain in information despite a smaller number of spatial units, which can be explained by a better consideration of the spatial variability of groundwater quality within the crystalline region. The results obtained should serve as the starting point for a revision of the GWBs for the island of Corsica based on groundwater quality, but also on lithology. As things stand at present, very small lithological units have been taken into account, and rightly so, in drawing up the GWBs, because of their specific nature. The calculation of groups of GWBs or sub-GWBs has confirmed their specificity, but they represent extremely small areas. On the other hand, the internal variability of the lithological units, which are much more extensive within crystalline Corsica, has not been taken into account, which could prove useful, particularly for health agencies, in developing a strategy for monitoring the quality of groundwater resources.

Author Contributions: Conceptualization, H.L., V.V. and I.K.; methodology, H.L., V.V. and L.B.; software, V.V. and I.M.; validation, A.T., M.A., I.M. and H.L.; formal analysis, H.L., V.V. and L.B.; investigation, H.L.; resources, É.G., F.H. and C.M.; data curation, A.-A.B., L.B. and I.M.; writing—original draft preparation, H.L. and V.V.; writing—review and editing, L.B.; visualization, L.B. and I.M.; supervision, I.K. and V.V.; project administration, I.K. All authors have read and agreed to the published version of the manuscript.

Funding: This research received no external funding.

Data Availability Statement: The data in this study are the property of Corsica Regional Health Agency and are included in the SISE-Eaux French database.

Acknowledgments: The authors would like to thank Patrice Grandjean (Corsica Regional Health Agency) for his help with data extraction.

Conflicts of Interest: The authors declare no conflict of interest.

References

1. Walker, D.B.; Baumgartner, D.J.; Gerba, C.P.; Fitzsimmons, K. Chapter 16—Surface Water Pollution. In *Environmental and Pollution Science*, 3rd ed.; Brusseau, M.L., Pepper, I.L., Gerba, C.P.B.T.-E., Eds.; Academic Press: Cambridge, MA, USA, 2019; pp. 261–292, ISBN 978-0-12-814719-1.
2. Li, P.; Karunanidhi, D.; Subramani, T.; Srinivasamoorthy, K. Sources and Consequences of Groundwater Contamination. *Arch. Environ. Contam. Toxicol.* **2021**, *80*, 1–10. [CrossRef] [PubMed]
3. Woldt, W.; Bogardi, I. Ground water monitoring network design using multiple criteria decision making and geostatistics. *JAWRA J. Am. Water Resour. Assoc.* **1992**, *28*, 45–62. [CrossRef]

4. Hudak, P.F.; Loaiciga, H.A. An optimization method for monitoring network design in multilayered groundwater flow systems. *Water Resour. Res.* **1993**, *29*, 2835–2845. [CrossRef]
5. Frapporti, G.; Hoogendoorn, J.H.; Vriend, S.P. Detailed Hydrochemical Studies as a Useful Extension of National Ground-Water Monitoring Networks. *Groundwater* **1995**, *33*, 817–828. [CrossRef]
6. Meyer, P.D.; Brill, E.D., Jr. A method for locating wells in a groundwater monitoring network under conditions of uncertainty. *Water Resour. Res.* **1988**, *24*, 1277–1282. [CrossRef]
7. Lee, J.-Y.; Yi, M.-J.; Yoo, Y.-K.; Ahn, K.-H.; Kim, G.-B.; Won, J.-H. A review of the National Groundwater Monitoring Network in Korea. *Hydrol. Process.* **2007**, *21*, 907–919. [CrossRef]
8. Esquivel, J.M.; Morales, G.P.; Esteller, M.V. Groundwater Monitoring Network Design Using GIS and Multicriteria Analysis. *Water Resour. Manag.* **2015**, *29*, 3175–3194. [CrossRef]
9. European Commission Directive 2006/118/EC of the European Parliament and of the Council of 12 December 2006 on the protection of groundwater against pollution and deterioration. *Off. J. Eur. Union* **2006**, *372*, 19–31.
10. Allan, I.J.; Vrana, B.; Greenwood, R.; Mills, G.A.; Knutsson, J.; Holmberg, A.; Guigues, N.; Fouillac, A.-M.; Laschi, S. Strategic monitoring for the European Water Framework Directive. *TrAC Trends Anal. Chem.* **2006**, *25*, 704–715. [CrossRef]
11. Martínez-Navarrete, C.; Jiménez-Madrid, A.; Sánchez-Navarro, I.; Carrasco-Cantos, F.; Moreno-Merino, L. Conceptual Framework for Protecting Groundwater Quality. *Int. J. Water Resour. Dev.* **2011**, *27*, 227–243. [CrossRef]
12. Masciale, R.; Amalfitano, S.; Frollini, E.; Ghergo, S.; Melita, M.; Parrone, D.; Preziosi, E.; Vurro, M.; Zoppini, A.; Passarella, G. Assessing Natural Background Levels in the Groundwater Bodies of the Apulia Region (Southern Italy). *Water* **2021**, *13*, 958. [CrossRef]
13. Frollini, E.; Preziosi, E.; Calace, N.; Guerra, M.; Guyennon, N.; Marcaccio, M.; Menichetti, S.; Romano, E.; Ghergo, S. Groundwater quality trend and trend reversal assessment in the European Water Framework Directive context: An example with nitrates in Italy. *Environ. Sci. Pollut. Res.* **2021**, *28*, 22092–22104. [CrossRef]
14. Goldscheider, N. Karst groundwater vulnerability mapping: Application of a new method in the Swabian Alb, Germany. *Hydrogeol. J.* **2005**, *13*, 555–564. [CrossRef]
15. Raihan, A.T.; Bauer, S.; Mukhopadhaya, S. An AHP based approach to forecast groundwater level at potential recharge zones of Uckermark District, Brandenburg, Germany. *Sci. Rep.* **2022**, *12*, 6365. [CrossRef]
16. Flindt Jørgensen, L.; Villholth, K.G.; Refsgaard, J.C. Groundwater management and protection in Denmark: A review of pre-conditions, advances and challenges. *Int. J. Water Resour. Dev.* **2017**, *33*, 868–889. [CrossRef]
17. Masses d'eau Souterraines-Métropole-Version État Des Lieux. 2019. Available online: https://geo.data.gouv.fr/fr/datasets/1a983edfe5ea441fef359a652e98217c9c3ce3c6 (accessed on 10 November 2023).
18. Tiouiouine, A.; Jabrane, M.; Kacimi, I.; Morarech, M.; Bouramtane, T.; Bahaj, T.; Yameogo, S.; Rezende-Filho, A.; Dassonville, F.; Moulin, M.; et al. Determining the relevant scale to analyze the quality of regional groundwater resources while combining groundwater bodies, physicochemical and biological databases in southeastern france. *Water* **2020**, *12*, 3476. [CrossRef]
19. Tiouiouine, A.; Yameogo, S.; Valles, V.; Barbiero, L.; Dassonville, F.; Moulin, M.; Bouramtane, T.; Bahaj, T.; Morarech, M.; Kacimi, I. Dimension reduction and analysis of a 10-year physicochemical and biological water database applied to water resources intended for human consumption in the provence-alpes-cote d'azur region, France. *Water* **2020**, *12*, 525. [CrossRef]
20. Jabrane, M.; Touiouine, A.; Bouabdli, A.; Chakiri, S.; Mohsine, I.; Valles, V.; Barbiero, L. Data Conditioning Modes for the Study of Groundwater Resource Quality Using a Large Physico-Chemical and Bacteriological Database, Occitanie Region, France. *Water* **2023**, *15*, 84. [CrossRef]
21. Jabrane, M.; Touiouine, A.; Valles, V.; Bouabdli, A.; Chakiri, S.; Mohsine, I.; El Jarjini, Y.; Morarech, M.; Duran, Y.; Barbiero, L. Search for a Relevant Scale to Optimize the Quality Monitoring of Groundwater Bodies in the Occitanie Region (France). *Hydrology* **2023**, *10*, 89. [CrossRef]
22. Mohsine, I.; Kacimi, I.; Abraham, S.; Valles, V.; Barbiero, L.; Dassonville, F.; Bahaj, T.; Kassou, N.; Touiouine, A.; Jabrane, M.; et al. Exploring Multiscale Variability in Groundwater Quality: A Comparative Analysis of Spatial and Temporal Patterns via Clustering. *Water* **2023**, *15*, 1603. [CrossRef]
23. De Graciansky, P.-C.; Roberts, D.G.; Tricart, P. Chapter Nine—The Tethyan Margin in Corsica. In *The Western Alps, from Rift to Passive Margin to Orogenic Belt*; De Graciansky, P.-C., Roberts, D.G., Tricart, P., Eds.; Elsevier: Amsterdam, The Netherlands, 2011; Volume 14, pp. 183–188, ISBN 0928-2025.
24. Chery, L.; Laurent, A.; Vincent, B.; Tracol, R. *Echanges SISE-Eaux/ADES: Identification Des Protocoles Compatibles Avec Les Scénarios D'échange SANDRE*; ONEMA: Vincennes/Orléans, France, 2011.
25. Gran-Aymeric, L. Un portail national sur la qualite des eaux destinees a la consommation humaine. *Tech. Sci. Méthodes* **2010**, *12*, 45–48. [CrossRef]
26. Steinskog, D.J.; Tjøstheim, D.B.; Kvamstø, N.G. A Cautionary Note on the Use of the Kolmogorov–Smirnov Test for Normality. *Mon. Weather Rev.* **2007**, *135*, 1151–1157. [CrossRef]
27. Feng, C.; Wang, H.; Lu, N.; Chen, T.; He, H.; Lu, Y.; Tu, X.M. Log-transformation and its implications for data analysis. *Shanghai Arch. Psychiatry* **2014**, *26*, 105–109. [PubMed]
28. Barbel-Périneau, A.; Barbiero, L.; Danquigny, C.; Emblanch, C.; Mazzilli, N.; Babic, M.; Simler, R.; Valles, V. Karst flow processes explored through analysis of long-term unsaturated-zone discharge hydrochemistry: A 10-year study in Rustrel, France. *Hydrogeol. J.* **2019**, *27*, 1711–1723. [CrossRef]

29. Helena, B.; Pardo, R.; Vega, M.; Barrado, E.; Fernandez, J.M.; Fernandez, L. Temporal evolution of groundwater composition in an alluvial aquifer (Pisuerga River, Spain) by principal component analysis. *Water Res.* **2000**, *34*, 807–816. [CrossRef]
30. Rezende Filho, A.; Furian, S.; Victoria, R.; Mascré, C.; Valles, V.; Barbiero, L. Hydrochemical variability at the upper paraguay basin and pantanal wetland. *Hydrol. Earth Syst. Sci.* **2012**, *16*, 2723–2737. [CrossRef]
31. Gleser, L.J. A Note on the Sphericity Test. *Ann. Math. Stat.* **1966**, *37*, 464–467. [CrossRef]
32. Achen, C.H. What Does "Explained Variance" Explain?: Reply. *Polit. Anal.* **1990**, *2*, 173–184. [CrossRef]
33. Miles, J. R Squared, Adjusted R Squared. In *Wiley StatsRef: Statistics Reference Online*; Wiley: Hoboken, NJ, USA, 2014; ISBN 9781118445112.

hydrology

MDPI

Review

Stream-Aquifer Systems in Semi-Arid Regions: Hydrologic, Legal, and Management Issues

Neil S. Grigg *, Ryan T. Bailey and Ryan G. Smith

Department of Civil and Environmental Engineering, Colorado State University, Fort Collins, CO 80523, USA; ryant.bailey@colostate.edu (R.T.B.); ryan.g.smith@colostate.edu (R.G.S.)
* Correspondence: neilg@engr.colostate.edu

Abstract: Integrated solutions to groundwater management problems require effective analysis of stream-aquifer connections, especially in irrigated semi-arid regions where groundwater pumping affects return flows and causes streamflow depletion. Scientific research can explain technical issues, but legal and management solutions are difficult due to the complexities of hydrogeology, the expense of data collection and model studies, and the inclination of water users not to trust experts, regulatory authorities, and in some cases, their management organizations. The technical, legal, and management issues are reviewed, and experiences with integrated management of stream-aquifer systems are used to illustrate how governance authorities can approach engineering, legal, regulatory, and management challenges incrementally. The situations in three basins of the State of Colorado with over-appropriated water resources are explained to identify modeling and control issues confronting regulators and managers of water rights. Water rights administration in the state follows the strict appropriation method and a workable technical-legal approach to establishing regulatory and management strategies has been developed. The explanations show how models and data management are improving, but the complexities of hydrogeology and institutional systems must be confronted on a case-by-case basis. Stream-aquifer systems will require more attention in the future, better data will be needed, model developers must prove superiority over simpler methods, and organizational arrangements will be needed to facilitate successful collective action amidst inevitable legal challenges. Continued joint research between technical, legal, and management communities will also be needed.

Keywords: groundwater; management; stream-aquifer; models; data; water rights

1. Introduction

The global urgency of groundwater resource management is increasing due to rising exploitation that causes declining levels in aquifers [1–6]. Another adverse trend is evident in regional waterlogging and flooding of cultivated fields, infrastructure, and transportation systems [7–10]. These problems have multiple causes, including lack of effective regulation, extreme weather events, inadequate drainage, and poor management of irrigation water.

Solutions to many of these problems require analysis of how groundwater and surface water are intrinsically linked and should be treated as a single resource. The solutions will require an assessment of the situations where groundwater management is a critical issue, such as in declining levels in alluvial and confined aquifers, depletion of surface water caused by well pumping, and, in some cases, exploitation of fossil water, subsidence [11,12], and degradation of water quality [13].

Regardless of the situation, integrated approaches to management are required to manage hydraulic connections in the stream-aquifer systems that can lead to significant water exchanges in influent or effluent streams [14–16]. The problem is especially important in irrigated semi-arid regions where river diversions and groundwater pumping create interdependencies involving groundwater return flows and streamflow depletion [17–21].

Such irrigated semi-arid regions occur throughout the world where crop yields increase with applied irrigation water, such as in parts of Iran, Central Asia, China, Africa, Australia, and western North America [22]. Depending on its level of development, governance, legal structure, and culture, each region has a unique approach to water rights administration. In some cases, little administration occurs, and water users must work out their problems locally, whereas in others, large-scale effects of groundwater pumping require more formal approaches. Compared to surface water laws, the status of groundwater law and administration of pumping rights is in a relative state of infancy because of the complexity of hydrogeology and the greater difficulty in measuring the short- and long-term effects of pumping. This gap occurs because, while to scientists and engineers, the relationship of groundwater to surface waters is known, to lawyers and courts it is not [23].

Irrigated regions must work out their unique issues according to physical laws and the needs of their water users. Most reports of experiences are research-based and reflect the perspectives of the analysts. For example, Foglia et al. [24] reported a modeling project in California that supported the sustainability of environmental goals in a basin. Other researchers propose methods of investigation of aquifer pumping effects on streams [25,26].

Based on the complexity of the issue and the status of the knowledge base, it is apparent that the problem of stream-aquifer water rights administration represents a new frontier for research in the combined arenas of hydrology and water resources management. This paper presents a review of the technical, legal, and management issues involved in the administration of stream-aquifer systems, and provides an example of how governance authorities have approached the issues in a semi-arid irrigated region in the United States, the State of Colorado, which now has more than 50 years of experience with integrated management of surface and groundwater. Success in this management approach is essential in the state's alluvial valleys where most of the population lives, as well as in important agricultural areas where the state must meet interstate compact requirements. The experiences gained in 50 years of management have included engineering, legal, and social adjustments, and regulatory and management approaches have relied heavily on hydrologic and engineering advances, with emphasis on data and modeling. They provide perspectives that may aid water managers and regulators in other regions to compare and adapt their approaches. The experiences are explained to include the setting and explanations of how the stream-aquifer issues emerged, the challenges they created, the impacts, the institutional and hydrologic responses, and the lessons learned.

2. Technical, Legal, and Management Issues of Stream-Aquifer Systems

The technical issues of stream-aquifer systems were explained in an extensive report by the United States Geological Survey, which traced the beginning of the systematic study to 1940 and the work by Theis, who investigated the problem of well-pumping [17]. After an extensive review of the studies of stream-aquifer analysis, the authors concluded that the major issues were identifying sources of water to a well, depletion of streamflow and its time response, effects of confining layers on streamflow depletion, and the distribution of streamflow depletion along stream reaches. Management-related issues stemming from these problems include analysis of the time rates of pumping and well shutdown and the difficult problem of basin-wide analysis, where hundreds, or even thousands, of wells may draw water from hydrologic systems.

The use of models to simulate pumping effects has focused on simplified analytical modeling methods, which cannot represent aquifer heterogeneity or complex stream and aquifer geometries. Numerical methods are better for handling the complexities of groundwater systems and accounting for the effects of pumping histories from large numbers of wells. Their disadvantages are, of course, based on their complexities, cost, and need for accurate data which may not always be available.

The legal complexities of stream-aquifer interconnections are embedded in the hydrogeology, the uncertain flows of groundwater, the difficulty of determining how pumping impacts streams, and other technical questions outlined above. As Dellapenna [23] explained, these legal issues are aligned with questions such as how to limit use to ensure sustainability and how much of the pumped water eventually will find its way back to an aquifer, even if not the aquifer of origin. The legal management issues are to regulate pumping to maintain water tables and return or substitute at least some of the extracted water. The author thought that regulations, while intended to sustain scarce and ambulatory resources, have not been entirely successful. He also observed that slow-developing public awareness about groundwater leads to little thought given to resource management and toward a "tragedy of the commons", where collective action fails to sustain a shared resource. The public awareness problem also saps trust in management institutions, as in a cited case involving local groundwater districts in eastern Colorado and Kansas where some 40% of irrigators doubted whether districts served their interests.

Management issues of stream-aquifer systems are derived from the technical and legal challenges involved, primarily that computation of the exact impacts of pumping on streams is inherently complex and that the legal system, together with its public opinion branch, does not trust experts or the regulatory system to protect their interests. This means that management institutions, such as groundwater districts or river basin organizations, face uphill battles to develop effective approaches and, even then, they must negotiate with the regulatory system, including the justice system, to gain approval for their methods.

3. Stream-Aquifer Regulation within the Water Rights System of Colorado

Colorado is a mid-continent state with an area of 104,094 square miles (269,603 square km) and a population of almost six million [27]. Based on the frequency of reports and coverage by the scientific research base, Colorado's experience with stream-aquifer management is more developed than in other western states. A national map of the United States would show the state just west of the center of the nation. Colorado has several distinct regions comprising the eastern plains, the Front Range, the mountains, and the western plateau (Figure 1). The general north-to-south line of the Continental Divide divides the state generally into east and west slopes. The average annual precipitation is 17 inches (432 mm), which varies markedly between the plains and mountain parts of the state [28]. Runoff from mountain snowpack provides much of the streamflow leaving the state on both sides of the Continental Divide.

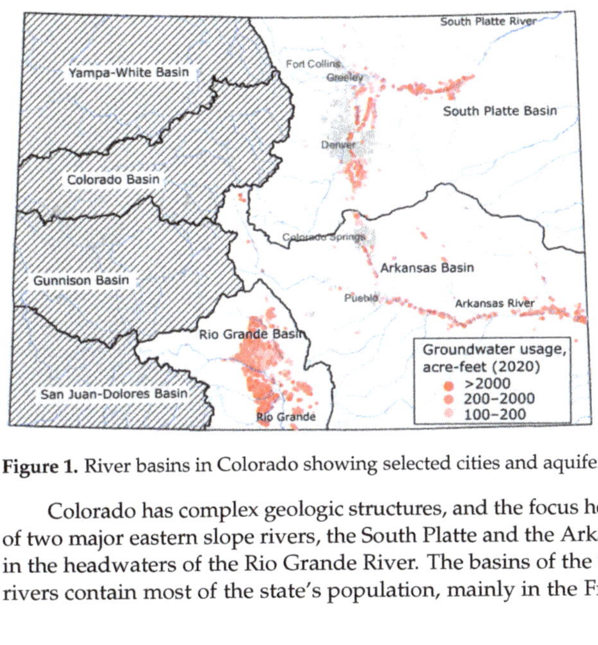

Figure 1. River basins in Colorado showing selected cities and aquifers.

Colorado has complex geologic structures, and the focus here is on the alluvial valleys of two major eastern slope rivers, the South Platte and the Arkansas, and a distinct region in the headwaters of the Rio Grande River. The basins of the South Platte and Arkansas rivers contain most of the state's population, mainly in the Front Range cities shown in

Figure 1. These two basins also contain most of the state's irrigated land and each river has an interstate water compact with neighboring states [29]. In addition to managing water quantities in the rivers, both have water quality issues. The Rio Grande basin is in the San Luis Valley, which is only sparsely populated but has extensive agriculture and water uses with regulatory controls to respond to interstate issues. Figure 1 illustrates the basins west of the Continental Divide as hatched areas and illustrates areas of heavy groundwater pumping in the South Platte, Arkansas, and Rio Grande basins.

The functions of water rights administration in Colorado are explained by authors such as Stenzel and Cech [30], who described the regulatory perspective. In the US, water rights are based primarily on state laws and Colorado follows the appropriation doctrine, sometimes called the "priority doctrine" with "first in time, first in right" governing entitlement to water. The doctrine is embodied in Article XVI of the Colorado Constitution, which was developed at the time of statehood in 1876. Variations of the doctrine are used throughout the drier parts of the western United States. Some states, such as California, use mixed water law systems because they have both arid and humid parts. Colorado's system is sometimes referred to as a pure appropriation doctrine because it applies the same across the state and is based on strict priorities. In 1879, the Colorado Legislature assigned adjudication of water rights to the state courts and created the first water districts and offices of water commissioners to oversee diversions. Soon thereafter, the Legislature created Division Engineers for each river basin and a State Engineer's Office to oversee the statewide system.

The legislation requiring analysis of stream-aquifer connections stems back to the first well developed in the South Platte basin in 1886 using a centrifugal pump driven by a steam engine. This led to the first lawsuit between ground and surface water users, with the court finding the well in violation of the doctrine of prior appropriation. Pumps installed during the period of rural electrification in the 1930s led to much more groundwater pumping. Prior to 1957, no permit was required for a well and there were few court decisions because hydraulic connections were not well understood. This began to change, and the Groundwater Management Act of 1965 affirmed that groundwater users must comply with the appropriation doctrine and required that the State Engineer's office could only permit a well if it would not injure senior rights. The Act also confronted the State Engineer with the difficult task of shutting off wells, although regulation was relatively flexible, at least initially.

The Colorado Legislature passed the more comprehensive Water Rights Determination and Administration Act in 1969 to clarify procedures and provide for improved water administration. This Act organized the seventy water districts and seven water divisions into the present system with a water court and division engineer in each river basin [31]. It defined the roles of water divisions, division engineers, water judges, referees, and water clerks and it set the intent to "integrate the appropriation, use, and administration of underground water tributary to a stream with the use of surface water, in such a way as to maximize the beneficial use of all of the waters of this state".

The 1969 Act provided a more serious approach to integrating wells with the surface water system than had existed before. Specifically, it required the user of tributary groundwater to provide replacement water at times when a water rights call is on the river. The complex system of aquifers requires different approaches to management. Tributary aquifers are considered hydrologically connected such that they influence streamflow directly. For non-tributary aquifers outside of designated basins, the requirement is that pumping will not within 100 years deplete streamflow at a rate greater than 0.1% of the annual rate of withdrawal.

There are eight designated basins with non-tributary aquifers in the eastern plains, which have little surface water. These are managed by 13 Ground Water Management Districts. Another category comprises the not non-tributary aquifers located in the Denver Basin, which is generally under the metropolitan area of the city. These do not influence streamflow as directly as tributary aquifers significantly, and they are regulated differently because of their importance to the economy. Regulation of the Denver Basin requires that at least 4% of pumped water be returned to surface streams. The State Engineer administers non-tributary and Denver Basin groundwater, and the Colorado Groundwater Commission and local groundwater management districts administer the Designated Basins [32].

The South Platte Basin was the first in the state to undergo intensive water development for farming, cities, and industry, and its population growth has been much greater than other basins in the state. The development of groundwater sources has been essential to supplement over-appropriated sources of surface water. In the basin, water is cycled from the river to the fields, to the aquifer, to the river, and then back again repeatedly from Denver to the Nebraska state line. Return flows are determined by farming practices, with high returns from flood irrigation and lower returns from the use of sprinklers. Groundwater pumping from the alluvial aquifer leads to the depletion of river water [17,21], and wells are either shut off or managed through augmentation plans, where recharge ponds mimic return flows to the river that would occur without well pumping. In some cases, canal seepage to the aquifer is used for augmentation credit [33], to allow pumping in subsequent years.

After the first well in the basin in 1886, a lawsuit was filed between ground and surface water users, and the court found that the well was in violation of the doctrine of prior appropriation. Despite this finding, prior to 1957 no permit was required for a well and there were few court decisions because hydraulic connections were not well understood. During the early and mid-20th Century, well development and availability of electric power in rural areas tapped more groundwater, and there was a total of 3200 wells pumping at least 100 GPM (379 L/min) by 1972. Only a fraction of the farms involved reported leasing surface water to compensate, and water right owners began to argue about depletions.

The Groundwater Management Act of 1965, which affirmed that tributary well users must comply with the appropriation doctrine, gave the State Engineer the difficult task of shutting off wells permanently, although initial regulation was flexible. In the 1967 Fellhauer case, the court ruled that the State Engineer had not developed rules and regulations to administer wells with the priority system, and a 1968 Engineering report found that wells were reducing stream flows.

These developments set the stage for the 1969 Act and signaled the beginning of the end of unrestricted groundwater pumping. In this period, the new rules were challenged, and various agreements were made. The legislature gave well owners the right to organize and up to three years to file a plan of augmentation. The State Engineer could continue using substitute water supply plans. Well owners were encouraged to form associations or conservancy districts to comply. In the 1970s two well augmentation groups were created to operate under the substitute water supply plan rule.

The period from 1980 to 2000 was wet, but a severe drought struck in 2000 and the State Engineer began to require replacement of out-of-priority depletions with augmentation plans as calculated by a stream depletion factor method. The record drought and a related lawsuit about augmentation plans led to a greater focus on stream-aquifer interactions and well shutdowns in the basin. As the drought worsened, river calls were earlier and lasted longer than in the past. Agreements were dismantled and some farmers with both wells and surface rights stopped using groundwater to reduce the amount of water they would need for augmentation plans. This resulted in irrigation ditches placing river calls even earlier and caused older patterns of river calls to move earlier in the season, which impacted recharge operations during dry periods [30].

The seeds were planted for discord as increased use of center pivot systems, improved irrigation efficiencies, and greater recapture of imported water began to reduce return flows to the river. The associations that had formed for augmentation started coming apart and some well owners filed individual plans of augmentation and withdrew from the common pool. The associations did not have enough substitute water supplies and had to cease operations.

The lawsuit filed in 2000 to address a situation in the Arkansas Basin affected water management in the South Platte in a major way. Litigation was between the Empire Lodge Homeowners Association and the Moyers family, and it addressed the State Engineer's authority to approve substitute water supply plans. The outcome, together with the drought led to the shutdown of about 2000 tributary groundwater irrigation wells. This had severe social and economic impacts as tens of thousands of acres (hectares) of productive agricultural lands reverted to dryland crops or weeds [34,35].

As evident in the South Platte Basin, the parallel evolution of water demands, drought, water rights regulation, and modeling technologies illustrates a hydro-legal approach to confronting stream-aquifer interactions. This approach is illustrated vividly in a major ruling by the US Supreme Court (SCOTUS) about water management in the Arkansas River Basin.

In the Arkansas River Basin, stream-aquifer interconnections are like those in the South Platte, but all water is managed under an interstate compact with Kansas. The basin has areas of high groundwater levels caused by inadequate drainage [36], but any change in irrigation practices from flood to sprinkler systems cannot modify the amount of river water delivered to Kansas.

The Arkansas River Compact between Colorado and Kansas signed in 1949 led to controversies over stream-aquifer modeling [37]. A long period of disputes began in 1902 with Kansas' complaints that Colorado was using excessive amounts of water. The SCOTUS ruled that no decision could be made without further evidence. Subsequently, the states tried different legal maneuvers and the lack of definite outcomes and high costs of litigation led to the 1949 Compact. In 1985, Kansas filed a complaint that Colorado had overused their compact share of water because groundwater pumps installed after 1949 and a Winter Water Storage Program had depleted river flows. In 1995 a special master ruled that groundwater pumping violated the compact, and the SCOTUS accepted his verdict. Colorado has assessed damages for the excess water use.

In subsequent rulings, the court decided in 2009 that a Hydrologic-Institutional Model (H-I Model) was required to determine depletions and accretions to flow into Kansas caused by groundwater pumping and replacement. The model simulates the hydrologic and institutional systems that influence flow from Pueblo, Colorado to the state line and estimates the flow of the Arkansas River that would have occurred in the absence of post-compact well pumping [38]. The surface water simulation includes native streamflow, transmountain deliveries, tributary inflows, reservoir operations, and irrigation diversions, while the groundwater component includes the effects of well pumping and recharge. The institutional system modeled includes rules for the diversion of surface water, operation of John Martin Reservoir, resolutions of the Arkansas River Compact Administration, operation of other off-channel reservoirs, and replacement operations.

There was considerable discussion of the model in the case. The Special Master explained that modeling the Arkansas River Basin is extraordinarily complex. Despite continuing refinements, the Special Master held that the model is not accurate on an annual basis and adopted a proposal from Colorado that results be applied over a ten-year period to average out errors. Also, he advised that continuing needs to update the model would be required, although the states might disagree on changes to be made. There was discussion of whether a River Master should be appointed to resolve modeling issues and make other decisions, but this did not occur. Water management and modeling decisions were left to the Arkansas River Compact Administration, although this was not considered an ideal solution due to the possibility of future disputes [37].

Water management in the Upper Rio Grande Basin in Colorado's San Luis Valley has focused on meeting agricultural needs and compliance with two interstate compacts. Compared to the South Platte and Arkansas basins, it has much less urban and industrial demands [39]. The Valley is a major potato-producing region, and water use has been ruled unsustainable. The Valley only receives 7 to 10 inches (178 to 254 mm) of precipitation per year, so surface runoff lacks an abundant supply, but a vast aquifer system is linked to the river. As well pumping increased, aquifer levels and river flows declined to the point that Colorado's ability to deliver compact water to New Mexico and Texas was threatened [40].

The Rio Grande Basin in the Valley has several productive alluvial aquifers with saturated thicknesses of several kilometers [41]. The confining units are as shallow as 20 m in some parts of the Valley. This shallow confined aquifer spurred the development of early artesian wells in the late 1800s for irrigation, and with electric power, many more were drilled later. The interstate issues of the basin date back to the Rio Grande Compact of 1939, which set required flows from Colorado to New Mexico and then to Texas. By 1965, Colorado had accumulated a water debt of roughly one million acre-feet of water, resulting in severe curtailment of water use in the valley. In response, the State Engineer placed moratoriums on new wells in the confined aquifer in 1972 and on those in the primary unconfined aquifer in 1981 [42]. In the 1990s, speculative proposals to pump water out of the valley to the front range began. The first proposal was denied in court because groundwater was considered tributary to natural streams. Later proposals have also failed, but water speculation efforts continue. The 2002 drought prompted further legislation to avoid depletion in the valley. Subdistricts were formed to remedy injurious groundwater depletions and replenish surface water to fulfill Colorado's compact obligation. Farmers have worked voluntarily since then to reduce groundwater use and restore Rio Grande surface flows by establishing fees with revenues used to retire farmland.

A groundwater model to quantify the impact of withdrawals on streamflow depletion was initiated during the 1990s [43]. New rules in 2004 led to the Rio Grande Decision Support System (RGDSS) study to model impacts on its artesian pressures. To support the RGDSS, the State Engineer released a Groundwater Measurement Rule in 2005 to require wells in Water Division 3 to meter water and report annual water pumped from large wells. In some parts of the Valley, the hydrogeology is not understood well enough to model accurately with the RGDSS [40,42].

4. Modeling Issues in Support of Water Rights Administration

While the State Engineer's office was required to administer interconnected stream-aquifer systems of water rights by the 1969 Act, it lacked hydrologic models and the databases were inadequate. The quantitative method used at the time was the analytical Glover-Balmer model [44], which estimates the time-dependent flow rate and volume of stream water that enters the aquifer due to pumping from a full-penetrating groundwater well. The model has been popular since its inception [45] due to its ease of use, as calculations can be implemented in spreadsheets or even by hand. The Glover-Balmer model has many simplifying assumptions, such as a straight-line river, homogeneous aquifer, fully penetrating well, fully penetrating river, and the same hydraulic conductivity for the aquifer and streambed. Later models were published to relax some of the assumptions with modifications [46,47]. These models can be used in spatial and temporal superposition to estimate streamflow depletion from multiple pumping wells and from a time series of pumping rates, including the amount of streamflow depletion that occurs after the pump has been shut off.

Due to inherent assumptions in their derivation, these analytical models are too simplistic for most situations, especially when pumping wells are more than a few kilometers from streams [48]. However, because of their ease of use, such as implementation in spreadsheet format for water managers to use, they have been accepted as "correct" models in water rights cases and augmentation plans. A typical application would be to use the analytical model in a spreadsheet to estimate the volume of streamflow depletion through time and then model where recharge ponds should be located to mimic return flows that would occur without pumping.

Researchers continued to develop new groundwater models based on solutions of partial differential equations using finite difference and finite element methods. A promising effort over a 15-year period in the 1970s and 1980s developed the Stream-Aquifer Model for Management by Simulation (Samson Model) [49] for the South Platte Basin. By the

1990s, modeling had advanced to become effective as a tool for management decisions. This was the beginning of the development of Decision Support Systems by the Colorado State Government for each river basin.

Numerical models such as MODFLOW and HydroGeoSphere are time-consuming to create, but they are more realistic for the simulation of physical processes than analytical models [9,33]. They have been applied in the South Platte Basin for research and via the SPDSS, as well as in the Arkansas River Basin [36] to investigate the impacts of management strategies on groundwater levels and return flows, and to quantify the impact of future climate change [50]. These physically based spatially distributed models can be used effectively to investigate groundwater-surface water interactions and management. However, defending complex numerical models in water courts can be a daunting task and, due to decades of use and acceptance, analytical models continue to be used for water rights cases and streamflow depletion calculations. In some venues, the prospect of developing numerical models is daunting due to their costs, and another challenge is to apply them to complex hydrogeological situations. This appears to be another situation where technology outpaces implementation, and gaps will probably widen when artificial intelligence and data-driven models are available.

The early stages of model development were limited by a lack of data, particularly groundwater withdrawals. Data to validate models and prepare augmentation plans were scarce and not well organized. Colorado State University digitized water rights records from the State Engineers Office, and this effort led eventually to the state's current HydroBase database [51]. Lawsuits prompted by a drought of 2002 led to mandatory metering of all non-exempt wells in the South Platte, Arkansas, and Rio Grande Basins, and resulted in high-quality groundwater data.

Data issues for testing of models involve groundwater levels, streamflow, and depletion volumes. Surface water data are from US Geological Survey (USGS) gages. Groundwater data include wells belonging to the USGS wells and Colorado Division of Water Resources wells. While these are increasing in number, more are always needed. Groundwater withdrawal data come via metering of use, which was limited until 2005–2010 when lawsuits from the 2002 drought led to their requirement.

Due to the limited data, consumptive use models were used to include the location of irrigated fields and crop water demand [52]. Stream depletion from these is difficult to estimate because the magnitude of pumped water is small compared to streamflow. A study of a short stretch of the South Platte River showed that, although depletion from high-capacity pumping wells occurs, the volumes can only be quantified confidently at lower river flows of the same order of magnitude as the pumping rates [53]. For larger rivers with high flows, the volume of pumping-induced streamflow depletion is hard to measure with field instruments.

5. Discussion, Conclusions, and Future Directions

Water interactions in stream-aquifer systems pose significant challenges in semi-arid areas like Colorado, and their complexities will become evident in many other places as groundwater management challenges are exacerbated by problems due to over-pumping and climate change. Confronting these challenges successfully will require effective management and regulatory systems that use advanced tools and valid data to inform decisions by managers and governance authorities.

The review of Colorado's experiences demonstrates how models and data management can be improved, but many obstacles remain in place. It shows how continued work is required to address the complexities of local and regional hydrogeology and institutional systems that are adapting to the challenges of water management. Issues to confront include adequate groundwater data, effective modeling strategies, organizational arrangements to facilitate collective action, building water management capacity, and legal challenges.

In Colorado, groundwater data are improving, but the complex hydrogeology and diverse types of aquifers continue to present formidable challenges. Data mandated in court settlements is especially useful when it is tested in adversarial legal proceedings, but research-based data can also be quite useful, especially when published in peer-reviewed venues. Compiling these data in ways to support management and regulatory systems presents organizational and financial challenges stemming from unclear responsibilities, fragmentation, and the cost of data management systems.

Stream-aquifer models keep improving and evolving decision support systems provide visual displays and tools to enable stakeholders to examine management scenarios. However, data uncertainties and the hidden nature of model coding create doubt in the minds of referees and judges and can explain why, in Colorado, the courts have not accepted advanced numerical models and continue to rely on older analytical models that are simpler but not able to simulate complex systems well.

Management districts, such as those organized in the San Luis Valley, provide a means to use common pool resource principles to manage water resources. In such local districts, the introduction of innovative tools such as aquifer-storage-recovery systems or management of well-pumping to stabilize stream flows may help. While this approach may work in relatively small areas, it faces challenges when stream-aquifer interactions extend over large distances in priority-based water right systems. The use of larger-scale models and coordinated management and regulatory approaches will be needed.

Colorado's experience with a court-based regulatory system connects water administrators in partnership with the legal system. Once a management plan is approved through the courts, the State Engineer's Office has a set of rules to follow and the complexity to administer water is reduced, but the approval process and the possibility of later legal challenges pose challenges to the reliability and trust in the regulatory system.

In Colorado, the historical legal controls on water management were established coincidentally with the appropriation doctrine of water rights. Subsequently, generations of water lawyers have become familiar with hydrology and water management strategies and engineers, modelers, and regulators have learned a great deal about water laws. The solution of the problems that block the acceptance of advanced model strategies will depend on continued joint work between the technical and legal communities. To support this work, researchers can demonstrate the superiority of newer models over older ones by using court case outcomes to demonstrate the impacts of errors in using outdated methods.

Looking to the future, water managers and their hydrologic support staff will face the technical, legal, and management quandaries outlined in the paper. Hydrogeologic diversity, lack of data, and modeling complexities and expenses will inhibit the application of the most advanced technologies to regulatory and management control schemes. The slow evolution of legal approaches to confront technical and environmental problems will mean that water managers cannot expect solutions from the justice system, and they must continue to innovate with new approaches without assurance of acceptance of their methods. Meanwhile, the older analytical approaches will not go away and may provide useful constructs to explain to the public about their choices and the limits of the regulatory system. Rapidly advancing data sources based on remote sensing, along with data methods based on machine learning and artificial intelligence may supplant some of the numerical models of today. As water becomes scarcer and our shared dependence on it intensifies, the methods and issues outlined in the paper will come increasingly into focus in the water management community.

Author Contributions: Conceptualization, N.S.G., R.T.B. and R.G.S.; methodology, N.S.G., R.T.B. and R.G.S.; software, NA; validation, NA; formal analysis, NA; investigation, NA; resources, NA; data curation, NA; writing—original draft preparation, N.S.G.; writing—review and editing, R.T.B. and R.G.S.; visualization, NA; supervision, NA; project administration, NA; funding acquisition, NA. All authors have read and agreed to the published version of the manuscript.

Funding: This research received no external funding.

Data Availability Statement: No data were developed in conjunction with the study.

Conflicts of Interest: The authors declare no conflict of interest.

References

1. Konikow, L.F.; Kendy, E. Groundwater depletion: A global problem. *Hydrogeol. J.* **2005**, *13*, 317–320. [CrossRef]
2. Famiglietti, J.S.; Lo, M.; Ho, S.L.; Bethune, J.; Anderson, K.J.; Syed, T.H.; Swenson, S.C.; de Linage, C.R.; Rodell, M. Satellites measure recent rates of groundwater depletion in California's Central Valley. *Geophys. Res. Lett.* **2011**, 38. [CrossRef]
3. Scanlon, B.R.; Faunt, C.C.; Longuevergne, L.; Reedy, R.C.; Alley, W.M.; McGuire, V.L.; McMahon, P.B. Groundwater depletion and sustainability of irrigation in the US High Plains and Central Valley. *Proc. Natl. Acad. Sci. USA* **2012**, *109*, 9320–9325. [CrossRef] [PubMed]
4. Konikow, L.F. Long-Term Groundwater Depletion in the United States. *Groundwater* **2015**, *53*, 2–9. [CrossRef]
5. Gorelick, S.M.; Zheng, C. Global change and the groundwater management challenge. *Water Resour. Res.* **2015**, *51*, 3031–3051. [CrossRef]
6. Perrone, D.; Jasechko, S. Deeper well drilling an unsustainable stopgap to groundwater depletion. *Nat. Sustain.* **2019**, *2*, 773–782. [CrossRef]
7. Cobby, D.; Morris, S.; Parkes, A.; Robinson, V. Groundwater flood risk management: Advances towards meeting the requirements of the EU floods directive. *J. Flood Risk Manag.* **2009**, *2*, 111–119. [CrossRef]
8. Abboud, J.; Ryan, M.; Osborn, G. Groundwater flooding in a river-connected alluvial aquifer. *J. Flood Risk Manag.* **2018**, *11*, e12334. [CrossRef]
9. Deng, C.; Bailey, R.T. Assessing causes and identifying solutions for high groundwater levels in a highly managed irrigated region. *Agric. Water Manag.* **2020**, *240*, 106329. [CrossRef]
10. Merchán-Rivera, P.; Geist, A.; Disse, M.; Huang, J.; Chiogna, G. A Bayesian framework to assess and create risk maps of groundwater flooding. *J. Hydrol.* **2022**, *610*, 127797. [CrossRef]
11. Smith, R.G.; Majumdar, S. Groundwater Storage Loss Associated with Land Subsidence in Western United States Mapped Using Machine Learning. *Water Resour. Res.* **2020**, *56*, e2019WR026621. [CrossRef]
12. Hasan, F.; Smith, R.; Vajedian, S.; Pommerenke, R.; Majumdar, S. Global land subsidence mapping reveals widespread loss of aquifer storage capacity. *Nat. Commun.* **2023**, *14*, 6180. [CrossRef] [PubMed]
13. Pauloo, R.A.; Fogg, G.E.; Guo, Z.; Harter, T. Anthropogenic basin closure and groundwater salinization (ABCSAL). *J. Hydrol.* **2021**, *593*, 125787. [CrossRef]
14. Fleckenstein, J.H.; Krause, S.; Hannah, D.M.; Boano, F. Groundwater-surface water interactions: New methods and models to improve understanding of processes and dynamics. *Adv. Water Resour.* **2010**, *33*, 1291–1295. [CrossRef]
15. Lewandowski, J.; Meinikmann, K.; Krause, S. Groundwater–Surface Water Interactions: Recent Advances and Interdisciplinary Challenges. *Water* **2020**, *12*, 296. [CrossRef]
16. Graaf, I.E.M.d.; Stahl, K. A model comparison assessing the importance of lateral groundwater flows at the global scale. *Environ. Res. Lett.* **2022**, *17*, 044020. [CrossRef]
17. Barlow, P.M.; Leake, S.A. *Streamflow Depletion by Wells: Understanding and Managing the EFFECTS of Groundwater Pumping on Streamflow*; US Geological Survey: Reston, VA, USA, 2012.
18. Condon, L.E.; Maxwell, R.M. Simulating the sensitivity of evapotranspiration and streamflow to large-scale groundwater depletion. *Sci. Adv.* **2019**, *5*, eaav4574. [CrossRef] [PubMed]
19. Liu, W.; Park, S.; Bailey, R.T.; Molina-Navarro, E.; Andersen, H.E.; Thodsen, H.; Nielsen, A.; Jeppesen, E.; Jensen, J.S.; Jensen, J.B.; et al. Comparing SWAT with SWAT-MODFLOW hydrological simulations when assessing the impacts of groundwater abstractions for irrigation and drinking water. *Hydrol. Earth Syst. Sci. Discuss.* **2019**, *2019*, 1–51.
20. Lapides, D.A.; Maitland, B.M.; Zipper, S.C.; Latzka, A.W.; Pruitt, A.; Greve, R. Advancing environmental flows approaches to streamflow depletion management. *J. Hydrol.* **2022**, *607*, 127447. [CrossRef]
21. Zipper, S.C.; Farmer, W.H.; Brookfield, A.; Ajami, H.; Reeves, H.W.; Wardropper, C.; Hammond, J.C.; Gleeson, T.; Deines, J.M. Quantifying Streamflow Depletion from Groundwater Pumping: A Practical Review of Past and Emerging Approaches for Water Management. *JAWRA J. Am. Water Resour. Assoc.* **2022**, *58*, 289–312. [CrossRef]
22. European Commission. World Atlas of Desertification. 2023. Available online: https://wad.jrc.ec.europa.eu/irrigations (accessed on 20 October 2023).
23. Dellapenna, J.W. Dellapenna, A Primer on Groundwater Law. *Idaho L. Rev.* **2013**, *49*, 265.
24. Foglia, L.; Neumann, J.; Tolley, D.G.; Orloff, S.G.; Snyder, R.L.; Harter, T. Modeling guides groundwater management in a basin with river–aquifer interactions. *Calif. Agric.* **2018**, *72*, 84–95. [CrossRef]
25. Davids, J.C.; Mehl, S.W. Sustainable Capture: Concepts for Managing Stream-Aquifer Systems. *Groundwater* **2014**, *53*, 851–858. [CrossRef] [PubMed]
26. Malama, B.; Pritchard-Peterson, D.; Jasbinsek, J.J.; Surfleet, C. Assessing Stream-Aquifer Connectivity in a Coastal California Watershed. *Water* **2021**, *13*, 416. [CrossRef]
27. Colorado State Demography Office. Demographic Summary. 25 February 2022. Available online: https://storymaps.arcgis.com/stories/e840d22c99334901a42c7cd1781183bf (accessed on 20 October 2023).
28. Colorado Climate Center. Climate of Colorado. 2023. Available online: https://climate.colostate.edu/co_cag/index.html (accessed on 20 October 2023).
29. Colorado Water Conservation Board. Technical Update to the Water Plan. 2023. Available online: https://cwcb.colorado.gov/colorado-water-plan/technical-update-to-the-plan (accessed on 20 October 2023).

30. Stenzel, R.; Cech, T.V. *Water: Colorado's Real Gold. A History of the Development of Colorado's Water, the Prior Appropriation Doctrine and the Division of Water Resources*; Richard Stenzel: Denver, CO, USA, 2013; ISBN 978-0-9887170-0-8.
31. Hobbs, G.J., Jr. Colorado's 1969 Adjudication and Administration Act: Settling In. *Univ. Denver Water Rev.* **1999**, *3*, 1–18. Available online: https://www.courts.state.co.us/userfiles/File/Court_Probation/Supreme_Court/Committees/Water_Court_Committee/Hobbs_1969_Act_CLE_January_20_2007.pdf (accessed on 20 October 2023).
32. Colorado State University. Groundwater Rights. 2023. Available online: https://waterknowledge.colostate.edu/water-management-administration/water-rights/groundwater-rights/ (accessed on 20 October 2023).
33. Deng, C.; Bailey, R.T. Assessing the Impact of Artificial Recharge Ponds on Hydrological Fluxes in an Irrigated Stream–Aquifer System. *Hydrology* **2022**, *9*, 91. [CrossRef]
34. Justia US Law. Empire Lodge Homeowners' Assn. v Moyer. 2023. Available online: https://law.justia.com/cases/colorado/supreme-court/2002/00sa211-0.html (accessed on 20 October 2023).
35. Cech, T. Environmental Reviews & Case Studies: Surface Water and Groundwater Interaction, Management, and Conflict in Colorado: Alarming Trends for the 21st Century. *Environ. Pract.* **2010**, *12*, 304–315. [CrossRef]
36. Morway, E.D.; Gates, T.K.; Niswonger, R.G. Appraising options to reduce shallow groundwater tables and enhance flow conditions over regional scales in an irrigated alluvial aquifer system. *J. Hydrol.* **2013**, *495*, 216–237. [CrossRef]
37. US Department of Justice. Original: Kansas v Colorado—Opposition. Brief for the United States in Opposition to the Exceptions of Kansas. 2023. Available online: https://www.justice.gov/osg/brief/original-kansas-v-colorado-opposition (accessed on 20 October 2023).
38. U.S. Supreme Court. Amended Appendix C.1. Hydrologic—Institutional Model Model Documentation. As Amended September 2011. Available online: https://www.supremecourt.gov/specmastrpt/Amended_Appendix_C-1_9_2011.pdf (accessed on 20 October 2023).
39. Colorado Water Conservation Board. Rio Grande Basin Implementation Plan. 2023. Available online: https://dnrweblink.state.co.us/cwcbsearch/0/edoc/216713/RioGrande_BIP_Volume1_2022.pdf (accessed on 20 October 2023).
40. Coleman, C. Hundreds of San Luis Valley Farm Wells at Risk as State Shortens Deadline to Repair Rio Grande River. 2020. Available online: https://www.watereducationcolorado.org/fresh-water-news/hundreds-of-san-luis-valley-farm-wells-at-risk-as-state-shortens-deadline-to-repair-rio-grande-river/ (accessed on 20 October 2023).
41. Bexfield, L.M.; Anderholm, S.K. *Section 10—Conceptual Understanding and Groundwater Quality of the Basin-Fill Aquifer in the San Luis Valley, Colorado and New Mexico*; US Geological Survey Professional Paper; US Geological Survey: Reston VA, USA, 2010; pp. 165–187.
42. King, N.; May, C.J.; Parker, M.; Sargent, V.; Schaffer, L. Water Exports and the San Luis Valley in Colorado: Understanding the History and Current Regulatory Framework. 2020. Available online: https://www.getches-wilkinsoncenter.cu.law/wp-content/uploads/2020/12/Water-Exports-and-the-San-Luis-Valley-in-Colorado.pdf (accessed on 20 October 2023).
43. Malers, S.A.; Bennett, R.R.; Nutting-Lane, C. Colorado's decision support systems: Data-centered water resources planning and administration. In *Watershed Management and Operations Management*; American Society of Civil Engineers: Reston, VA, USA, 2001; pp. 1–9. [CrossRef]
44. Glover, R.E.; Balmer, G.G. River depletion resulting from pumping a well near a river. *Trans. Am. Geophys. Union* **1954**, *35*, 468–470.
45. Zipper, S.C.; Gleeson, T.; Kerr, B.; Howard, J.K.; Rohde, M.M.; Carah, J.; Zimmerman, J. Rapid and Accurate Estimates of Streamflow Depletion Caused by Groundwater Pumping Using Analytical Depletion Functions. *Water Resour. Res.* **2019**, *55*, 5807–5829. [CrossRef]
46. Hantush, M.S. Wells near streams with semipervious beds. *J. Geophys. Res.* **1965**, *70*, 2829–2838. [CrossRef]
47. Hunt, B. Unsteady Stream Depletion from Ground Water Pumping. *Groundwater* **1999**, *37*, 98–102. [CrossRef]
48. Sophocleous, M.; Koussis, A.; Martin, J.L.; Perkins, S.P. Evaluation of Simplified Stream-Aquifer Depletion Models for Water Rights Administration. *Groundwater* **1995**, *33*, 579–588. [CrossRef]
49. Illangasekare, T.H.; Morel-Seytoux, H.J. A discrete kernel simulation model for conjunctive management of a stream—Aquifer system. *J. Hydrol.* **1986**, *85*, 319–338. [CrossRef]
50. Aliyari, F.; Bailey, R.T.; Arabi, M. Appraising climate change impacts on future water resources and agricultural productivity in agro-urban river basins. *Sci. Total. Environ.* **2021**, *788*, 147717. [CrossRef]
51. Colorado Division of Water Resources. Data & Information. 2023. Available online: https://dwr.colorado.gov/services/data-information (accessed on 20 October 2023).
52. Colorado's Decision Support Systems. Consumptive Use (StateCU). 2023. Available online: https://cdss.colorado.gov/modeling-data/consumptive-use-statecu (accessed on 20 October 2023).
53. Flores, L.; Bailey, R.T.; Kraeger-Rovey, C. Analyzing the Effects of Groundwater Pumping on an Urban Stream-Aquifer System. *JAWRA J. Am. Water Resour. Assoc.* **2020**, *56*, 310–322. [CrossRef]

hydrology

MDPI

Article

Integrated Exploitation of Rainwater and Groundwater: A Strategy for Water Self-Sufficiency in Ca Mau Province of the Mekong Delta

Dang Hoa Vinh [1], Dung Duc Tran [2,*], Dao Dinh Cham [3,4], Phan Thi Thanh Hang [3,4], Duong Ba Man [1], Danh Mon [1], Luu Hai Tung [1], Le Van Kiem [5], Thien Duc Nguyen [2] and Duong Thi Ngoc Tuyen [6]

1 HCMC Institute of Resources Geography, Tay Nguyen Institute of Scientific Research, Vietnam Academy of Science and Technology, Ho Chi Minh City 700000, Vietnam; dhvinh@hcmig.vast.vn (D.H.V.); dmon@hcmig.vast.vn (D.M.); lhtung@hcmig.vast.vn (L.H.T.)

2 Center of Water Management and Climate Change, Institute of Environment and Natural Resources, Vietnam National University Ho Chi Minh City, Ho Chi Minh City 700000, Vietnam; thienduc295@gmail.com

3 Graduate University of Science and Technology, Vietnam Academy of Science and Technology, Hanoi 100000, Vietnam; ddcham@ig.vast.vn (D.D.C.); hangphanvn95@gmail.com (P.T.T.H.)

4 Institute of Geography, Vietnam Academy of Science and Technology, Hanoi 100000, Vietnam

5 Southern Institute for Water Resources Research, Ho Chi Minh City 700000, Vietnam; kiemsiwrr@gmail.com

6 Department of Natural Resources and Environment of Ca Mau Province, Ca Mau 970000, Vietnam; tuyensotnmt@gmail.com

* Correspondence: dungtranducvn@yahoo.com

Abstract: Groundwater sources have been exploited excessively for numerous purposes worldwide, leading to increasingly severe depletion. However, the replenishment of groundwater sources has not usually been a focus in economically and socially underdeveloped countries and regions. In coastal provinces of the Vietnamese Mekong Delta (VMD), rural areas are facing difficulties in accessing fresh water due to shortages from the water supply plant and excessive use of groundwater, highlighting an urgent need for sustainable development solutions. Our study first conducted interviews with 200 households in Ca Mau Province of the VMD to identify the current situation and the challenges and obstacles of rainwater harvesting and to find sustainable and proactive solutions. We then analyzed daily rainfall data from 10 meteorological stations to construct four scenarios of the water balance method: (i) potential rainwater harvesting based on existing roof area; (ii) optimal scale of storage tank and catchments for different levels of water usage; (iii) tank scale utilizing rainwater entirely during the rainy season and basic needs during the dry season; and (iv) integrated water supply between rain and groundwater. The results showed that using rainwater entirely for domestic water supply requires large storage tank capacities, making these scenarios difficult to achieve in the near future. Our research introduces a novel integrated water supply approach to storing rain and groundwater that has demonstrated high effectiveness and sustainability. With existing tank capacities (0.8 m^3 per person), rainwater could only meet over 48% (14 m^3 per year) of the water demand while requiring 14.8 m^3 of additional groundwater extraction. With a tank capacity of 2.4 m^3 per person, ensuring rainwater harvesting meets basic demand, harvested rainwater could satisfy 64% of the demand, with artificial groundwater supplementation exceeding 1.79 times the required extraction, while excess rainwater discharge into the environment would be minimal. Our research results not only provide potential solutions for rainwater and groundwater collection to supplement sustainable domestic water sources for Ca Mau but also serve as an example for similar regions globally.

Keywords: rainwater; groundwater; harvesting; water balance; Mekong

1. Introduction

Clean, fresh water is invaluable for human life and development. Hence, it is essential to safeguard human society from the impacts of water scarcity (i.e., droughts) or excess (i.e., floods), which could contaminate water quality and change its quantity [1]. The process of urbanization, coupled with population explosions, is driving up the increasing demand for clean water globally [2–4]. Meanwhile, climate change has increased its impacts by altering rainfall patterns, increasing temperatures, and negatively affecting water sources, leading to serious consequences for water security and causing severe livelihood disruptions, particularly for impoverished populations in particularly rural and coastal areas [5]. Despite the attention and research on water scarcity or excess and clean water issues [2,6–8], the consequences are escalating, demanding effective solutions, especially in poorer and developing countries and particularly in rural and coastal rural areas facing challenges such as climate change and sea-level rise driving environmental issues like salinity intrusion [9–11].

Groundwater (GW) is one of the most vital water sources in rural and coastal areas where accessing freshwater sources poses challenges [12]. However, GW extraction has been continuously increasing, reaching a rate of 1500 km^3 per year across countries worldwide [13], exceeding natural replenishment levels and causing significant global-scale impacts such as GW level depletion, land subsidence, and the contamination of aquifers [14–16]. Meanwhile, rainwater harvesting is perceived as a sustainable solution to obtain clean water at low costs and with minimal energy consumption [17,18], yet it remains an underappreciated water source. For instance, rainwater harvesting has been commonly applied in Australia and countries across Asia (i.e., Japan, China) and Africa (i.e., Botswana, Ethiopia, Kenya) [19]. Due to the unpredictable nature of rainfall in terms of onset, duration, and withdrawal, rainwater harvesting often faces challenges in meeting humans' water demands [20]. Therefore, GW and rainwater are considered important backup water sources with untapped potential, mostly during the dry season, and require management, protection, and renovation solutions.

Recognizing the importance of GW to meet future needs, the management, protection, and replenishment of GW resources are seriously considered [21]. Among management techniques, Managed Aquifer Recharge (MAR) has been globally implemented [22,23], treating GW as a reserve water bank and tightly managing aquifers to ensure replenishment. Several studies have outlined the benefits of MAR, not only for cost-effective water use but also for ensuring the secure stability of future water sources [24–26]. Water bank models, achievements, and challenges in MAR implementation have been introduced by Niazi et al. [27] and Arshad et al. [28], who suggested that GW utilization through MAR is more economical than surface water reservoirs, although the treatment costs are relatively high in cases of insufficient clean water sources for supplementation.

Ca Mau is the southernmost coastal province of the Vietnamese Mekong Delta (VMD) (Figure 1), where access to surface water sources during the dry season is challenging [29]. In the province, the primary water source for domestic needs is groundwater; however, it is currently facing severe depletion, evidenced by land subsidence (0.5 to 4 cm annually), declining water levels, and aquifer contamination [30,31]. In addition, the freshwater shortage has been exacerbating due to salinity intrusion caused by sea level rise driven by climate change [32]. According to Bruno et al. [33], climate change-induced sea level rise would inundate a substantial portion of the province, encompassing roughly 43% to 75% of its land area in the years 2030 and 2050, respectively. Thus, the abundant amount of rainfall within Ca Mau Province is considered a valuable resource that assists people in fulfilling their water needs. Rainfall on rooftops is deemed adequate to supply domestic water needs [20]; however, rainwater harvesting is currently limited across the province and does not fully utilize the available potential.

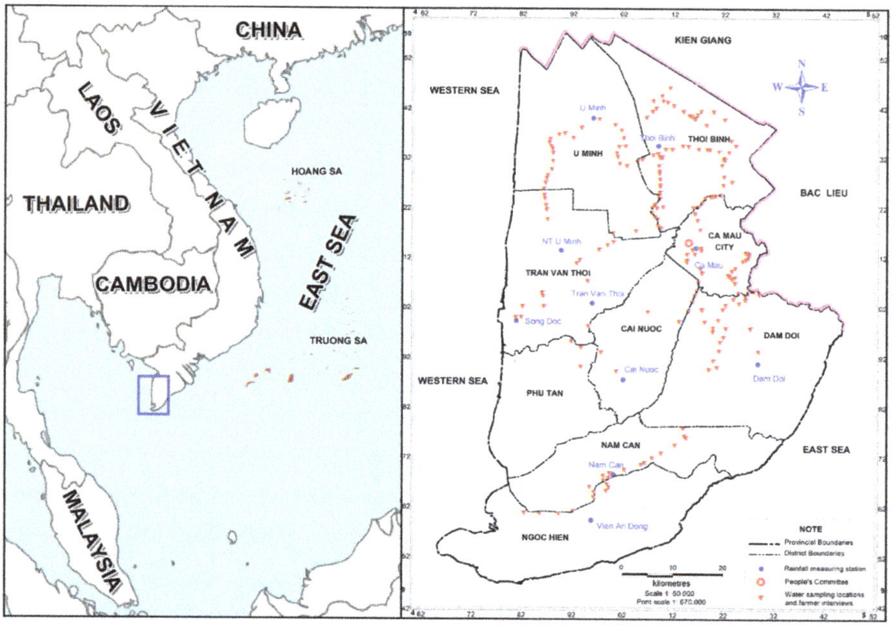

Figure 1. Map depicting the research area location of Ca Mau Province within the Mekong Delta, Vietnam (VMD) (**left**) and Ca Mau Province (**right**), illustrating provincial boundaries, districts, the locations of 10 rainfall monitoring stations, and 200 interviewing locations.

Recent studies have investigated the negative impacts of groundwater overexploitation, such as land subsidence in Ca Mau Province, and the need to sustain this resource [34,35]. Restricting groundwater usage while improving rainwater harvesting could alleviate depletion effects, but it remains underexplored in the province and most other coastal provinces in the VMD. In addition, our interdisciplinary approach, comprising a survey of 200 households and water balance calculations using rainfall data analysis from 10 meteorological stations in Ca Mau, aims to devise an integrated rainwater and groundwater strategy. This strategy considers rainwater (a renewable resource) as the primary water source, while groundwater becomes a backup source for cases of freshwater shortage [36]. The backup water source is protected and rejuvenated by artificially recharging groundwater, such as through MAR, using excess rainwater from its harvesting system. We hence address three main objectives, including (i) analyzing and evaluating the potential for rainwater harvesting (catchment and storage capacity), identifying obstacles, and assessing the willingness of the community to enhance rainwater harvesting; (ii) calculating rainwater harvesting options, implementation methods, and technical requirements; and (iii) quantifying excess rainwater in the system for artificial groundwater recharge. Our study offers a practical method and evidence for constructing an effective water storage strategy through rainwater harvesting to address freshwater shortages driven by salinity intrusion in the VMD provinces and similar regions globally, especially during the dry season.

2. Materials and Methods

2.1. Study Area and Data

Ca Mau covers nine (9) districts and is one of the seven (7) coastal provinces (i.e., Kien Giang, Bac Lieu) located in the Vietnamese Mekong Delta (Figure 1), with a total natural area of 5242.5 km^2, equivalent to 13.1% of the delta region's area and 1.58% of the national territory. It has a coastline stretching over 254 km, bordering both the East Sea and the West Sea, making it an advantageous location for seaport development. With a population of 1,194,476 people (2019), 73% settle in rural areas engaged in year-round brackish aquaculture (mainly shrimp) and paddy rice cultivation during the rainy season.

Annually, the dry season typically spans from January to May, while the rainy season extends from June to December. Fortunately, the province experiences an average annual rainfall of approximately 2247 mm. Nevertheless, many areas contend with a shortage of clean and fresh water during the dry season, leading many households to extract freshwater from underground aquifers. Located in a coastal zone, salinity usually intrudes further inland to affect aquaculture and agricultural activities across the province [32,33].

Daily rainfall data from 10 monitoring stations located within Ca Mau were provided by the Southern Regional Hydro-meteorological Center in Vietnam. To ensure consistency over time, the study selected a calculation timeframe for the entire dataset from 1989 to 2017. All the data used in the study were checked for reliability using the T-test method and tested for homogeneity using the one-sign analysis of variance [37]. After that, a statistical descriptive method was applied to determine the characteristics of the rainfall data. Details of the methods can be found in a previous study by Dang et al. [20].

2.2. Methods

2.2.1. Interview Survey

A survey of 200 households was conducted in April 2020 to mainly gather information on the current status of rainwater harvesting systems, the groundwater exploitation situation, and the obstacles they have faced (Figure 1, right map). We used stratified random sampling to interview households' heads. During the survey, three social groups, including poor households, middle-income households, and relatively affluent households, were selected as the main interview subjects. The purpose of dividing the households into the three various categories was to objectively understand the use of domestic water and rainwater collection practices within the communities. In addition, information about the perceptions of people with different economic capacities was expected to provide evidence supporting the feasibility of implementing a new rainwater harvesting strategy.

The interview questionnaires focused on clarifying key points, including general household information, domestic water usage and access to each water source, the current operation and maintenance status of rainwater harvesting systems, residents' perceptions and concerns about rainwater quality, and residents' readiness to increase rainwater harvesting capacity. These points were investigated in each interview, which took about 30 min to make sure they could address the research questions. Survey data were then input into Excel software 2016 version and processed for aggregation. Descriptive statistical methods were also employed to analyze the social characteristics of the surveyed group, water usage status, rainwater usage practices, perceptions of rainwater quality, and characteristics of rainwater harvesting systems. Our questionnaire can be found in the Supplementary Material.

2.2.2. Water Balance Calculations

Water balance calculations were used to assess the potential of rainwater and determine the amount of rainwater that can be harvested annually through existing catchment capabilities. Additionally, the minimum required rainwater catchment area and minimum tank capacity to meet usage needs were quantified. Subsequently, the excess rainwater in the specific calculated rainwater harvesting scenarios was determined to artificially refill groundwater.

The water balance equation is typically represented as a function of the amount of water in and out [38]:

$$W^t{}_i = W^t{}_{i-1} + W^c{}_i - W^o{}_i \ (m^3) \quad (1)$$

where

$W^t{}_i$ and $W^t{}_{i-1}$: the volume of water stored in the reservoir (m^3) in a given time period i, i − 1; The initial calculation period $W^t{}_1 = 0$;

$W^t{}_i \leq W^t$, where W^t is the capacity of the tank;
$W^c{}_i$: water volume entering the tank at time i (m^3);
$W^o{}_i$: volume of water discharged from the tank at time i (m^3).

Scenario 1: Evaluating the Potential for Rainwater Harvesting through the Existing Roof Area

This scenario aims to determine the amount of usable water that rainwater can provide through the existing rainwater catchment area. Here, the potential for rainwater harvesting is specified based on two conditions: (i) the storage tank capacity is unlimited and (ii) the catchment area is the existing average roof area (F) available to residents, which is 21.4 m^2 as per the Population and Housing Census Steering Committee of Vietnam [39].

The variables in Equation (1) include the following:

$W^o{}_i$: the water outflow from the storage tank or the amount of water used over the calculation period (day); $W^o{}_i$ = D/1,000, where D represents the water demand (L/person-day);

$W^c{}_i$ (the water volume entering the tank at time i) = $W^m{}_i$; the amount of rainwater collected through a catchment area during the period i (m^3) is determined via the following equation:

$$W^m{}_i = (X_i - A) \times F, \quad (2)$$

in which the variables are as follows:

X_i: the amount of rainfall at time i;
A: the amount of rainwater at the beginning of the rainfall event that needs to be discharged; in this study, A = 3 mm based on experience.

The water demand, D, was gradually tested to determine the maximum level at which there are no instances of the water reservoir running dry ($W^t{}_i \leq 0$). Afterward, the minimum capacity of the storage tank was re-evaluated to meet the previously determined maximum water demand (W^t—m^3).

Scenario 2: Determining the Minimum Roof Area/Size and Tank Capacity Requirements to Meet Water Demand Levels

The variables in Equation (1) at this time are as follows:
Water outflow from the storage tank:

$$W^o{}_i = W^d{}_i + W^x{}_i; \quad (3)$$

$W^d{}_i$: the water amount used during the calculated time period (day); $W^d{}_i$ = D/1000
$W^x{}_i$: the water amount discharged or overflowed during the calculated time period.
Equation (1) becomes the following:

$$W^t{}_i = W^t{}_{i-1} + W^c{}_i - W^d{}_i - W^x{}_i \ (m^3) \quad (4)$$

When $W^t{}_i \leq W^t$, then no water is discharged or overflowing, or $W^x{}_i = 0$;
When $W^t{}_i > W^t$, then the water will proceed to be discharged or overflow ($W^x{}_i > 0$).

The input information included the daily rainfall data from the observation stations; the calculated levels of water demand (D) are as follows: 3, 5, 15, 20, 30, 40, 50, 60, 70, 80, 90, and 100 L/person-day; the minimum roof area was determined as $F_{min} = D \times 365/X_{year}$, where X_{year} is the total average annual rainfall; and the minimum storage tank capacity was calculated using $W^t{}_{min} = D \times NRY$, where NRY is the maximum average number of consecutive non-rainy days in the year.

The roof area (F) and storage tank capacity (W^t) were calculated incrementally from small to large to ensure meeting the water demand without any instances of the storage tank running out of water ($W^t{}_i \leq 0$). The calculation resulted in the optimal roof area and storage tank capacity.

Scenario 3: Determining Tank Requirements for Maximum Use of Rainwater in the Rainy Season and Essential Water Needs in the Dry Season

This scenario utilized rainwater to fully meet the water demand throughout the year, requiring a relatively large storage tank capacity. The essential water demand was 15 L/person-day [39]. To accurately identify the transition from the rainy season to the dry season, which is challenging for most residents, our study used the criterion of water level in the storage tank to recognize the change in water usage demand; for example, when the water level in the tank was above 75%, it was used for full demand, and conversely, when the water level in the tank was below 75%, an essential water demand was set.

Equation (4) is applied, with the difference being the determination of the variable $W^d{}_i$: when $W^t{}_i > W^t \times 0.75$, then $W^d{}_i = D/1000$ (m^3), and when $W^t{}_i \leq W^t \times 0.75$, then $W^d{}_i = 0.015$ (m^3).

In this scenario, the roof area (F) is also 21.4 m^2/person. The calculation steps are similar to scenario 2.

Scenario 4 Integrating Rain and Groundwater

In this scenario, the tank capacity was specified from the results of scenario 3. Our objective was to maximize rainwater exploitation, just use groundwater when there was a shortage, and artificially supplement groundwater from excess rainwater. In case the tank ran out of water ($W^t{}_i \leq 0$), the tank was then replenished from the supplementary water source (groundwater or other water source), and the excess rainwater source during the rainy season was used to maximize artificial groundwater recharge.

Whenever the storage tank ran dry, it was replenished with an amount of water equal to 50% of the tank's capacity (this result has been experimentally calculated across scenarios of 25%, 50%, and 75% replenishment, indicating that replenishing 50% each time the tank runs dry was suitable).

The artificial groundwater recharge method was applied as follows. During the permissible period for artificial groundwater recharge, when the water level in the storage tank exceeded 50% of the tank's capacity, the water in the tank automatically flowed into the artificial groundwater recharge well. The artificial recharge period started on 1 May each year and ended on 31 October, about 20 days before the end of the rainy season.

The following equation is considered:

$$W^c{}_i = W^m{}_i + W^{GW}{}_i \tag{5}$$

$W^{GW}{}_i$: the amount of water needed to be extracted from the groundwater when the storage tank runs dry ($W^t{}_i \leq 0$).

$$W^o{}_i = W^d{}_i + W^{AR}{}_i + W^{OW}{}_i \tag{6}$$

$W^{AR}{}_i$: the amount of water artificially replenished into the groundwater during the period i;

$W^{OW}{}_i$: the amount of water that needs to be discharged or overflowed during the period i;

Equation (1) is applied:

$$W^t{}_i = W^t{}_{i-1} + W^m{}_i + W^{GW}{}_i - W^d{}_i - W^{AR}{}_i - W^{OW}{}_i \ (m^3) \tag{7}$$

This scenario aims to determine the amount of harvested rainwater used for domestic purposes, the amount of water that needs to be supplemented when there is a shortage, and the amount of rainwater needed for artificial groundwater recharge. We developed this scenario considering not only a technical assessment but also social and economic aspects.

In this scenario, the system consists of six main components (Figure 2): (i) a roof catchment for collecting rainwater, including gutters, debris swept/strainers, first flush diverters, and conduits with a bottom valve; (ii) a first storage tank for storing untreated water; (iii) a water treatment filtration system; (iv) a storage tank for treated water; (v) a groundwater extraction and recharge system with an inverted U-shaped structure, with a height equal to half of the tank height, and a groundwater extraction well integrated; and (vi) a groundwater extraction system, including extraction pumps and pipelines transferring water to the first storage tank. Rainwater collected from the roof catchment is stored in the first storage tank, treated through the filtration system, and transferred to the storage tank for distribution to users. Additional branches diverting water into the groundwater extraction well are installed along the pipeline leading water out of the storage tank. The inverted U-shaped structure ensures that the water level in the storage tank is maintained at 50%. During the permitted time, the check valve of the groundwater extraction and recharge system remains open and closes during the dry season. The groundwater extraction well is designed to also serve as a recharge well. Finally, the groundwater extracted from the well is transferred to the first storage tank for treatment and then sent to the storage tank for usage.

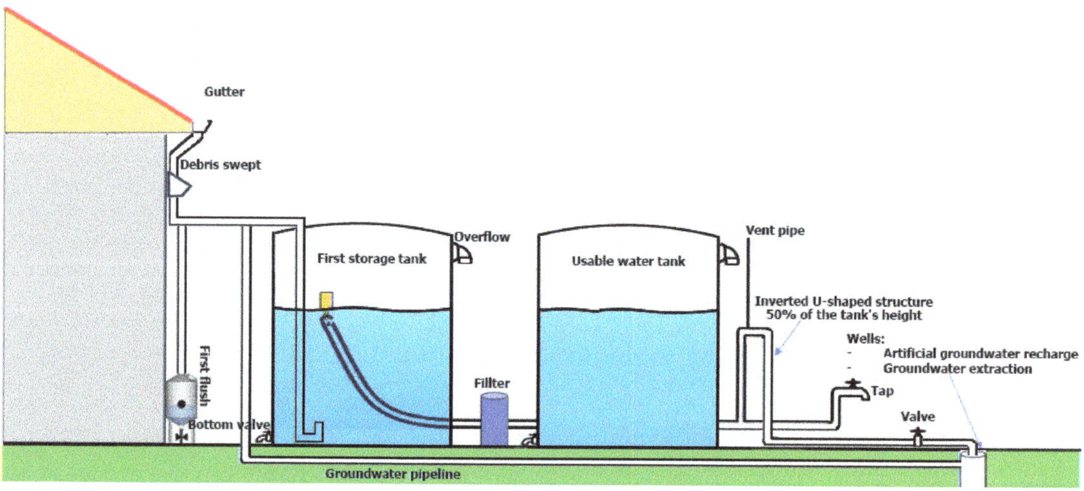

Figure 2. Rainwater harvesting system developed in the study, including artificial groundwater recharge and groundwater extraction, adopted from the website https://content.ces.ncsu.edu/mosquito-control-for-rainwater-harvesting-systems assessed on 31 December 2021.

3. Results and Discussion

3.1. Rainwater Harvesting for Domestic Use in Ca Mau Province

3.1.1. Interview Survey

The survey results show that residents in rural areas of Ca Mau Province use five different water sources, including tap water, groundwater, surface water, rainwater, and bottled water (Figure 3). On average, each household uses 2–3 different water sources.

Tap water in Ca Mau Province is supplied from water treatment plants under the Center of Clean Water and Environmental Sanitation in Rural Areas or from water supply companies. All these plants extract water from groundwater sources where the quality is good and abundant. Due to the uneven distribution of these plants and their placement in different areas, only 8% of households have access to this water source. Meanwhile, surface water is extracted from household ponds, lakes, or canals in areas with freshwater sources, but the quality is very poor, so only 2% of households can use it. Bottled water is quite commonly used by 53% of households due to its low cost. Our findings are similar to those in other regions of developing countries where surface freshwater is limited or affected by droughts and salinity intrusion [40,41].

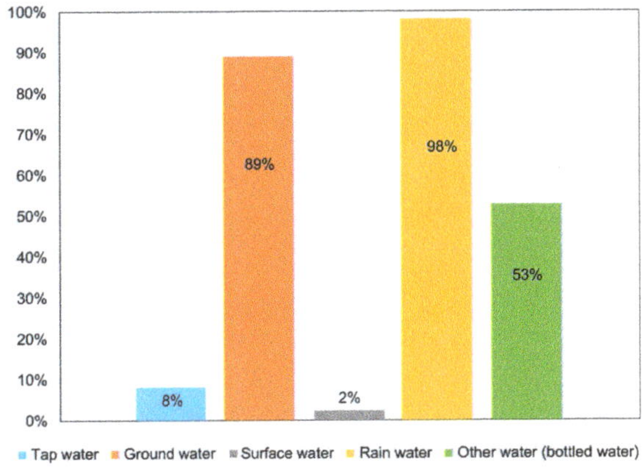

Figure 3. Five water resources for domestic usage in Ca Mau Province.

Rainwater and groundwater are the main sources of water for household purposes, which people harvest themselves on-site. Almost all residents (98%) use rainwater. Due to the convenience of groundwater, it is the preferred water source for 89% of the population, and households that do not use groundwater simply cannot access it.

Rainwater and bottled water are the main sources for drinking purposes, with rates of 59.1% and 31.8%, respectively, while other sources are less common (Figure 4). Rainwater is predominantly used for cooking, accounting for 49.3% of households, followed by groundwater at 47.1%. People rarely use rainwater for other purposes, mainly relying on groundwater for tasks such as washing food and kitchen utensils. Our survey results on water sources and usage patterns also indicate that rainwater and groundwater are the primary sources for household use. It is suggested that increasing the use of rainwater contributes to decreasing groundwater extraction.

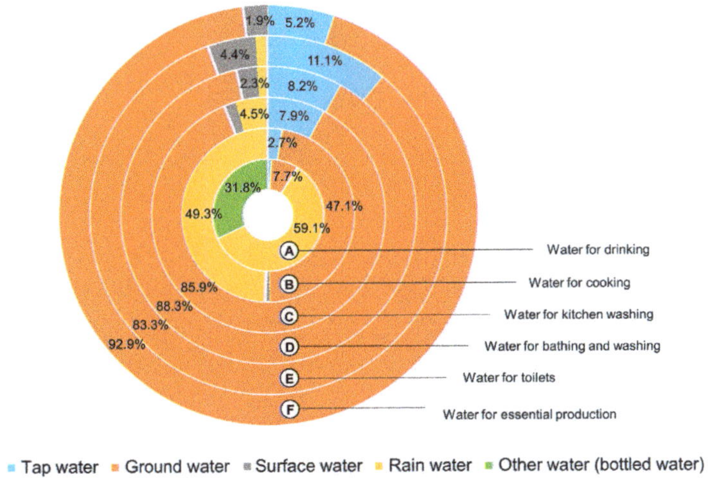

Figure 4. Percentage of residents using water sources for different purposes.

3.1.2. Existing Context of Rainwater Harvesting for Ca Mau Residents

In Ca Mau Province as well as others across the VMD, there are two common types of roof catchments, namely corrugated metal roofs (50.9%) and fiber cement roofs (49.7%), with nearly equal proportions (Figure 5). The average roof area per household is 77.24 m^2. Household types with roof areas ranging from 50 to 100 m^2 per household are prevalent (68.42%), while very small roofs (<30 m^2) or very large roofs (>200 m^2) are less common. On average, each household typically has around 3-4 individuals, resulting in an average roof catchment area of approximately 19.3 to 25.7 m^2 per person. The survey data are consistent with the results from the 2019 Population and Housing Census Steering Committee in Vietnam [39].

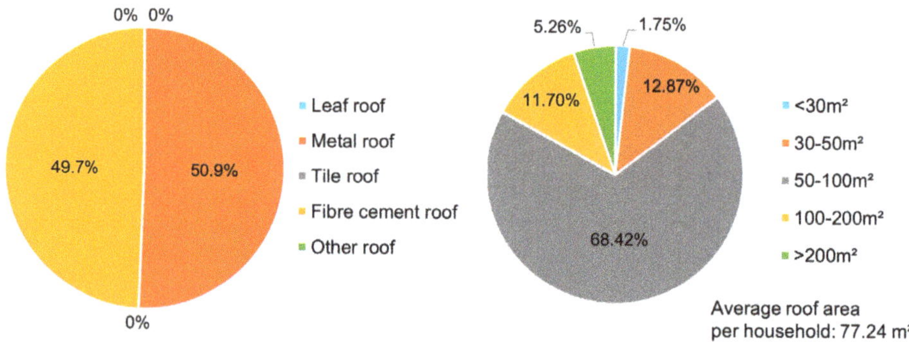

Figure 5. Characteristics of rainwater roofs in terms of material (**left**) and area/size (**right**).

The average storage tank capacity is 2.56 m^3 per household, with an average of about 0.64 to 0.85 m^3 of storage tank capacity per person (Figure 6). Commonly, households have storage tank capacities ranging from 1 to 3 m^3 (50.6%). The number of households with larger storage capacities (5 to 10 m^3) accounts for 5%, while those with very large capacities (>10 m^3) only make up 1.19%. The majority of residents primarily use water pots (48.52%) and prefabricated (44.52%) tanks.

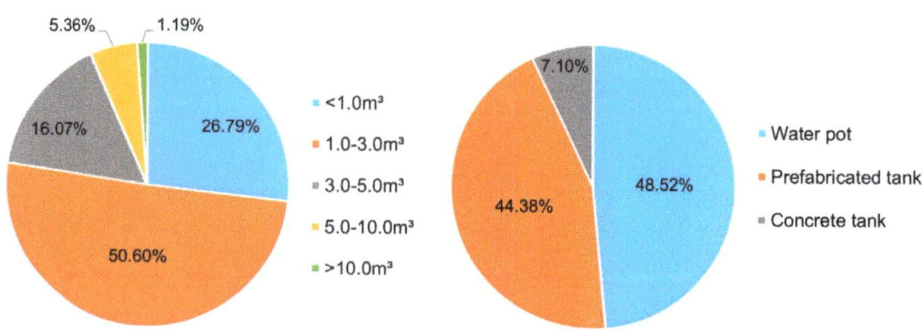

Figure 6. Characteristics of rainwater tanks in terms of capacity (**left**) and materials (**right**).

The primary purpose of rainwater harvesting is mainly for drinking and cooking, so residents do not have much motivation to increase rainwater storage when they already have enough water to use. Up to 84% of households responded that they do not want to increase their rainwater storage capacity. Among them, the main reasons are a lack of necessity (82%), a lack of funds (13%), and a lack of space for construction (5%).

Overall, rainwater is commonly used by residents in Ca Mau Province. However, the high installation costs of rainwater harvesting systems, along with the low direct economic efficiency [42–44], coupled with the ease of accessing piped water supply, have limited the increase in rainwater usage. Therefore, residents only harvest enough rainwater to meet their highest needs for drinking and cooking, with an estimated demand of about 2.5 to 3.0 L/person-day. Other needs with much larger volumes are mainly met from groundwater sources, which are currently under considerable pressure due to depletion.

3.2. *Rainwater Potential and Rainwater Harvesting Scenarios*

3.2.1. Rainwater Characteristics Related to Harvesting Capacity

Table 1 presents the statistical characteristics describing the annual rainfall and the greatest number of consecutive rainless days (or days with rainfall < 5 mm) based on data collected from 10 rainfall monitoring stations in Ca Mau Province. The rainfall amount characteristic reflects the potential, while the number of rainless days indicates limitations affecting rainwater harvesting.

Table 1. Statistical descriptive characteristics of annual rainfall data at 10 monitoring stations.

Station / Characteristic	Ca Mau	Cai Nuoc	Dam Doi	Nam Can	NT U Minh	Song Doc	Tran Van Thoi	Thoi Binh	U Minh	Vien An Dong	Average
Amount of Rainfall (mm)											
Average	2378	2148	1932	2329	2455	2353	2237	2363	2428	2149	2247
Median	2304	2115	1893	2250	2415	2381	2180	2336	2429	2114	2199
Max	3549	3151	2516	2964	3258	3406	3306	2885	3291	3385	2906
Min	1932	1554	1408	1781	1894	1695	1542	2014	1668	1024	1908
Standard deviation	306	365	314	316	365	359	381	252	360	467	236
Cv	0.13	0.17	0.16	0.14	0.15	0.15	0.17	0.11	0.15	0.22	0.11
Cs	0.72	0.28	0.37	0.75	0.33	0.23	0.45	0.32	0.01	0.22	1.43
Continuous days without rain (days)											
Average	77	96	107	101	87	84	86	82	82	108	91
Median	62	81	113	101	78	79	78	72	68	96	83
Max	153	174	171	187	162	150	158	138	171	185	141
Min	23	35	32	34	18	29	30	30	21	29	39
Standard deviation	39	45	38	44	41	38	36	36	41	39	33
Cv	0.50	0.47	0.36	0.43	0.47	0.45	0.42	0.43	0.50	0.36	0.36
Cs	1.18	1.03	0.46	0.02	0.64	0.36	0.69	0.83	1.05	0.90	0.78

The average annual rainfall in Ca Mau is 2247 mm/year and has remained stable over the years. With a natural area of 5242.5 km^2, the total rainfall in Ca Mau amounts to 11.78 billion m^3/year. The total population of Ca Mau in 2019 was 1,194,476 people, resulting in over 9600 m^3 of rainfall per person per year, which is close to the average in Vietnam (9434 m^3/person) and significantly higher than the average in Southeast Asian countries (4900 m^3/person) [45]. Even in the year with the least rainfall (1908 mm/year), there was still over 10 billion m^3 of rainfall (8158 m^3/person). This amount is 6.60 times higher than the total freshwater demand in Ca Mau [46]. The coefficient of variation (C_v) is low (0.11 to 0.22), indicating relatively stable annual rainfall in the province. Although we did not calculate the above values considering the impact factor of climate change, our results imply that climate change did not have much influence on the total rainfall amount, excluding the changes in rainfall patterns [9].

The average greatest number of consecutive rainless days or days with less than 5 mm of rainfall (NRY) across the entire province is 91 days (with a deviation of 33 days). The station with the shortest NRY is Ca Mau with 77 days, while the longest is Vien An Dong with 108 days. The coefficient of variation (C_v) is relatively high (0.36 to 0.50), indicating significant fluctuations in the NRY over time. The difference between the highest and lowest NRY values is substantial.

Overall, in the research area, the amount of rainfall can meet usage needs, but the challenge lies in the large and unstable NRY. The instability of the NRY makes it difficult to determine the required storage tank capacity.

3.2.2. Potential of Rainfall Harvesting

Table 2 shows that the amount of rainfall collected from existing rooftops can fully meet the water usage needs for household purposes in rural areas under the current conditions (60 L/person-day) as well as in the 2030 planning scenario (80 L/person-day) following Decision No. 2140/QD-TTg signed in 2016 by the Vietnam Prime Minister [47].

Table 2. Rainfall harvesting availability and requirement of tank volume based on current capacity of catchment roofs.

Station / Capacity	Ca Mau	Cai Nuoc	Dam Doi	Nam Can	NT U Minh	Song Doc	Thoi Binh	Tran Van Thoi	U Minh	Vien An Dong
Rainfall availability for harvesting (L/person-day)	130.0	113.3	106.5	110.7	127.5	114.9	130.1	125.1	134.0	122.8
Tank volume required (m^3)	68.3	43.9	63.3	33.2	33.2	47.9	44.6	49.8	42.8	75.9

The potential for rainwater harvesting can be expanded by increasing the roof catchment area. This expansion is achievable through various means: (i) repurposing the roofs of water storage tanks as additional catchment areas; (ii) installing awnings or canopies to extend the catchment surface; and (iii) utilizing mobile awnings for water collection. Consequently, the ability to increase rainwater harvesting is still possible. However, a significant obstacle to rainwater harvesting is the demand for large storage tank capacities (33.2–75.9 m^3 per household). The survey data indicate that the average storage tank capacity is 2.56 m^3 per household, corresponding to 0.64–0.85 m^3 per person, which is too small compared to the required capacity. Our results imply a positive policy change in rainwater harvesting for not only the VMD but also for any nation that encourages people to collect rainwater for storage through all means and provides robust programs for collecting rainwater at a larger scale for rural communities, especially those that have been strongly impacted by droughts and salinity intrusion.

3.3. *Scale Requirements for Storage Tanks and Roof Catchments for Rainwater Harvesting Systems*

3.3.1. Storage Tank and Roof Catchment Requirements for Different Water Supply Levels

Figure 7 illustrates the requirements for the storage tank capacity and roof catchment area to meet water supply needs. At the level of water demand for drinking, the requirements for the roof catchment area and storage tank capacity are very low, averaging 0.78 m^3 per tank and 1.27 m^2 for a demand of 5 L/person-day. With an average storage tank capacity of 2.56 m^3 per household (averaging 3–4 persons per household), rainwater harvesting can meet drinking water needs.

To ensure water supply for other basic needs such as basic sanitation and food and utensil washing, with a demand of about 15 L/person-day, the required storage tank capacity is 3.1 m^3 per person, averaging 9–12 m^3 per household. These types of tanks are typically made of concrete, with a cost of approximately 1–1.2 million VND per m^3 (equivalent to USD 40—45), a cost that is acceptable for households with moderate incomes. The roof catchment area requirement for cases exceeding 15 L/person-day is still relatively low, at only 3.58 m^2 per person, which is much smaller than the existing roof catchment area.

The existing roof catchment area of 21.4 m^2 per person can meet water supply demands, exceeding 80 L/person-day. This indicates that people do not currently need to increase their roof catchment areas for rainwater harvesting. However, to meet the demand for 80 L/person-day, a very large storage tank capacity is required, around 16 m^3 per person, corresponding to 48–64 m^3 per household. When following traditional solutions, many households cannot afford this.

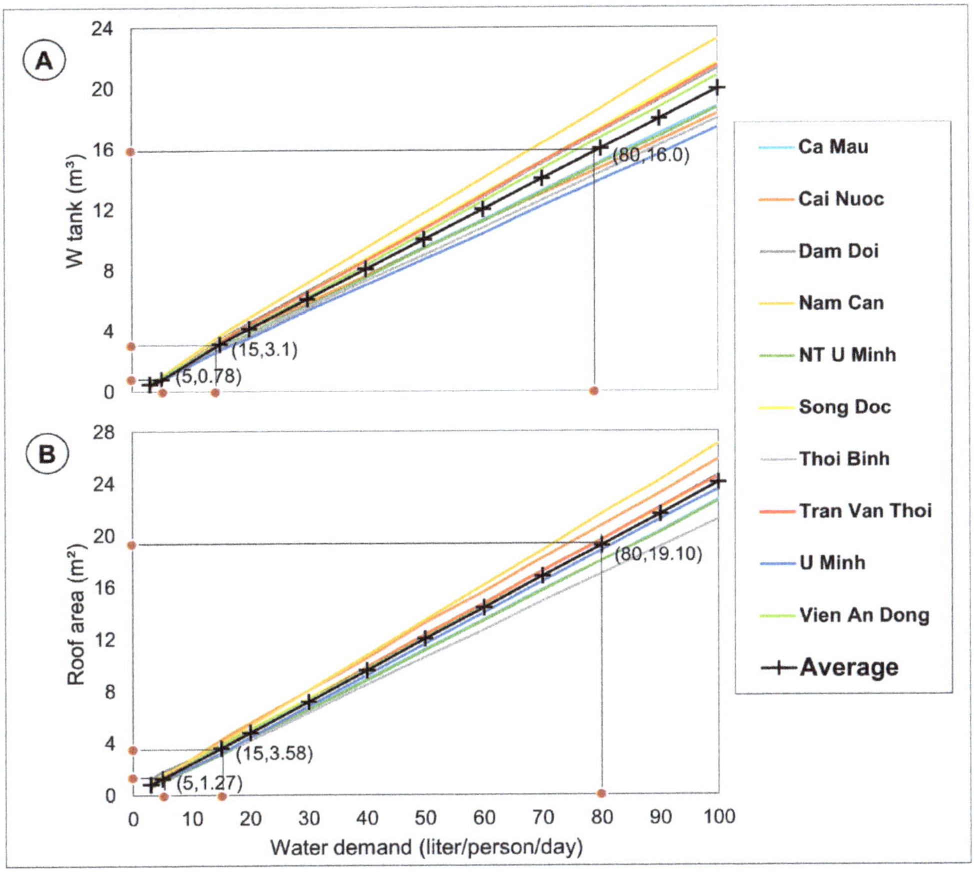

Figure 7. The requirement for the tank capacity (**A**) and the rainfall catchment area (**B**) to meet the water supply needs calculated based on the rainfall of 10 meteorological stations. The values from the red dots are the averages from 10 stations.

In general, the main difficulty is the requirement for large storage tank capacities in rainwater harvesting. Increasing storage tank capacities for rainwater use may present several obstacles, such as community needs, budget constraints, and space requirements for installation. The current average storage tank capacity only meets the water demand for drinking purposes at 5 L/person-day. However, a survey conducted in 2020 showed that about 5% of households have storage tank capacities ranging from 5 to 10 m^3, and 1% have capacities exceeding 10 m^3, which could enable rainwater harvesting to meet demands of 7 to 15 L/person-day. Therefore, rainwater harvesting systems should be divided into two types, respectively: (i) Rainwater harvesting for essential needs (maximum demand of 15 L/person-day), which requires a separate catchment area that is easy to clean to ensure high-quality water collection. Traditional storage tanks should be used, minimizing the risk of contamination during storage. (ii) Rainwater harvesting for additional needs (45 to 65 L/person-day) that utilizes all types of catchment areas and may involve improved storage tanks in the form of excavated ponds combined with waterproofing materials such as waterproof membranes. This is considered a long-term solution for rainwater harvesting systems, such as the Pond Harvesting System (PHS) and MAR, which are practiced in many countries [45].

One of the major reasons for reluctance to increase rainwater harvesting is that residents only perceive the need for enough water for drinking and cooking. Therefore, there is a need for awareness campaigns to encourage residents to increase their use of rainwater for other purposes. Additionally, there should be financial and technical support measures provided to help people increase their storage tank capacities.

3.3.2. Storage Tank Scale Required to Meet Basic Household Needs in the Dry and Rainy Seasons

The results indicate that the average storage tank capacity requirement is 2.4 m^3 per person, equivalent to 7.2–9.6 m^3 per household, to meet the essential water needs (Table 3). Increasing the demand to 80 L/person-day during the rainy season and 15 L/person-day during the dry season, the storage tank capacity requirement becomes 3.3 m^3 per person, equivalent to 10–13 m^3 per household. At these scales, residents are more likely to accept increasing the scale of rainwater harvesting. In reality, 6% of households already have storage tank capacities that can meet this demand. Hence, costs should be carefully analyzed to increase the possibility and willingness of residents to upscale their tanks under an integrated rainfall harvesting system, although the technical and social benefits are pointed out by our study. To make this level of storage tank more widely accepted and further disseminated, additional support from the government as well as various social organizations is needed in various forms, such as loans and capital subsidies for impoverished households.

Table 3. Tank volume required to meet essential water demand during the dry season and sufficient supply during the rainy season—the whole time series (holistic) regulation method (m^3/person).

Water Demand (Liter/Person/Day)	15	30	60	80	100
Station					
Ca Mau	2.5	3.0	3.4	3.5	3.6
Cai Nuoc	2.4	3.0	3.3	3.4	3.4
Dam Doi	2.9	3.4	3.8	3.9	4.0
Nam Can	2.8	3.3	3.6	3.7	3.8
NT U Minh	2.3	2.9	3.2	3.3	3.4
Song Doc	2.1	2.5	2.9	2.9	2.9
Thoi Binh	2.2	2.7	2.9	3.0	3.1
Tran Van Thoi	2.0	2.4	2.8	2.9	2.9
U Minh	2.1	2.5	2.9	3.0	3.0
Vien An Dong	2.8	3.3	3.6	3.8	3.8
Average	2.4	2.9	3.2	3.3	3.4

3.4. Integration of Rainwater Harvesting with Groundwater Usage and the Potential for Artificial Groundwater Recharge

This scenario is a promising solution which needs to be evaluated in terms of technical, economic, and social aspects. Building upon this, we technically examined the following: (i) the rainwater harvesting potential at different storage tank scales; (ii) the additional groundwater extraction required to meet the shortfall; and (iii) the potential use of surplus rainwater in the rainwater harvesting system to replenish groundwater. Specifically, the existing storage tank scales (0.8 m^3 per person), a storage tank meeting basic needs (2.4 m^3 per person), and a storage tank meeting maximum needs (3.4 m^3 per person) were analyzed. The excess water discharge, harvested rainwater, additional groundwater extraction needed, and rainwater available for groundwater replenishment were calculated for each scenario (Table 4).

Table 4. Synthesis of calculation results for tank scales (m^3/year).

Characteristic	Excess Rainfall Runoff			Rainwater Harvesting Volume			Groundwater Extraction Volume			Artificial Groundwater Recharge Volume		
Tank Volume (m^3)	0.8	2.4	3.4	0.8	2.4	3.4	0.8	2.4	3.4	0.8	2.4	3.4
Station												
Ca Mau	12.5	2.9	1.5	15.1	19.6	20.8	13.6	9.4	6.8	15.1	20.2	20.3
Cai Nuoc	12.8	3.4	1.8	13.0	17.6	18.6	15.8	11.4	11.2	13.4	18.2	18.7
Dam Doi	9.1	1.8	1.0	13.0	17.4	18.4	15.9	11.8	12.4	12.9	15.8	15.6
Nam Can	11.5	2.8	1.6	13.3	18.1	19.2	15.5	11.0	10.8	12.9	16.7	16.8
NT U Minh	14.8	3.6	1.6	14.8	18.9	20.4	14.0	10.2	8.0	15.5	22.6	23.1
Song Doc	11.2	2.4	1.2	13.4	18.3	19.6	15.3	10.7	12.4	12.6	16.5	16.4
Thoi Binh	13.1	2.6	1.1	14.6	19.5	20.7	14.2	9.6	11.2	15.4	21.1	21.4
Tran Van Thoi	11.8	2.5	1.1	13.8	18.5	19.7	15.0	10.5	9.6	14.5	19.0	19.2
U Minh	16.0	4.3	2.2	14.4	19.4	21.0	14.4	9.7	10.0	15.1	21.9	22.3
Vien An Dong	9.2	2.6	1.4	14.2	17.8	18.7	14.7	11.4	9.6	15.5	18.5	18.8
Average	12.2	2.9	1.5	14.0	18.5	19.7	14.8	10.6	10.2	14.3	19.0	19.3

In the case of the existing storage tank scale (0.8 m^3 per person), an average of 14 m^3 of rainwater is harvested, requiring 14.8 m^3 of additional groundwater extraction. Approximately 14.3 m^3 of rainwater can be used to replenish the groundwater, while the surplus rainwater amounts to 12.2 m^3 discharged into the environment. Overall, the integrated water usage solution has proven highly effective. Rainwater can meet more than 48% of the water demand, and the surplus rainwater available for groundwater replenishment is nearly equal to the additional groundwater extraction needed. This case demonstrates high efficiency under current conditions, but a limitation is the significant amount of surplus rainwater discharged into the environment. Additionally, the essential water supply needs to rely on groundwater of low quality during the dry season. Overall, this solution is beneficial for households lacking space to increase the scale of their storage tanks.

In the case of a storage tank capacity of 2.4 m^3 per person on average, approximately 18.5 m^3 of rainwater is harvested, requiring 10.6 m^3 of additional groundwater extraction. About 19.0 m^3 of rainwater can be used to replenish the groundwater, while the surplus rainwater amounts to 2.9 m^3 discharged into the environment. The harvested rainwater can meet 64% of the water demand, and the surplus rainwater available for groundwater replenishment is more than 1.79 times the additional groundwater extraction needed. The amount of surplus rainwater discharged into the environment is relatively small. These results indicate that this solution is highly effective and economically sustainable, representing a step towards achieving water self-sufficiency in rural areas.

In the case of maximum storage tank capacity (3.4 m^3 per person on average), approximately 19.7 m^3 of rainwater is harvested, corresponding to 10.2 m^3 of additional groundwater extraction. About 19.3 m^3 of rainwater can be used to replenish the groundwater, resulting in only 1.5 m^3 of surplus rainwater being discharged into the environment. This scenario is efficient in terms of utilizing rainwater and increasing the surplus rainwater available for groundwater replenishment while reducing the amount of additional groundwater needed. However, a storage tank capacity ranging from 2.4 to 3.4 m^3 per person is deemed appropriate because increasing the scale further does not enhance efficiency.

The scenario/method of integrating rainwater and groundwater has been implemented and tested by a research group at Tri Phai Primary School in Tri Phai commune, Thoi Binh district of Ca Mau Province. The system, which is illustrated in Figure 2, consists of two storage tanks with a capacity of 25 m^3 each, totaling 50 m^3, and a roof catchment area of 200 m^2. The groundwater extraction is conducted via a 49 mm diameter well reaching the Pleistocene aquifer at a depth of 100 m. This system provides water for the school and its

600 students. During the rainy season, the system adequately supplies water for the school. In the dry season, the system only provides drinking water for students (after treatment), while other needs such as sanitation and irrigation are met from the groundwater source. The flow rate of groundwater extraction reaches 900 L/h, and during the rainy season of 2022, the amount of extracted groundwater was 215 m^3, with no observed overflow from the system. The system considers rainwater as the primary water source prioritized for maximum utilization. Groundwater serves as a reserve source used in cases of water shortage, particularly during the dry season and drought years.

4. Conclusions

Millions of residents in the Vietnamese Mekong Delta (VMD), especially in Ca Mau Province, have relied on various water sources ingrained in their culture. Despite ample rainfall, efficient rainwater harvesting in Ca Mau falls short of meeting demands due to inadequate storage capacity and challenges like low awareness, high construction costs, and limited space. Our study uses data from 10 meteorological stations and interviews with 200 households to propose a sustainable water harvesting scenario. Emphasizing rainwater's importance, especially during dry periods, we stress the need to maintain large tank capacities for effective harvesting. Additionally, we advocate for integrating rainwater and groundwater systems to enhance water resource management and self-sufficiency. Based on the findings, two main conclusions are drawn:

- To satisfy the basic water supply requirement (15 L/person-day) from rainwater, each person needs an average tank capacity of 3.1 m^3 and a roof catchment area of 3.6 m^2. The roof catchment area requirement is not overly large but allows for adaptive operation and maintenance to ensure the harvesting of good-quality rainwater. In the scenario where all available roof catchment areas are utilized, a tank capacity of 2.4 m^3 per person can meet the basic water supply demand, and approximately 6% of households have achieved this. The scenario of utilizing rainwater entirely during the rainy season and meeting basic needs during the dry season demonstrates that the demand of 80 L/person-day during the rainy season and 15 L/person-day during the dry season can be met with a tank scale of 3.4 m^3 per person. These options could be flexibly selected for different areas with different economic conditions and water demands.
- The integrated water supply scenario of rain and groundwater demonstrates the highest effectiveness, even with the current rainwater tank size, nearly achieving water balance. At a tank scale of 2.4 m^3 per person, equivalent to 7.2 to 9.6 m^3 per household, 64% of rainwater can be utilized for domestic use. The supplementary artificial groundwater recharge exceeds the amount of groundwater extraction by 1.79 times, resulting in minimal surplus water discharge into the environment. The artificial groundwater recharge system is relatively simple and not expensive, involving the addition of inverted U-shaped structures and the adaptation of groundwater extraction wells to serve both extraction and recharge purposes. This approach offers sustainability and potential water independence across rural areas of the VMD. However, many challenges exist, such as the willingness of people and financial constraints, especially for the poor, to apply or transform this approach, which may prolong its widespread acceptance. This approach can also be applied in urban areas, directly saving water costs and indirectly mitigating urban flooding while replenishing groundwater, but it needs strong support from local authorities via long-term official programs.

Supplementary Materials: The following supporting information can be downloaded at https://www.mdpi.com/article/10.3390/hydrology11040055/s1: Questionnaire.

Author Contributions: Conceptualization, D.H.V.; methodology, D.H.V. and D.D.T.; investigation, D.H.V., D.D.C. and P.T.T.H.; data curation, D.B.M., D.M., L.H.T., L.V.K. and D.T.N.T.; writing—original draft preparation, D.H.V.; writing—review and editing, D.H.V. and D.D.T.; visualization,

D.H.V., D.B.M. and T.D.N.; supervision, D.D.C. and P.T.T.H. All authors have read and agreed to the published version of the manuscript.

Funding: The authors express gratitude to the Vietnam Academy of Science and Technology (VAST) for funding support under the project number VAST06.04/21-22.

Data Availability Statement: Data are provided upon a reasonable request.

Acknowledgments: Many thanks to colleagues, particularly at VAST, who have made time for discussion during the research.

Conflicts of Interest: The authors declare no conflicts of interest.

References

1. Jaramillo, P.; Nazemi, A. Assessing urban water security under changing climate: Challenges and ways forward. *Sustain. Cities Soc.* **2018**, *41*, 907–918. [CrossRef]
2. Vorosmarty, C.J.; Green, P.; Salisbury, J.; Lammers, R.B. Global water resources: Vulnerability from climate change and population growth. *Science* **2000**, *289*, 284–288. [CrossRef] [PubMed]
3. Ercin, A.E.; Hoekstra, A.Y. Water footprint scenarios for 2050: A global analysis. *Environ. Int.* **2014**, *64*, 71–82. [CrossRef] [PubMed]
4. De Fraiture, C.; Wichelns, D. Satisfying future water demands for agriculture. *Agric. Water Manag.* **2010**, *97*, 502–511. [CrossRef]
5. Allan, C.; Xia, J.; Pahl-Wostl, C. Climate change and water security: Challenges for adaptive water management. *Curr. Opin. Environ. Sustain.* **2013**, *5*, 625–632. [CrossRef]
6. Oki, T.; Agata, Y.; Kanae, S.; Saruhashi, T.; Yang, D.; Musiake, K. Global assessment of current water resources using total runoff integrating pathways. *Hydrol. Sci. J.* **2001**, *46*, 983–995. [CrossRef]
7. Wolfe, S.; Brooks, D.B. Water scarcity: An alternative view and its implications for policy and capacity building. *Nat. Resour. Forum* **2003**, *27*, 99–107. [CrossRef]
8. Rijsberman, F.R. Water scarcity: Fact or fiction? *Agric. Water Manag.* **2006**, *80*, 5–22. [CrossRef]
9. Thuc, T.; Van Thang, N.; Huong, H.T.L.; Van Khiem, M.; Hien, N.X.; Phong, D.H. *Climate Change and Sea Level Rise Scenarios for Vietnam*; Ministry of Natural Resources and Environment: Hanoi, Vietnam, 2016.
10. Loc, H.H.; Low Lixian, M.; Park, E.; Dung, T.D.; Shrestha, S.; Yoon, Y.-J. How the saline water intrusion has reshaped the agricultural landscape of the Vietnamese Mekong Delta, a review. *Sci. Total Environ.* **2021**, *794*, 148651. [CrossRef]
11. Duc Tran, D.; Thi Bich Thuc, P.; Park, E.; Thi Thanh Hang, P.; Ba Man, D.; Wang, J. Extent of saltwater intrusion and freshwater exploitability in the coastal Vietnamese Mekong Delta assessed by gauging records and numerical simulations. *J. Hydrol.* **2024**, *630*, 130655. [CrossRef]
12. Chen, Y.; Li, Z.; Fan, Y.; Wang, H.; Deng, H. Progress and prospects of climate change impacts on hydrology in the arid region of northwest China. *Environ. Res.* **2015**, *139*, 11–19. [CrossRef] [PubMed]
13. Döll, P.; Hoffmann-Dobrev, H.; Portmann, F.T.; Siebert, S.; Eicker, A.; Rodell, M.; Strassberg, G.; Scanlon, B.R. Impact of water withdrawals from groundwater and surface water on continental water storage variations. *J. Geodyn.* **2012**, *59*, 143–156. [CrossRef]
14. Abidin, H.Z.; Andreas, H.; Gumilar, I.; Fukuda, Y.; Pohan, Y.E.; Deguchi, T. Land subsidence of Jakarta (Indonesia) and its relation with urban development. *Nat. Hazards* **2011**, *59*, 1753–1771. [CrossRef]
15. Chen, C.-T.; Hu, J.-C.; Lu, C.-Y.; Lee, J.-C.; Chan, Y.-C. Thirty-year land elevation change from subsidence to uplift following the termination of groundwater pumping and its geological implications in the Metropolitan Taipei Basin, Northern Taiwan. *Eng. Geol.* **2007**, *95*, 30–47. [CrossRef]
16. Dong, S.; Samsonov, S.; Yin, H.; Ye, S.; Cao, Y. Time-series analysis of subsidence associated with rapid urbanization in Shanghai, China measured with SBAS InSAR method. *Environ. Earth Sci.* **2014**, *72*, 677–691. [CrossRef]
17. Kim, Y.; Han, M.; Kabubi, J.; Sohn, H.-G.; Nguyen, D.-C. Community-based rainwater harvesting (CB-RWH) to supply drinking water in developing countries: Lessons learned from case studies in Africa and Asia. *Water Sci. Technol. Water Supply* **2016**, *16*, 1110–1121. [CrossRef]
18. Nguyen, D.C.; Dao, A.D.; Kim, T.; Han, M. A sustainability assessment of the rainwater harvesting system for drinking water supply: A case study of Cukhe Village, Hanoi, Vietnam. *Environ. Eng. Res.* **2013**, *18*, 109–114. [CrossRef]
19. Campisano, A.; Butler, D.; Ward, S.; Burns, M.J.; Friedler, E.; DeBusk, K.; Fisher-Jeffes, L.N.; Ghisi, E.; Rahman, A.; Furumai, H.; et al. Urban rainwater harvesting systems: Research, implementation and future perspectives. *Water Res.* **2017**, *115*, 195–209. [CrossRef] [PubMed]
20. Dang, V.H.; Tran, D.D.; Cham, D.D.; Hang, P.T.; Nguyen, H.T.; Truong, H.V.; Tran, P.H.; Duong, M.B.; Nguyen, N.T.; Le, K.V.; et al. Assessment of Rainfall Distributions and Characteristics in Coastal Provinces of the Vietnamese Mekong Delta under Climate Change and ENSO Processes. *Water* **2020**, *12*, 1555. [CrossRef]
21. Megdal, S.B.; Dillon, P. Policy and economics of managed aquifer recharge and water banking. *Water* **2015**, *7*, 592–598. [CrossRef]
22. Dillon, P. Future management of aquifer recharge. *Hydrogeol. J.* **2005**, *13*, 313–316. [CrossRef]
23. Bouwer, H. Artificial recharge of groundwater: Hydrogeology and engineering. *Hydrogeol. J.* **2002**, *10*, 121–142. [CrossRef]
24. Maliva, R.G. Economics of managed aquifer recharge. *Water* **2014**, *6*, 1257–1279. [CrossRef]

25. Megdal, S.B.; Dillon, P.; Seasholes, K. Water banks: Using managed aquifer recharge to meet water policy objectives. *Water* **2014**, *6*, 1500–1514. [CrossRef]
26. Fernandez Escalante, E.; Calero Gil, R.; San Miguel Fraile, M.Á.; Sánchez Serrano, F. Economic assessment of opportunities for managed aquifer recharge techniques in Spain using an Advanced Geographic Information System (GIS). *Water* **2014**, *6*, 2021–2040. [CrossRef]
27. Niazi, A.; Prasher, S.O.; Adamowski, J.; Gleeson, T. A system dynamics model to conserve arid region water resources through aquifer storage and recovery in conjunction with a dam. *Water* **2014**, *6*, 2300–2321. [CrossRef]
28. Arshad, M.; Guillaume, J.H.A.; Ross, A. Assessing the feasibility of managed aquifer recharge for irrigation under uncertainty. *Water* **2014**, *6*, 2748–2769. [CrossRef]
29. Joffre, O.M.; Bosma, R.H.; Bregt, A.K.; van Zwieten, P.A.M.; Bush, S.R.; Verreth, J.A.J. What drives the adoption of integrated shrimp mangrove aquaculture in Vietnam? *Ocean Coast. Manag.* **2015**, *114*, 53–63. [CrossRef]
30. Van Linh, D. *Investigate and Evaluate Modern Geodynamics to Complete Climate Change Scenarios and Propose Adaptation Solutions in the Vietnamese Mekong Delta (Điều tra, đánh giá địa động lực hiện đại để hoàn thiện kịch bản biến đổi khí hậu và đề xuất các giáp pháp thí)*; The Vietnam Geological Department: Ho Chi Minh, Vietnam, 2015.
31. Vuong, B.T.; Chan, N.D.; Nam, L.H. *Assessing the Impact of Climate Change on Groundwater Resources in the Mekong Delta, Proposing Solutions to Respond (Đánh giá tác động cùa biến đôi khí hậu đến tài nguyên nước dưới đất vùng ĐBSCL, đề xuất các giải pháp ứng phó)*; a National Report; Hanoi, Vietnam, 2013.
32. Thi Nhung, T.; Le Vo, P.; Van Nghi, V.; Quoc Bang, H. Salt intrusion adaptation measures for sustainable agricultural development under climate change effects: A case of Ca Mau Peninsula, Vietnam. *Clim. Risk Manag.* **2019**, *23*, 88–100. [CrossRef]
33. Giusto, B.D.; Le, T.M.; Nguyen, T.T.; Nguyen, T.T.; Vu, N.U.; Lavallee, J.P. Development versus Adaptation? Facing Climate Change in Ca Mau, Vietnam. *Atmosphere* **2021**, *12*, 1160. [CrossRef]
34. Karlsrud, K.; Vangelsten, B.; Frauenfelder, R. Subsidence and Shoreline Retreat in the Ca Mau Province—Vietnam Causes, Consequences and Mitigation Options. *Geotech. Eng.* **2017**, *48*, 26–32.
35. Minderhoud, P.S.J.; Erkens, G.; Pham, V.H.; Bui, V.T.; Erban, L.; Kooi, H.; Stouthamer, E. Impacts of 25 years of groundwater extraction on subsidence in the Mekong delta, Vietnam. *Environ. Res. Lett.* **2017**, *12*, 64006. [CrossRef] [PubMed]
36. Dang, H.V.; Tran, D.D.; Pham, B.T.; Khoi, N.D.; Tran, H.P.; Nguyen, T.N. Exploring Freshwater Regimes and Impact Factors in the Coastal Estuaries of the Vietnamese Mekong Delta. *Water* **2019**, *11*, 782. [CrossRef]
37. Thanh, N.D.; Tan, P. Van Non-parametric testing of changing trends of some meteorological factors for the period 1961–2007 (Kiểm nghiệm phi tham số xu thế biến đổi của một số yếu tố khí tượng cho giai đoạn 1961–2007). *Vietnam J. Sci. Technol.* **2012**, *28*, 129–135.
38. Planners, C.E. *Improved Watershed Management (Decentralized Level), Local Governance and Capacity Building-Guidelines for Estimating Water Balance*; LDK Consultants Engineers & Planners SA: Kifisia, Greece, 2019.
39. Central Population and Housing Census Steering Committee Vietnam, Hanoi. *Population and Housing Census at 0:00 on April 1, 2019—Implementation Organization and Preliminary Results (Tổng điều tra dân số và nhà ở thời điểm 0 giờ ngày 1 tháng 4 năm 2019—tổ chức thực hiện và kết quả sơ bộ)*; Statistical Publishing House: Hanoi, Vietnam, 2019.
40. Shaffril, H.A.M.; Krauss, S.E.; Samsuddin, S.F. A systematic review on Asian's farmers' adaptation practices towards climate change. *Sci. Total Environ.* **2018**, *644*, 683–695. [CrossRef] [PubMed]
41. Sultana, P.; Thompson, P.M. Adaptation or conflict? Responses to climate change in water management in Bangladesh. *Environ. Sci. Policy* **2017**, *78*, 149–156. [CrossRef]
42. Hafizi Md Lani, N.; Yusop, Z.; Syafiuddin, A. A review of rainwater harvesting in Malaysia: Prospects and challenges. *Water* **2018**, *10*, 506. [CrossRef]
43. Hajani, E.; Rahman, A. Reliability and cost analysis of a rainwater harvesting system in peri-urban regions of Greater Sydney, Australia. *Water* **2014**, *6*, 945–960. [CrossRef]
44. Kim, K.; Yoo, C. Hydrological modeling and evaluation of rainwater harvesting facilities: Case study on several rainwater harvesting facilities in Korea. *J. Hydrol. Eng.* **2009**, *14*, 545–561. [CrossRef]
45. World Bank Group. *Vietnam: Toward a Safe, Clean, and Resilient Water System*; World Bank Group: Washington, DC, USA, 2019.
46. Vinh, D.H. *Analyze Water Demand and Investigate and Evaluate the Current Status of Water Sources in the Coastal Provinces of the Vietnamese Mekong Delta (Phân tích nhu cầu dùng nước và điều tra, đánh giá thực trạng các nguồn nước đến các tỉnh ven biển ĐBSCL)*; 2019.
47. Minister of Construction. *Water Supply Planning for the Vietnamese Mekong Delta until 2030, a Vision in 2050 (In Vietnamese: Quy hoạch cấp nước vùng Đồng Bằng sông Cửu Long đến năm 2030, tầm nhìn đến năm 2050)*; Minister of Construction: Hanoi, Vietnam, 2016.

hydrology

MDPI

Article

Estimation of Groundwater Recharge in a Volcanic Aquifer System Using Soil Moisture Balance and Baseflow Separation Methods: The Case of Gilgel Gibe Catchment, Ethiopia

Fayera Gudu Tufa [1,2,*], Fekadu Fufa Feyissa [2], Adisu Befekadu Kebede [1,2], Beekan Gurmessa Gudeta [1,2], Wagari Mosisa Kitessa [1,2], Seifu Kebede Debela [1,2], Bekan Chelkeba Tumsa [1,2], Alemu Yenehun [1], Marc Van Camp [1] and Kristine Walraevens [1]

[1] Laboratory for Applied Geology and Hydrogeology, Ghent University, 9000 Ghent, Belgium; adisubefekadu.kebede@ugent.be (A.B.K.); beekan.gudeta@ugent.be (B.G.G.); wagarimosisa.kitessa@ugent.be (W.M.K.); seifukebede.debela@ugent.be (S.K.D.); bekan.chelkeba@ju.edu.et (B.C.T.); alemuyenehun.beyene@ugent.be (A.Y.); marc.vancamp@ugent.be (M.V.C.); kristine.walraevens@ugent.be (K.W.)

[2] Faculty of Civil and Environmental Engineering, Jimma University, Jimma P.O. Box 378, Ethiopia; fekadu.fufa@ju.edu.et

* Correspondence: fayeragudu.tufa@ugent.be or fayerag2@gmail.com

Abstract: Understanding the recharge–discharge system of a catchment is key to the efficient use and effective management of groundwater resources. The present study focused on the estimation of groundwater recharge using Soil Moisture Balance (SMB) and Baseflow Separation (BFS) methods in the Gilgel Gibe catchment where water demand for irrigation, domestic, and industrial purposes is dramatically increasing. The demand for groundwater and the existing ambitious plans to respond to this demand will put a strain on the groundwater resource in the catchment unless prompt intervention is undertaken to ensure its sustainability. Ground-based hydrometeorological 36-years data (1985 to 2020) from 17 stations and satellite products from CHIRPS and NASA/POWER were used for the SMB method. Six BFS methods were applied through the Web-based Hydrograph Analysis Tool (WHAT), SepHydro, BFLOW, and Automated Computer Programming (PART) to sub-catchments and the main catchment to estimate the groundwater recharge. The streamflow data (discharge) obtained from the Ministry of Water and Energy were the main input data for the BFS methods. The average annual recharge of groundwater was estimated to be 313 mm using SMB for the years 1985 to 2020 and 314 mm using BFS for the years 1986 to 2003. The results from the SMB method revealed geographical heterogeneity in annual groundwater recharge, varying from 209 to 442 mm. Significant spatial variation is also observed in the estimated annual groundwater recharge using the BFS methods, which varies from 181 to 411 mm for sub-catchments. Hydrogeological conditions of the catchment were observed, and the yielding capacity of existing wells was assessed to evaluate the validity of the results. The recharge values estimated using SMB and BFS methods are comparable and hydrologically reasonable. The findings remarkably provide insightful information for decision-makers to develop effective groundwater management strategies and to prioritize the sub-catchments for immediate intervention to ensure the sustainability of groundwater.

Keywords: groundwater recharge; catchment hydrology; soil moisture balance; baseflow separation; sustainability assessment

1. Introduction

Groundwater, a crucial element of the hydrologic cycle, plays a pivotal role in regulating surface water flow, sustaining ecosystems, and providing life-sustaining water resources [1]. Approximately 2.5 billion people across the world rely exclusively on groundwater resources to meet their basic daily water needs [2]. Its historical use for domestic water supply is long-standing, with increasing significance anticipated due to factors like population growth, surface water pollution, and unreliable surface water sources for irrigation [3–5]. Globally, groundwater allocation stands at 69% for agriculture, 22% for domestic use, and 9% for industrial purposes, with figures in Africa being 65%, 32%, and 2%, respectively [6]. In Ethiopia, groundwater plays a substantial role, providing 90% of domestic use, 95% of industrial need, and 1% of irrigation water supply, with an ambitious plan to develop large groundwater-based irrigated agriculture [7].

Despite its critical importance, groundwater constitutes less than 2% of the world's total water resources and often suffers from inadequate understanding, undervaluation, mismanagement, and over-exploitation [1]. Densely populated regions and established agricultural areas witness intensive groundwater development and depletion, underscoring the urgent need for responsible groundwater management [5]. The Gilgel Gibe catchment currently maintains extra inflows compared to outflows [8]. However, impending challenges like population growth, land cover changes, and increasing water demands raise concerns about potential future groundwater stress.

Takala's [9] analysis of land use and land cover (LULC) changes in the upper Gilgel Gibe catchment reveals an increase in agricultural land and a significant rise in urban areas between 1987 and 2010. A recent LULC map of the catchment extracted from the global Sentinel-2 LULC product with a 10 m resolution indicates that approximately 69% of the catchment is now classified as cropland. The catchment's population is projected to rise from 2.07 million in 2020 to 3.65 million by 2050 [10], leading to a dramatic surge in water demand. In 2020, average daily water demand, including both domestic and non-domestic usage, amounted to 70,244 m^3/day, with projections indicating a three-fold increase by 2050.

To address the escalating water demands, several deep and shallow wells are under construction. This includes deep wells for the Jimma Industrial Park and Jimma University, along with additional wells in various stages of construction. Additionally, seven shallow wells are being built in the Kersa district to meet local domestic water needs.

Given these developments, proactive policies and management strategies are crucial to ensure effective groundwater utilization in the catchment. The accurate estimation of groundwater recharge and identification of potential sources play a central role in guiding policy decisions and determining sustainable extraction limits from the aquifer [11]. So far, a limited number of studies have been conducted in the Gilgel Gibe catchment regarding groundwater recharge. Mengistu [12] characterized the recharge potential area of the upper Gilgel Gibe catchment based on proxy data using GIS and analytical hierarchy processes (AHP). However, he did not quantify the groundwater recharge of the catchment. Abebe [8] evaluated the groundwater potential of the upper Gilgel Gibe catchment based on the limited recorded data using a water balance model (WTRBLN). The model was run on a monthly basis and the annual groundwater recharge and abstraction rates were estimated to be 298 and 3.5 mm, respectively. This study does not cover the whole Gilgel Gibe catchment, and the spatial and temporal distribution of groundwater recharge was not discussed. Shimelis [13] applied SMB and BFS methods to the Bulbul catchment to estimate the annual groundwater recharge and found them to be 351 and 406 mm, respectively.

In the present study, SMB and BFS methods were applied on daily time steps to the Gilgel Gibe catchment. The SMB method, relying on historical climatological data, serves as a valuable tool for estimating groundwater recharge [14]. Recent advancements in satellite-based climatic data, which can well represent spatial and temporal variations, expand its applicability and enable upscaling to watershed or regional scales. The SMB method has been commonly applied in Ethiopia, particularly in the Upper Blue Nile River Basin [15–18]. Previous studies have validated its effectiveness against alternative methodologies, further affirming its reliability in groundwater recharge estimation [16–18]. Baseflow separation methods, especially in regions with readily available streamflow data, play a pivotal role in estimating groundwater recharge [8,11,13,16,19–25]. The method is reportedly efficient in humid and sub-humid regions [11,23]. Various techniques, ranging from geochemical to analytical and graphical approaches, have been developed and are applied globally, attesting to their versatility and applicability [26,27].

The graphical method, in particular, relies on specific assumptions to delineate baseflow from streamflow dynamics [28–30] and it is applicable in regions like the Gilgel Gibe basin where predefined catchment indices like alpha may be lacking. Noteworthy developments in this domain include the HYSEP computer program, which incorporates algorithms for graphical baseflow separation methods, enhancing efficiency [30]. Efforts have also been made to address user-friendliness concerns through integration with other software, such as BFI + 3.0, and web-based tools like WHAT [31]. The recent development of SepHydro, integrating 11 distinct baseflow separation methods including graphical methods, offers both user-friendliness and a platform for method intercomparison [28].

Several recursive digital filter methods are also available for baseflow separation including Chapman [32], Boughton [33], Chapman and Maxwell [34], Szilagyi and Parlange [35], BFLOW [11], Furey and Gupta [26], Eckhardt [36], and EWMA [37]. BFLOW is a computer program developed by Arnold and Allen [11] for a one-parameter recursive digital filter based on the Lyne and Hollick method. The method filters the high-frequency flow component (runoff) from the low-frequency flow component (baseflow) based on signal analysis and processing [38] and has been used by several scholars [11,23,39]. The method has no true physical bases but is shown to estimate a realistic value of baseflow when an appropriate filter parameter is used [11,23,31,39].

As the observed data acquisition is difficult for cross-validation, the application of two or more methods is convenient for groundwater recharge estimation. This study aims to employ the SMB method and BFS techniques to accurately estimate groundwater recharge in the Gilgel Gibe catchment, providing critical insights for the formulation of policies and management strategies to ensure sustainable groundwater utilization, in light of escalating water demands and potential future stressors. The BFS techniques applied include graphical methods through SepHydro, Eckhardt through WHAT, BFLOW, and PART.

2. Materials and Methods

2.1. The Study Area

The study area encompasses the Gilgel Gibe catchment, which is a sub-basin within the larger Omo Gibe river basin, located in the southwestern region of Ethiopia. Its centroid is positioned approximately 260 km away from Addis Ababa City and 70 km from Jimma City (Figure 1). The total land area of the Gilgel Gibe catchment is about 5145 km^2, delineated at the confluence of the primary Omo-Gibe River. Predominantly, 95% of the catchment area is situated within the Oromia National Regional State, specifically in the Jimma zone, while the remaining 5% is located in the Central Ethiopia Regional State, specifically in the Gurage and Yem zones. The terrain of the catchment is characterized by undulating topography, comprising mountains, hills, plains, deep V-shaped valleys, and flat river terraces along the main river course. This geographical layout facilitates the natural processes of both groundwater recharge and discharge [40]. The catchment's drainage system displays characteristics of dendritic, parallel, and subparallel patterns. The elevation of the study area exhibits a range from 3329 to 1080 m above sea level.

The Gilgel Gibe catchment is situated within a humid tropical region, predominantly characterized by a monomodal pattern of rainfall distribution. However, specific areas within dissected gorges exhibit traits of a semi-arid climate [41]. The principal source of water within the catchment emanates from rainfall [12]. While rainfall is distributed across the year, the catchment experiences a peak in rainfall during July and August. From March through October, the region receives an average monthly rainfall ranging from 92 mm to over 259 mm, whereas from November to January, the average monthly rainfall ranges from 29 mm to 42 mm based the data from CHIRPS for the period from 1985 to 2020. The annual average rainfall varies between 1193 mm and 1716 mm. The average temperature within the catchment is 19 °C.

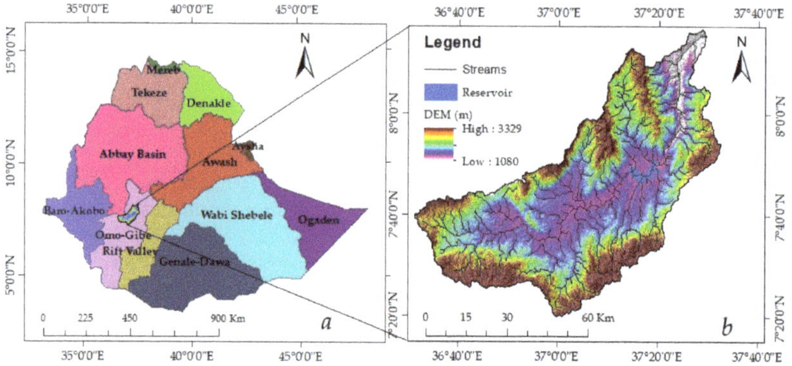

Figure 1. Location of Gilgel Gibe catchment: (**a**) Ethiopian river basins and (**b**) Gilgel Gibe catchment.

As delineated by the Land Use and Land Cover (LULC) map of 2018 derived from ESA Sentinel-2 imagery at 10 m resolution [42], cropland covers approximately 69% of the catchment area, while tree/forest occupies 13%, the built-up area covers 10%, rangeland encompasses 7%, and the remaining 1% is covered by water body, bare land, and regularly flooded vegetation. LULC stands as a pivotal determinant impacting groundwater recharge, surface runoff, and evapotranspiration processes.

The catchment is predominantly inhabited and cultivated by a substantial population of smallholding farmers, as noted by Teklu and Talema [43]. The natural vegetation in the region comprises dense, towering, broad-leaved trees, alongside evergreen trees, and expanses of open grassland [41]. Within the quaternary alluvial area, one can observe a transition from grassland to boggy terrain. In certain highland regions, bamboo forests are sporadically interspersed with trees, particularly along minor river courses. Additionally, apart from the indigenous vegetation, the cultivation of various tree species such as Eucalyptus, Coffee, and Inset is prevalent in specific localities [41].

It is important to note that the catchment area is subject to extensive cultivation, with LULC changes driven primarily by agricultural expansion, often at the expense of grassland, shrub-covered areas, and tree-dotted regions, as highlighted by Jillo [44]. The flatter, inadequately drained lowlands within the catchment typically remain uncultivated, instead serving as areas for dry season grazing and the establishment of Eucalyptus tree plantations. Figure 2a provides a detailed depiction of the LULC types and their distribution across the study area.

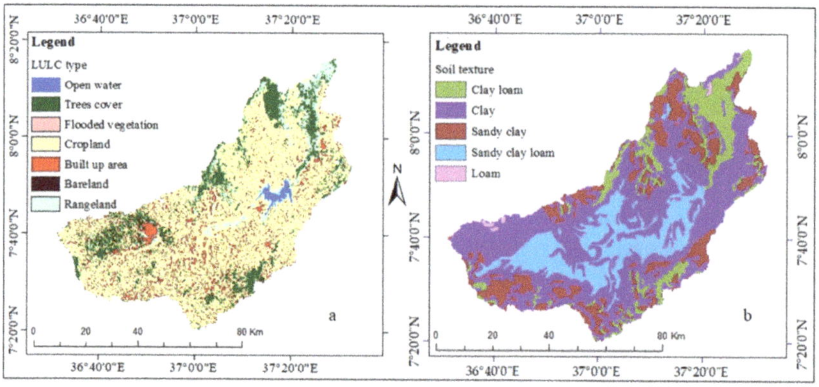

Figure 2. (**a**) LULC type and (**b**) Soil type of Gilgel Gibe catchment.

The study area predominantly features soils formed from tertiary and quaternary volcanic rocks [45–47]. In 2013, the Ministry of Water and Energy (MoWE) of Ethiopia undertook comprehensive field soil surveys and laboratory analyses, leading to the development of a soil map at a scale of 1:250,000. This initiative aimed to formulate a strategic plan for irrigation development across major river basins, notably the Omo-Gibe region [45]. Attempts have also been made by researchers to develop more detailed soil maps of specific localities in the catchment [46]. For this study, the MoWE's map was employed because the map covers the whole study area, despite the existence of other specific soil types in certain areas of the study region.

According to the MoWE's soil map, eight distinct soil types are identified within the catchment. However, we grouped them into five classes (Figure 2b) based on their textural type, which determines soil water-holding capacity and infiltration for recharge estimation. Notably, clay soil predominates, encompassing 50% of the study area. It is primarily concentrated in the central, northern upstream, and southern downstream segments of the catchment (Figure 2b). Sandy clay and sandy clay loam soils cover 35% of the total area. Sandy clay occupies the northern and southern extremities of the catchment, while sandy clay loam soils are predominantly situated in the middle section (Figure 2b). In addition, clay loam and loam soils collectively account for 15% of the catchment area. These soils are primarily located downstream, aligning with the main river valley, and extend to higher elevations in the northern and southern catchment regions along the ridges (Figure 2b).

2.2. Geology and Hydrogeology

Over time, various regional geological investigations have been conducted in and around the study area, as documented by Alemu [40], Basalfew [48], Davidson [49], Haro [41], Merla et al. [50], and Tefera et al. [51]. However, mapping specifically focused on the geology of the Gilgel Gibe catchment has been comparatively limited, with studies by Regassa et al. [52], Tadesse [53], and Van Daele [54] providing notable contributions.

Merla et al. [50] initially mapped the geology of Ethiopia, representing the Gilgel Gibe catchment with Jimma volcanic flows that erupted during the Oligocene to Miocene periods. Subsequently, Tefera [51] provided a more detailed geological classification, incorporating Jimma volcanic formations (ranging from the late Eocene to late Oligocene), Makonnen Basalts (Oligocene to Miocene in age), and Nazret Series (Miocene to Pliocene in age). Recent geological surveys conducted by the Ethiopian Geological Survey have further refined our understanding of the country's geological formations [40].

The predominant rock types within the catchment encompass basalt flows, trachyte flows, rhyolite flow, pyroclastic flows, welded ignimbrite, alluvial deposits, and stratified tuffs (Figure 3). This comprehensive geological framework provides a critical foundation for understanding the hydrogeological characteristics of the Gilgel Gibe catchment.

The geological formation of the Gilgel Gibe catchment can be categorized into four main units. At the base lies lower-tertiary volcanic flow, dating back to pre-Oligocene times, comprising lower basalt flows overlain by lower pyroclastic flow. Above this, upper-tertiary volcanic flows from the Oligocene era are characterized by middle basalt flows topped by Trachyte flows. Further up, upper-tertiary volcanic formations from the Miocene to Pliocene epochs include upper basaltic flows, rhyolites, upper pyroclastic material, and Setema basalt. Lastly, there are quaternary stratified tuff and alluvial deposits [41].

Lower basalt flows, covering the downstream of the Gilgel Gibe River valley and extending from the reservoir to the main outlet, are characterized by aphanitic and slightly weathered basalt, containing plagioclase and pyroclastics with trachyte plugs [40,41]. These lower basalt flows are underlain by Omo trachyte flow and overlain by a fine-grained basalt layer. Lower pyroclastic deposits extend around the Gilgel Gibe dam, encompassing areas from Asendabo to Dimtu and from Bore to Serbo, with ignimbrite, tuffs, and lithic tuffs containing trachytic obsidian lenses. Middle basalt flows, found in the northwest of Jimma and northern Dimtu, comprise fine-grained dark gray basalt, interspersed with paleo soils and, in some places, basaltic pyroclasts with varying sizes of particles cemented by fine-grained materials and scoriaceous basalt. Trachyte flows dominate the upstream area of the catchment, spanning Dedo, Jimma, and Sarbo, with mostly massive rock formations occasionally displaying fractures and intercalations of ignimbrite, basalt, and paleo soil. Downstream, upper basalt flows with minor trachyte and tuff layers overlay the trachyte flows and directly underlie the lower basalt along the Gilgel Gibe gorge, also extending into the southwestern catchment area south of Dedo. Quaternary alluvial deposits are prevalent along the main rivers in plateau regions, where streams flow and drain Tertiary to Pliocene volcanic flows [41]. Around the Gilgel Gibe dam reservoir, the Asendabo area is flat, with small hills and depressed areas that are filled with younger quaternary tuffs and alluvium. The depression extends to Jimma Town [55].

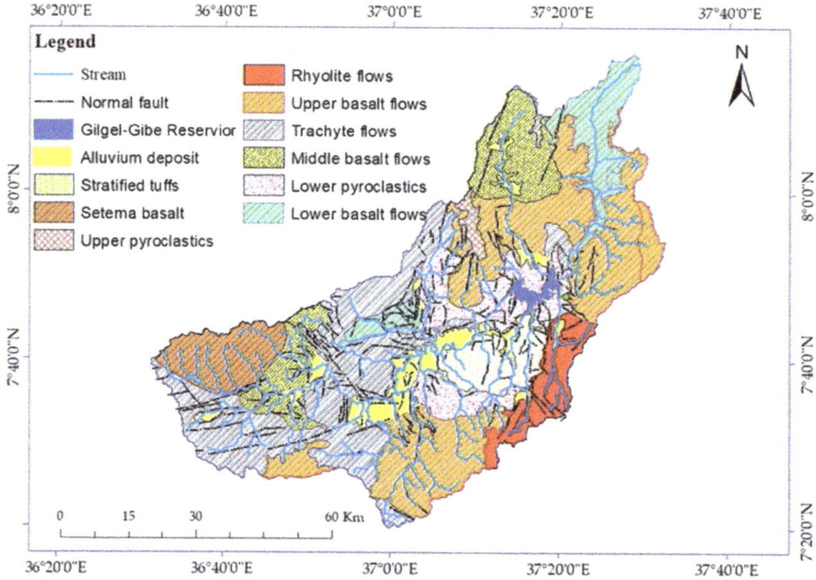

Figure 3. Geological map of Gilgel Gibe catchment.

Hydrogeologically, the aquifer/aquitard systems encompass quaternary alluvial deposits, stratified tuffs, basalts, trachyte, sedimentary, and pyroclastic rocks [55]. Different hydrogeological units are identified: (1) porous permeable aquifers, primarily consisting of alluvial sediments, which are concentrated along major rivers and the Asendabo area; (2) fissured permeable aquifers, located in the eastern part of Jimma, are predominantly composed of basalts with trachyte and rhyolites, featuring open faults and fault systems that serve as significant groundwater flow pathways; (3) mixed fissured and porous permeable aquifers, consisting of trachyte lava flows, ignimbrites, and rhyolite intercalated with porous tuffs and pumice. These aquifers can be both unconfined and confined, with varying thickness. The topographic configuration plays a pivotal role in defining potential recharge and discharge areas within the aquifers system.

2.3. Data Collection and Analysis

A 30 m resolution Digital Elevation Model (DEM) from the Shuttle Radar Topographic Mission (SRTM) "https://earthexplorer.usgs.gov (accessed on 15 February 2022)" was employed in conjunction with ArcGIS to analyze the catchment and its sub-catchments. Utilizing LULC and soil maps (Figure 2a,b), surface runoff was estimated using a modified SCS-CN method. Additionally, a geological map (Figure 3) was referenced to gain a comprehensive understanding of the aquifer system within the catchment.

The availability and reliability of hydrometeorological data are the key for groundwater recharge estimation and runoff assessment through the SMB method. In many African regions, traditional ground-based observations have been the primary source of such data [56]. Nevertheless, these methods are prone to uncertainties and offer limited spatial coverage. Even in well-monitored areas, temporal inconsistencies and prolonged periods of missing data are common [44,57]. Within the Gilgel Gibe catchment, there are 15 meteorological stations along with 2 neighboring stations (Figure 4), each exhibiting varying temporal coverage and substantial data gaps, particularly in rainfall records. Only the Jimma station consistently maintained reliable hydrometeorological data, with a mere 1% data deficit from 1985 to 2020. This underscores the inadequacy of relying solely on ground-based measurements for groundwater recharge estimation in this catchment.

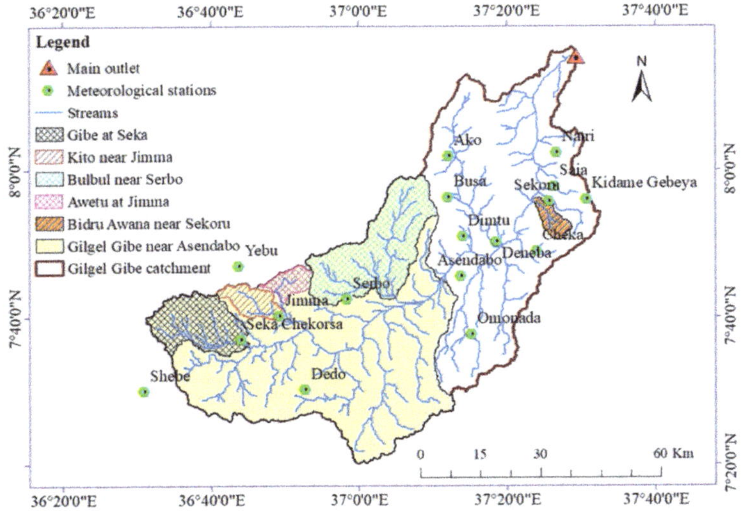

Figure 4. Map of gauged sub-basins and locations of meteorological stations.

To mitigate these challenges, satellite-based data, exhibiting good correlation with the available ground-based measurements in Ethiopian basins in general and the Gilgel Gibe basin in particular, were reviewed [16,56,58]. Accordingly, Climate Hazards Group InfraRed Precipitation with Stations (CHIRPS) with $0.05° \times 0.05°$ spatial resolution demonstrated a robust correlation with observed data in the catchment. While CHIRPS provides rainfall data from 1981 to the present at daily, monthly, and annual temporal scales [59], additional sources were needed for temperature, relative humidity, wind speed, and solar radiation. The National Aeronautics and Space Administration/Prediction of Worldwide Energy Resources (NASA/POWER), which emerged as a valuable data source offering satellite and model-based solar and meteorological data [60–62], was used. These parameters have demonstrated sufficient accuracy in providing reliable data [62]. NASA/POWER supplies daily meteorological data on a global grid with varying spatial resolutions, ranging from 1° latitude by 1° longitude for radiation datasets to 0.5° latitude by 0.625° longitude for meteorological datasets. These data are accessible from 1981 to near real-time. The rainfall data obtained from both CHIRPS and NASA/POWER were corroborated with available ground-based measurements (Table 1), revealing a strong agreement. However, the available ground-based measurements are insufficient and inconsistent to evaluate the correlation of other weather parameters. This combined dataset, encompassing rainfall from CHIRPS and comprehensive hydrometeorological data from NASA/POWER, was then utilized for groundwater recharge estimation.

Table 1. Correlation of CHIRPS and NASA/POWER with ground-based measured monthly rainfall.

Stations	Year	Number of Years	Correlation Coefficient	
			CHIRPS	NASA/POWER
Ako	2019	1	0.82	0.76
Asendabo	1995–2003	9	0.97	0.93
Busa	1994–2002	9	1.00	1.00
Cheka	1998–2005	8	0.97	0.93
Chekorsa	2000–2005	6	0.95	0.90
Dedo	1985–1990	6	0.95	0.90
Deneba	1994–2000	7	0.78	0.66
Dimtu	2004–2007	9	0.96	0.94
Jimma	1985–2017	33	0.98	0.92
Kidame Gebeya	1988–1993	6	0.96	0.93
Natri	1989–1990	2	0.85	0.76
Omonada	1999–2005	7	0.93	0.78
Saja	1985–1995	11	1.00	0.98
Sekoru	1985–1991	7	0.99	0.95
Serbo	1999–2005	7	0.98	0.84
Shebe	1985–1993	9	0.98	0.97
Yebu	1985–1992	8	0.95	0.96

Stream flow data from six gauging stations of different duration were collected from the Ministry of Water and Energy (Table 2). However, there is no gauging station at the main outlet of the catchment (Gilgel Gibe station). Stream flow for the Gilgel Gibe station was obtained by using an appropriate regionalization method. The modified linear interpolation [63] is an appropriate method for regionalization if the ungauged catchment is located either upstream or downstream of the gauged catchment. Asendabo, a gauged station that drains about 60% of the Gilgel Gibe catchment, is located upstream of the Gilgel Gibe outlet. The location of gauged sub-catchments and meteorological stations is depicted in Figure 4.

Table 2. Streamflow gauging stations.

Gauging Station	Longitude	Latitude	Area (km^2)	Year
Awetu at Jimma (Awetu)	36.83	7.68	72	1985–2007
Bidru Awana near Sekoru (Bidru Awana)	37.40	7.92	41	1985–2014
Gibe Near Seka (Seka)	36.75	7.60	280.4	1985–2014
Gilgel Gibe Near Asendabo (Asendabo)	37.18	7.75	2966	1985–2019
Kito Near Jimma (Kito)	36.83	7.70	85	1985–2005
Bulbul Near Serbo (Bulbul)	37.08	7.57	526	1986–2018
Gilgel Gibe at main outlet (Gilge Gibe)	37.49	8.26	5145.41	1985–2003

Asendabo station was selected for the regionalization method because it drains a larger area compared to other sub-catchments. The stream flow at the main outlet was determined from stream flow measured at the Asendabo station using the modified linear interpolation equation (Equation (1)).

$$Qu = \frac{Au * Pu}{Ag * Pg} * Qg \tag{1}$$

where Qu and Qg are the stream flow of ungauged and gauged catchment, respectively (m^3/s);

Au and Ag are the area of gauged and ungauged catchment, respectively (km^2);

Pu and Pg are the mean annual rainfall in the drainage area of the ungauged and gauged catchments, respectively (mm).

The stream flow data observed at sub-catchments and determined at the main outlet using Equation (1) were used to estimate the recharge using BFS methods for the respective sub-catchments. The catchments and sub-catchments area with the duration of recorded data are described in Table 2.

2.4. Recharge Estimation Methods

2.4.1. Soil Moisture Balance Method

The SMB method (Equation (2)) was developed by Thornthwaite [64], further revised in 1955 [65], and edited in 1957 by Thornwhaite and Mather [14]. The method estimates the groundwater recharge indirectly as a residual of water balance components of the soil layer. The method allows us to compute the groundwater recharge on hourly, daily, monthly, or annual timesteps using available climate data [66]. It estimates the diffuse recharge from rainfall and does not take into account the recharge from preferential flows. The required input data for the SMB method are rainfall, PET, runoff coefficient, PAW, and initial soil moisture content [67], allowing us to estimate the soil water balance components:

$$R = P - Ea + \Delta S - Ro \quad (2)$$

where R = recharge; P = precipitation; Ea = actual evapotranspiration; ΔS = change in soil water storage; and Ro = runoff.

The SMB method estimates groundwater recharge using a point scale. However, runoff is usually estimated for land areas of varying scales. The USDA Natural Resources Conservation Service, previously called the Soil Conservation Service (SCS-CN), curve number method [68] which was developed for small catchments [33,69–72], was applied in this study, creating a Thiessen polygon for 17 hydrometeorological stations. The area represented by each station and the estimated runoff coefficient are described in Table 3.

Table 3. Meteorological stations and respective PAW and runoff coefficients.

No	Stations	Area (km^2)	Average RD (cm)	(FC-PWP) (mm)	PAW (mm)	Runoff Coefficient (%)
1	Ako	365	156	216	338	23
2	Asendabo	303	77	209	161	35
3	Busa	319	114	203	231	26
4	Cheka	193	74	186	138	18
5	Seka Chekorsa	486	122	208	254	20
6	Dedo	704	79	211	167	26
7	Deneba	162	74	204	151	27
8	Dimtu	169	81	205	166	33
9	Jimma	264	140	212	297	18
10	Kidame Gebeya	81	94	204	192	10
11	Natri	392	160	192	307	12
12	Omonada	635	88	211	186	20
13	Saja	88	124	199	247	14
14	Sekoru	141	110	200	220	16
15	Serbo	646	92	216	199	25
16	Shebe	130	117	204	238	19
17	Yebu	67	118	186	220	32

The SCS-CN method is most widely used because of its ease of use and stability, and it considers the most important watershed characteristics such as soil, LULC, hydrologic condition, and antecedent moisture [69]. The depth of runoff is determined from rainfall using an empirical equation (Equation (3)) [69,73].

$$Q = \frac{(P - 0.2S)^2}{P + 0.8S} \text{ for } P > 0.2S, \text{ else } Q = 0 \quad (3)$$

where Q—surface runoff (mm), S–potential maximum retention (mm) and P—precipitation (mm).

Potential maximum soil storage is given by

$$S = \frac{25,400}{CN} - 254; CN \text{ is the curve number.} \tag{4}$$

Rainfall-runoff processes are highly affected by soil, LULC, and Antecedent Moisture Conditions (AMC). AMC refers to the degree of soil saturation before rainfall events and is categorized into 3 types (AMCI, AMCII, and AMCIII) [69,70]. The associated curve numbers, CNI, CNII, and CNIII, are required to determine runoff from rainfall. The curve number for AMCII (CNII) for soil and LULC complex was adopted from the Ethiopian Road Authority's Drainage Design Manual [71]. The land surface slope is another important factor that significantly affects the rainfall-runoff process but was not considered in the original curve number method. The land slope correction factor developed by Huang and Gallichand [72] (Equation (5)) was applied to improve the accuracy of the method to estimate the surface runoff from the catchment.

$$CNII\alpha = \frac{322.79 + 15.63\alpha}{\alpha + 323.52} \tag{5}$$

The slope α was generated from DEM using ArcGIS and Zonal statistics in the ArcGIS environment; this was applied to calculate the average value of α for each Thiessen polygon corresponding to meteorological stations.

The curve number for dry and wet conditions, *CNI* and *CNIII*, were derived from modified *CNII* using Equations (6) and (7), respectively [69,70].

$$\text{For dry conditions}: CNI = \frac{4.2CNII}{10 - 0.058CNII} \tag{6}$$

$$\text{For wet condition}: CNIII = \frac{23CNII}{10 + 0.13CNII} \tag{7}$$

The most widely used approach, the 5-day antecedent precipitation index (API), was used to determine the daily runoff from daily rainfall data.

Evapotranspiration was estimated using the modified Penman–Monteith method [74,75]. This method is widely accepted and has the advantage of considering the effect of standard climatological parameters including temperature, relative humidity, solar radiation, and wind speed [75]. Moreover, the method has the advantage of adapting local parameters of specific locations.

To estimate the plant's available water (PAW), the essential parameters including field capacity (FC), permanent wilting point (PWP), bulk density (BD), and rooting depth (RD) of the plants were obtained from both laboratory analyses and the literature. The FC was estimated in the soil laboratory of the Jimma Institute of Technology, Jimma University, using the gravimetric and oven drying techniques. A purposive sampling technique was used at seven points near the weather stations (Seka, Dedo, Jimma, Dedo, Asendabo, Dimtu, and Serbo and Mazoria near Yebu), representing major soil types including clay, loam, and sandy clay loam, which were the selected types. Two samples for each station at depths of both 0–30 cm and 30–60 cm were taken for the analysis. The samples were taken using an auger and transported to the laboratory using plastic bags. Then, they were saturated in the laboratory and subjected to gravity for 24 h to drain the excess water and consequently maintain the soil water content at FC. The weight of the sample was measured and oven-dried at 105 °C for 24 h. The oven-dried sample was reweighted, and the difference was calculated to obtain the gravimetric water content at FC. For the determination of BD, undisturbed samples were collected using core samplers of 38 mm in diameter. The samples were taken to the laboratory and extruded from the samplers using a hydraulic extruder. The volume of the sample was determined and oven-dried to obtain the dry mass of the sample. Then, the BD was determined by dividing the dry mass of the sample by the volume. The estimated gravimetric soil water content was converted to volumetric soil water content by multiplying it with the BD [76]. The PWP and FC of loam, clay loam, and sandy clay were adopted from previous studies [77–80].

The major LULC types of the catchment are cropland, tree covers, grassland, and shrubs (Figure 2a). The major crops grown in the area include teff, wheat, barley, beans, sorghum, and maize [81]. The average rooting depth of these crops is 70 cm [82]. The major vegetation in the catchment is evergreen trees [40,41] and the average rooting depth is 336 cm. The average rooting depths of shrubs and grassland are 350 and 140 cm, respectively [83].

To determine the plant available water (PAW), a Thiessen polygon was created for 17 meteorological stations and overlaid with soil and LULC maps to determine the weighted average of the root depth and water-holding capacity (FC-PWP) of the soil in the polygons. The initial soil moisture content is another input parameter required for the SMB method. Since the initial time step was set to the first day of January, the average soil moisture content measured at the Jimma and Sekoru stations on that day was used for the stations located upstream and downstream, respectively.

2.4.2. Baseflow Separation Methods

Baseflow separation techniques can provide valuable insight into the contribution of groundwater to rivers [27,84] as a proxy of groundwater recharge in areas where groundwater withdrawal is negligible. To estimate the groundwater recharge in the Gilgel Gibe catchment, algorithms from SepHydro, BFLOW, and WHAT were utilized.

1. SepHydro

Among eleven baseflow separation algorithms incorporated in SepHydro [28], fixed interval, sliding interval, and local minimum methods were used in the present study. The methods conceptually consider the line connecting the low points of the hydrograph to define the baseflow hydrograph [30]. The baseflow hydrograph is drawn for consecutive specified widths or windows. The length of the width is equal to 2N but the nearest integer is between 3 and 11. N can be determined from the catchment area using an empirical equation suggested by Pettyjohn and Henning [28] (Equation (8)). These methods are simple to use and are able to reproduce consistent results that can replace manual separation.

$$N = A^{0.2} \tag{8}$$

N is the number of days after which runoff ceases;
A is drainage area (km^2).

a. Fixed interval method

All baseflow values in a given interval or window are set to a minimum stream flow value in that interval. The first interval starts with the first day of the streamflow record. The method can be visualized as a moving window of 2N days, considering the width or window. The window moves horizontally by 2N days and the process is repeated until the last interval to complete the baseflow separation for the entire streamflow data is recorded [28].

b. Sliding interval method

The value of baseflow at the middle of each interval or window is set to the minimum streamflow value in the respective interval. The window moves vertically upward until it intersects with the lowest point of the stream flow hydrograph in the respective interval and the window then slides to the next day. The process is repeated until the last interval to complete the complete baseflow separation for the entire streamflow data is recorded [28].

c. Local minimum method

The streamflow value of each day in a given interval is checked whether it is the lowest value in that interval or not. If it is, then it is considered a local minimum and is joined with adjacent local minimums using a straight line. The resulting line, across the whole dataset, represents the baseflow hydrograph [28].

2. BFLOW

The algorithm of BFLOW [11] was developed based on the one-parameter recursive digital filter developed by Lyne and Hollick [38] and used by Nathan and McMahon [23] and Arnold, Allen [11]. The separation process is based on the wave of the hydrograph [31] and the filter passes over the entire stream flow three times (forward, backward, and forward) [11,39]. BFLOW algorithm uses the filter equation described in Equation (9).

$$qt = \beta q(t-1) + \frac{(1+\beta)}{2} * (Qt - Q(t-1)) \tag{9}$$

qt—filtered surface runoff (m^3/s) at the t time step (day);
Q—original streamflow (m^3/s);
β—filter parameter.

Once the runoff component (quick response) is determined using Equation (9), the baseflow component (bt) can be computed using Equation (10).

$$bt = Qt - qt \tag{10}$$

The value of the filter parameter was determined by Nathan and McMahon [23]. According to Nathan and McMahon [23], based on visual inspection, the baseflow index falls in an acceptable range when the value of the filter parameter is between 0.9 and 0.95. Nathan and McMahon [23] applied three values (0.9, 0.925, and 0.95) to the whole datasets of 186 catchments and identified the most appropriate value filter value, which was 0.925. The results from the digital filter were compared with those of smoothed minima and manual methods for all catchments. The digital filter provided a more stable baseflow index (BFI) than that of smoothed minima method with a 0.97 coefficient of determination. The filter parameter determined by Nathan and McMahon [23], 0.925, was used in this study.

3. WHAT

WHAT is a web-based hydrograph analysis tool developed by Lim [31] and is currently integrated with three baseflow separation methods including a two-parameter recursive digital filter, the Eckhardt method [36]. Eckhardt [36] developed the two-parameter recursive digital filter based on a one-parameter recursive digital filter [38], introducing an additional parameter called the Maximum Baseflow Index (*BFImax*) (Equation (11)). The filter separates the direct runoff from baseflow based on the frequency signals. Low-frequency signals are associated with baseflow, and high-frequency signals are associated with runoff [31].

$$bk = \frac{[(1 - BFImax) * \alpha b_{k-1} + (1-\alpha) * BFImax * yk]}{1 - \alpha * BFImax} \tag{11}$$

where

bk: baseflow at time step k;
b_{k-1}: baseflow at time step k − 1;
yk: total streamflow at time step k;
$BFImax$: Maximum Baseflow Index;
α: filter parameter.

The accuracy of the method depends on the quality of stream flow data and filter parameter used [85]. *BFImax* is the maximum value of a long-term ratio of baseflow total stream flow. Eckhardt [36] estimated the value of *BFImax* for different hydrological and hydrogeological conditions. According to Eckhardt [36], the *BFImax* for perennial streams with porous aquifers is 0.8, for ephemeral streams with porous aquifers it is 0.50, and it is 0.25 for perennial streams with hard rock aquifers. The values were validated by comparing them with the values obtained from other research under similar conditions. Lim, Park [31] applied the Eckhardt method in a small catchment with Eckardt's filter value (*BFImax* = 0.8 and α = 0.98) to filter the direct runoff components from streamflow. The result was highly correlated with that of BFLOW with a unit coefficient of determination. Eckhardt's filter parameter values were applied in the Gilgel Gibe catchment.

4. PART

PART uses streamflow partitioning to estimate baseflow from a daily streamflow record. The program scans the streamflow data series for days that fit a requirement of antecedent recession and designates baseflow to be equal to streamflow on these days, then linearly interpolates on the days that do not fit the requirement of antecedent recession [86]. Although the program uses daily streamflow data, Rutledge [87] recommended that results should be reported at a larger time scale, at least monthly. The program is recommended for the estimation of baseflow in aquifers characterized by diffuse areal recharge to the groundwater table and groundwater discharge to the stream [86]. The program is appropriate for the basin in which natural groundwater flow is maintained and gauging stations located downstream at the end of the basin that capture the most outflow. The only groundwater recharge source in the Gilgel Gibe catchment has been rainfall.

3. Results and Discussions

3.1. Soil Moisture Balance Method

Meteorological data highly influence groundwater recharge estimated using the SMB method. Rainfall and potential evapotranspiration play a major role. Rainfall is a primary source of groundwater recharge while PET adversely affects the recharge. Figure 5 describes the spatial distribution of rainfall and potential evapotranspiration in the catchment. The spatial distribution of annual rainfall shown in Figure 5a shows that rainfall decreases from west to east. The potential evapotranspiration was estimated using the Penman–Monteith method at each meteorological station and interpolated using the Inverse Distance Weighted (IDW) method in the GIS environment, as shown in Figure 5b. The map shows PET is increasing in the downstream direction, from a southwest to northeast direction. PET varies from 1171 to 1434 mm with a mean value of 1285 mm, whereas rainfall varies from 1193 to 1716 mm with a mean value of 1539 mm based on CHIRPS data. To develop the groundwater management strategies in the catchment, the spatial distribution of rainfall and PET along with the estimated groundwater recharge, offer valuable insights.

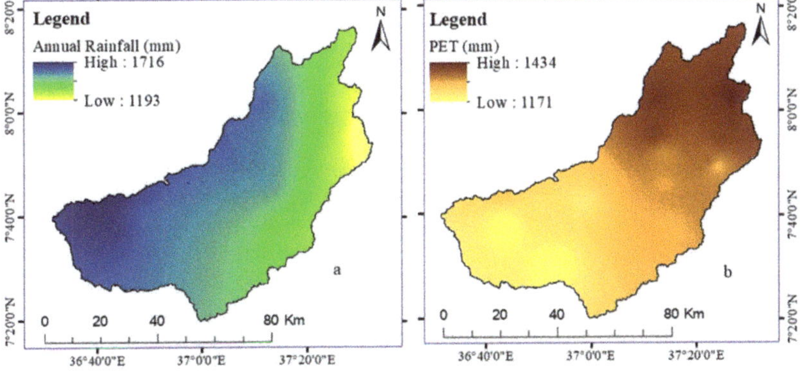

Figure 5. The map of (**a**) rainfall and (**b**) potential evapotranspiration (PET).

In addition to the daily rainfall and PET data, PAW, runoff coefficient, and initial soil moisture content are required in the SMB method. The method needs a single value for each parameter for each station representing the Thiessen polygon. The SMB method uses the initial soil moisture content to start the simulation and the parameter influences only the initial time steps. The model iteratively calculates the soil moisture content for the entire time steps and the influence of this parameter on the annual recharge is insignificant (Figure 6). The PAW, rooting depth, and runoff coefficient values determined at each meteorological station are described in Table 3, with respective area coverage as delineated by Thiessen polygons based on the soil and LULC matrix.

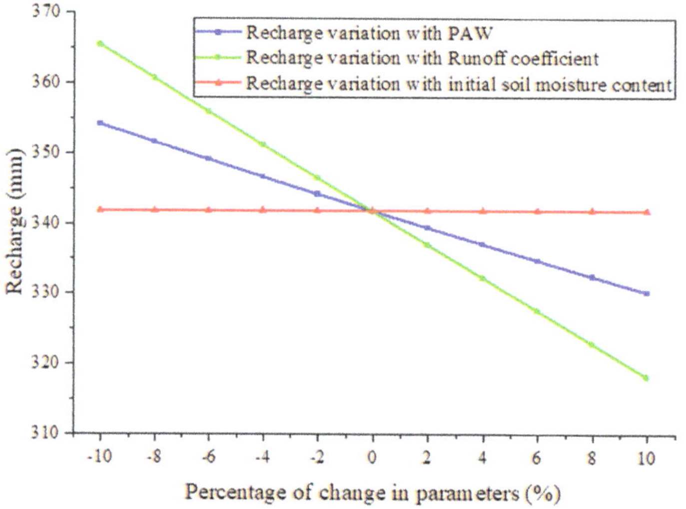

Figure 6. The effect of PAW, runoff coefficient, and initial soil moisture on the annual recharge.

A sensitivity analysis was carried out to identify the sensitivity of the lumped input parameters (PAW, runoff coefficient, and initial soil moisture content). The sensitivity analysis was conducted for the Jimma area to provide insight into the potential errors that might be incorporated, due to parameter value estimations. The meteorological station located in the area, Jimma station, provides better recorded data compared to the others. The analysis was carried out by changing a parameter at a certain time, keeping the remaining parameters constant (Figure 6).

The estimated value of PAW, runoff coefficient, and initial soil moisture content at Jimma station are 297 mm, 18%, and 144 mm, respectively, and the corresponding estimated annual recharge is 342 mm.

Figure 6 illustrates that the annual recharge does not vary with the initial soil moisture content, indicating that initial soil moisture is an insensitive parameter to the recharge. The increment of PAW by 2% decreases the annual recharge by 2.39 mm, while the increment of runoff coefficient by 2% decreases the recharge by 4.72 mm (Figure 6). The sensitivity analysis shows that the runoff coefficient is more sensitive than PAW, and the initial soil moisture is an insensitive parameter to the annual recharge. Among the lumped input parameters used in SMB method, the annual recharge is highly influenced by the runoff coefficient. This implies that to improve the groundwater recharge in the catchment, for instance, one can apply water management practices that reduce the surface runoff to be more efficient and effective.

The daily groundwater recharge was estimated for Thiessen polygons corresponding to each meteorological station, but the result is reported on a monthly, seasonal, and annual basis. Table 4 shows the annual rainfall (RF), potential evapotranspiration (PET), actual evapotranspiration (AET), and groundwater recharge estimated at each meteorological station corresponding to each Thiessen polygon. Based on the data obtained from CHIRPS, the catchment gains 1538 mm of water annually from rainfall and loses 862 mm through evapotranspiration, as estimated using the SMB method. The annual groundwater recharge estimated using the SMB method is 313 mm, which is 20% of the annual rainfall.

Table 4. Annual RF, PET, AET, and recharge.

Meteorological Stations	RF (mm)	PET (mm)	AET (mm)	Recharge (mm)
Ako	1657	1414	959	312
Asendabo	1569	1312	786	233
Busa	1573	1393	887	274
Cheka	1373	1300	775	350
Seka Chekorsa	1700	1171	914	442
Dedo	1585	1171	820	351
Deneba	1373	1349	774	228
Dimtu	1569	1321	797	253
Jimma	1588	1220	956	342
Kidame Gebeya	1129	1391	806	209
Natri	1304	1434	919	225
Omonada	1387	1300	827	282
Saia	1368	1402	892	282
Sekoru	1368	1382	884	264
Serbo	1582	1214	847	337
Shebe	1699	1190	926	446
Yebu	1758	1313	905	287
Weighted Average	1539	1285	862	313

Figure 7 shows the recharge distribution in the catchment as estimated using the SMB method per each Thiessen polygon. The Seka Chekorsa area, where the Gilgel Gibe rises, has the highest annual recharge of the catchment followed by the Dedo, Jimma, Serbo, Ako, and Cheka areas, which receive moderately high recharge. The annual recharge value estimated at these stations is greater than the average catchment recharge of 313 mm. High rainfall occurs with vegetated LULC characteristics favoring the recharge in these areas of the catchment. However, the Natri, Kidame Gebeya, Deneba, and Asendabo areas receive the relatively lowest annual recharge of the catchment. The annual recharge value estimated for these is less than the average recharge value of the catchment. The Omonada, Busa, Dimtu, Sekoru, and Saja areas cover the middle part of the catchment and receive relatively moderate annual recharge. Based on the SMB method, the annual recharge estimated at Kidame Gebeya is lower than that of the Seka Chekorsa station by 53%, where the discrepancy in the annual rainfall is 34%.

The temporal distribution of average annual rainfall and recharge for the Seka Chekorsa, located upstream, and the Sekoru station, located downstream, for the period from 1985 to 2020, is shown in Figure 8a,b, respectively. For both stations, the highest annual rainfall occurred in 2019, which is 2213 and 1739 mm at the Seka Chekorsa and Sekoru stations, respectively. The annual maximum groundwater recharge also occurred in 2019, which is 882 mm and 567 mm at the Seka Chekorsa and Sekoru stations, respectively. The minimum annual recharge values estimated at Seka Chekorsa and Sekoru were 210 and 64 mm, occurring in 2002 and 2000, respectively. The annual recharge percentage of rainfall varied from 14% in the year 1995 to 40% in 2019 at the Seka Chekorsa station and ranged from 5% (2000) to 33% (2019) at the Sekoru station. In general, groundwater recharge is highly varied with time, not only at the Seka Chekorsa and Sekoru stations (Figure 8a,b), but also for all the other stations. The recharge increases with rainfall but the trend is not a one-to-one correspondence (Figure 8). As a result of fluctuating annual rainfall, Figure 8a,b demonstrate that while PET varies little, recharge varies significantly. This indicates that the variation in recharge is more affected by the rainfall magnitude and intensity. The SMB estimates the recharge on a daily time step and the daily rainfall, and its frequency controls the recharge temporal variability. Annual recharge is calculated as a percentage of annual rainfall, and does not take the temporal variability of rainfall and recharge into account.

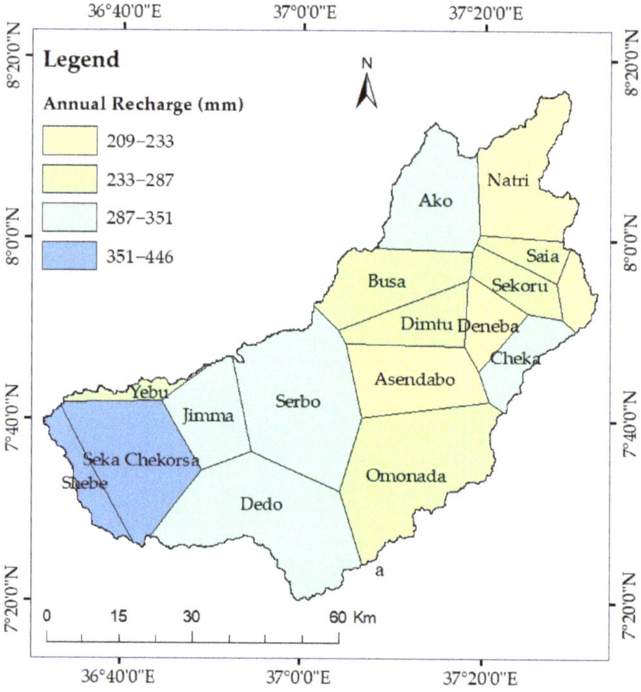

Figure 7. Annual recharge of the catchment estimated for each Thiessen polygon.

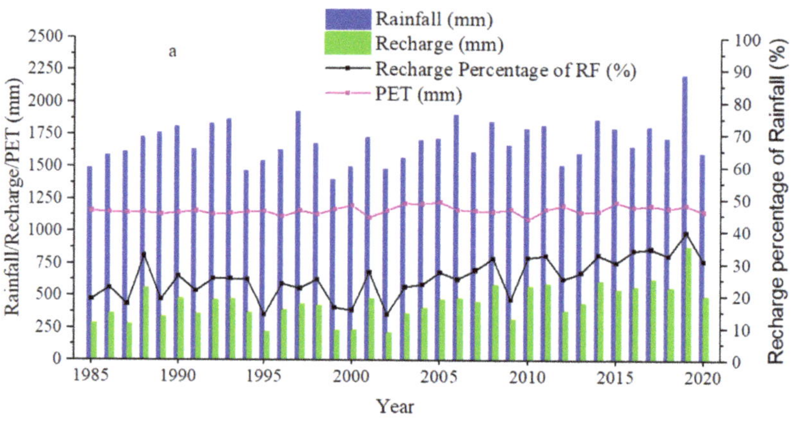

Figure 8. *Cont.*

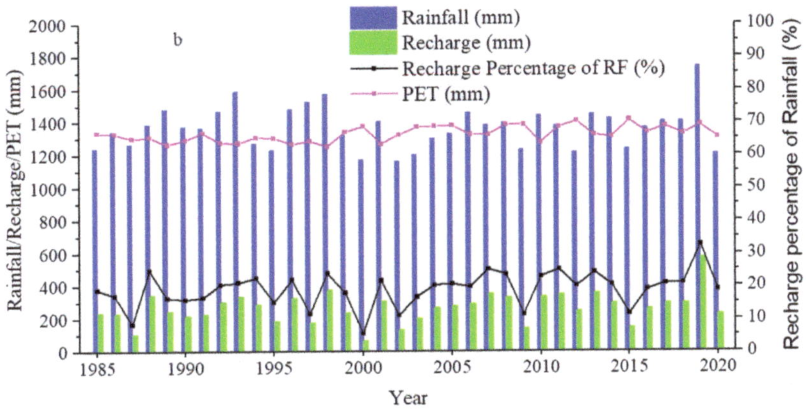

Figure 8. Recharge percentage of annual rainfall and annual recharge trend: (**a**) Seka Chekorsa station and (**b**) Sekoru Station.

Based on the SMB estimates, the seasonal groundwater recharge of the catchment was estimated to be 4 mm, 237 mm, 72 mm, and 0 mm in spring (March–April–May), summer (June–July–August), autumn (September–October–November) and winter (December–January–February), respectively. The groundwater recharge in summer accounts for 76% of the annual recharge of the catchment. Summer and autumn together account for 99% of the annual recharge, whereas spring receives only 1% of the annual groundwater recharge. Based on the monthly time scale, the groundwater recharge of the catchment varies from 0 to 105 mm (Table 5). The catchment receives significant monthly recharge from June to October. Even though some rainfall occurs, the recharge is nil in the remaining months because of the soil moisture deficit and high evapotranspiration demand during these months. Comparing the amount of recharge in May and November, the recharge percentage of rainfall in November is higher than that of May.

Table 5. Average monthly recharge and rainfall of Gilgel Gibe catchment (1985–2020).

Months	Jan.	Feb.	Mar.	Apr.	May	Jun.	Jul.	Aug.	Sep.	Oct.	Nov.	Dec.
RF (mm)	30	42	92	122	168	225	259	245	184	100	42	29
R (mm)	0	0	0	0	4	31	101	105	51	19	2	0
Recharge % of RF	0	0	0	0	3	14	39	43	28	19	5	0

Figure 9 describes monthly rainfall, evapotranspiration, and recharge for the Seka Chekorsa and Sekoru stations. These stations are situated in the upstream and downstream parts of the catchment (Figure 4). Even though they do not represent the entire upstream and downstream part of the catchment, the results can be compared and cross-validated with the estimates from the BFS method of corresponding sub-catchments, Seka and Bidru Awana. The PET is more or less higher for all months running from September to May for both stations.

The significant difference in the recharge percentage of rainfall in May and November could be due to soil moisture difference. Since May is preceded by dry months and November is preceded by wet months, the significant soil moisture deficit in May decreased the contribution of rainfall to groundwater recharge. From June and August, there is an abundance of rainfall, soil moisture reaching field capacity, and a decrease in PET, all of which contribute to increased groundwater recharge. For the Seka Chekorsa station, which is located in upstream of the catchment, the recharge occurs from May to November (Figure 9a).

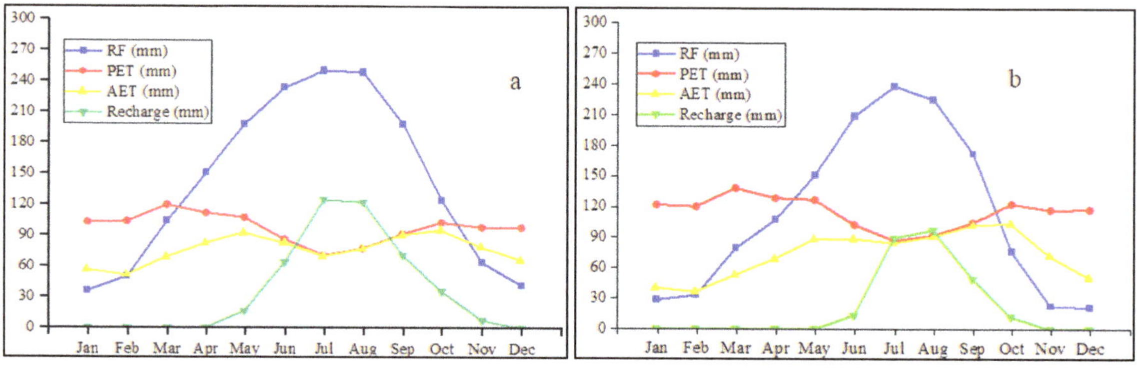

Figure 9. Monthly RF, PET, AET, and recharge, (**a**) Seka Chekorsa and (**b**) Sekoru.

However, for Sekoru station, which is located downstream of the catchment, recharge occurs for relatively a shorter period, from June to October (Figure 9b). Considering the results from all stations, from June to October, rainfall potentially recharges the groundwater through the catchment (Table 5).

In general, the estimate using the SMB method is reasonably acceptable. Previous studies conducted in the upper Gilgel Gibe catchment showed that the mean annual recharge was 298 mm, as estimated using the water balance method [8] which is comparable with the present result, 312 mm. Another study exclusively conducted in the Bulbul sub-catchment showed that the annual groundwater recharge as estimated using the water balance method was 351 mm [13].

3.2. Baseflow Separation (BFS)

Six BFS techniques were applied to the main catchment and six sub-catchments to estimate the annual groundwater recharge for the respective sub-catchments. However, the annual baseflow estimated at Awetu station is extremely large (600 mm), even greater than the maximum annual value estimated using the SMB method. The reason could be the measurement error and/or contribution of wastewater released from Jimma City to the Awetu River. The river flows through Jimma City from the northeast to the southwest. During field observations, several drainage outlets that feed wastewater into the Awetu River were noticed. The Kito River also drains a part of the city and is exposed to similar problems. For this reason, the results from both stations were intentionally rejected and only four sub-catchments and the main catchment were considered for the groundwater recharge estimation (Table 6). The seasonal and annual groundwater recharge of the main catchment and four sub-catchments, as estimated using six BFS techniques, is presented in Table 6. The average value was adopted for groundwater recharge estimation. The analysis period varies from station to station based on the availability of streamflow data.

The highest annual recharge was estimated for the Seka sub-catchment, followed by Bulbul and Asendabo (Table 6). However, the Asendabo sub-catchment encompasses Seka, Bulbul, Kito, Awetu, and other ungauged catchments (Figure 4), and the result shows the aggregated result of these sub-catchments. Since the catchment area of Asendabo station is much larger than that of Awetu and Kito, the aforementioned problems can be neglected.

The average annual recharge of the overall catchment was computed from the baseflow estimated at the main outlet of the catchment (Gilgel Gibe), which is 314 mm (Table 6). A previous study showed that the annual groundwater recharge estimated at the upper Gilgel Gibe catchment using the BFS method was 338 mm [8], which is comparable with the present estimate.

The aquifer system in the Seka sub-catchment consists of alluvial deposits, trachyte, pyroclastic, and fractured basaltic rocks, with relatively good water-bearing capacity. A 200 m well drilling was observed during the field campaign and the samples of well logging consist of alluvial deposits of mainly river gravels and sands. The well data collected from the Jimma Water and Energy Office indicate that the water-yielding capacity of the available wells located in the Seka sub-catchment ranges from 10 to 23 L/s. The Seka sub-catchment has good tree cover and soils, with a moderate infiltration capacity which is favorable for the recharge process. The higher annual groundwater recharge estimated for the Seka sub-catchment agrees with the hydrogeological condition of the sub-catchment. Hence, the result obtained from the BFS would be realistic.

Table 6. Seasonal and annual recharge estimated using BFS methods for the main catchment and sub-catchments.

Gauging Stations	Period	Fixed Interval	Sliding Interval	Local Minimum	Eckhardt	BFLOW	PART	Average
Asendabo (1985–2019)	Spring	22	22	22	20	22	18	21
	Summer	183	182	172	156	175	158	171
	Autumn	132	132	125	135	144	132	133
	Winter	18	18	18	19	20	19	19
	Annual	355	355	337	329	361	327	344
Bidru Awana (1985–2014)	Spring	20	20	20	18	19	18	19
	Summer	91	90	84	79	80	68	82
	Autumn	62	62	60	59	61	58	60
	Winter	20	20	21	17	20	20	20
	Annual	194	192	185	173	179	165	181
Bulbul (1986–2018)	Spring	13	13	12	10	11	11	12
	Summer	222	221	203	170	187	182	197
	Autumn	172	172	161	158	168	165	166
	Winter	14	14	14	13	14	14	14
	Annual	420	420	390	352	380	371	389
Gilgel Gibe (1985–2003)	Spring	19	18	17	17	18	17	18
	Summer	166	167	161	142	159	146	157
	Autumn	123	122	111	124	132	126	123
	Winter	16	16	16	17	18	18	17
	Annual	324	323	307	300	328	306	314
Seka (1985–2014)	Spring	44	43	42	37	40	38	41
	Summer	197	198	190	169	183	174	185
	Autumn	152	152	146	145	153	144	148
	Winter	37	37	37	35	38	37	37
	Annual	429	430	414	386	414	393	411

Likewise, the present estimate of the 389 mm annual recharge for the Bulbul sub-catchment is hydrogeologically plausible. The aquifer system of the catchment is porous and fissured, consisting of volcanic and sedimentary rocks. The sub-catchment also has good LULC and soil for infiltration. It receives high rainfall based on the data derived from CHIRPS. However, there are no ground-based rainfall measurements in the area to verify at a local level. The nearby station is Serbo station, but it is located in a lower area and the data from this station may not represent the Bulbul sub-catchment which is situated at a higher elevation. Well data are also not available to verify the result. However, previous studies revealed that the catchment receives high annual groundwater recharge, with a value of 406 mm as estimated using the BFS method [13].

Lower annual groundwater recharge was estimated via the BFS methods for the Bidru Awana sub-catchment, which is located in the southwestern part of the Gilgel Gibe catchment. Based on the estimates from BFS methods, the sub-catchment receives 56% less annual recharge than the Seka sub-catchment. However, the annual rainfall variation between the two stations is only 20%. The higher recharge variation could be due to the geologic characteristics of the localities. The aquifer system of the Bidru Awana sub-catchment consists of slightly fractured basaltic rocks with lower well yields ranging from

0.5 to 5 L/s, which is attributed to a low baseflow contribution observed at the Bidru Awana gauging station.

Figure 10 describes the seasonal baseflow contribution of the main catchment and sub-catchments based on the average estimates of BFS methods for the period of observed data depicted in Table 6. The estimated seasonal baseflow shows that the sub-catchments contribute 45 to 51% of the annual baseflow during summer, 33 to 43% during autumn, 4 to 11% during winter, and 3% to 11% during spring (Figure 10). Asendabo and Bulbul sub-catchments contribute a higher percentage of annual baseflow during summer and autumn and a lower percentage during spring and winter compared to other sub-catchments. Bidru Awana and Seka contribute a lower percentage in summer and autumn but a higher percentage in spring and winter compared to the other sub-catchments.

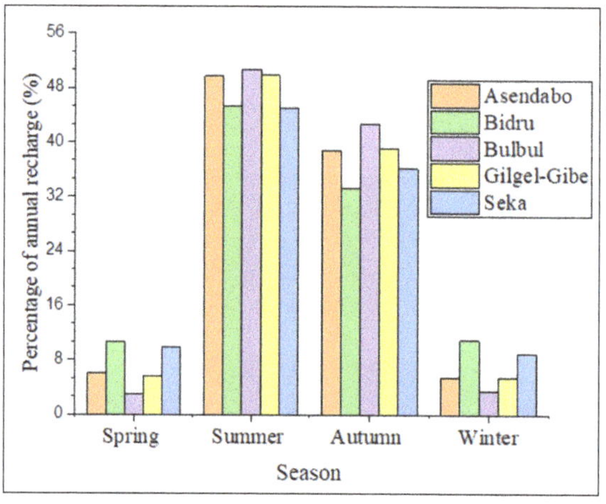

Figure 10. Percentage of annual recharge in catchments.

The seasonal baseflow estimated using BFS methods at the Gilgel Gibe catchment was comparable with the average value estimated at the sub-catchments (Figure 10). However, most of the gauged streams are located in the northwestern part of the catchment (Figure 4). This area is characterized by porous geological formations and receives higher annual rainfall (ranging from 1582 to 1758 mm). The gauged streams located in this region, including the Bulbul and Seka stations experienced higher baseflow (Table 6). The result of hydrograph analyses also shows that the BFI of the Bulbul and Seka stations are higher, 0.76 and 0.77, respectively. On the contrary, the southeastern part of the catchment, specifically, the Deneba, Kidame Gebeya, Sekoru, Saja, and Natri areas, are geologically less permeable due to the presence of massive basaltic layers and receive low annual rainfall ranging from 1129 to 1373 mm. The hydrograph analysis at Awana Dawa, the only gauged station in this area, shows low baseflow and BFI (181 mm and 0.56, respectively). Therefore, the average recharge estimated by averaging the baseflow obtained from only gauged stations would be higher than the actual recharge because there is only one gauged station representing the southeastern part of the catchment where the baseflow is expected to be lower.

The trend in average annual groundwater recharge estimated using the BFS methods for the main catchment and sub-catchments with their respective baseflow index (BFI) is shown in Figure 11. The groundwater recharge varies in response to rainfall variability and other climatic parameters, whereas the baseflow index (BFI) decreases with time for all catchments. This shows that the groundwater contribution to streamflow is decreasing, which could be attributed to the groundwater abstraction increasing in response to demand. The Bulbul sub-catchment has a relatively higher BFI than other sub-catchments with an average value of 78%. On the other hand, the Bidru Awana sub-catchment has a lower BFI, with an average value of 56%.

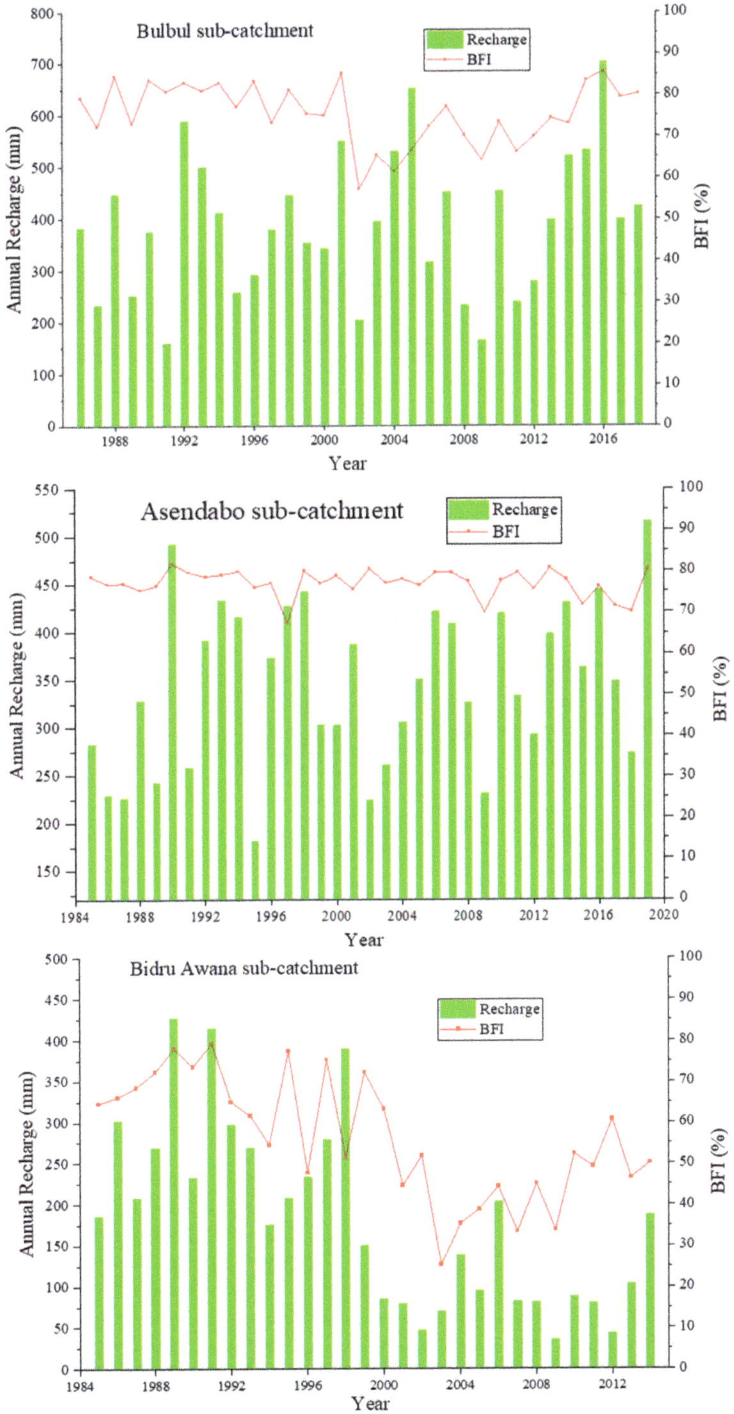

Figure 11. *Cont.*

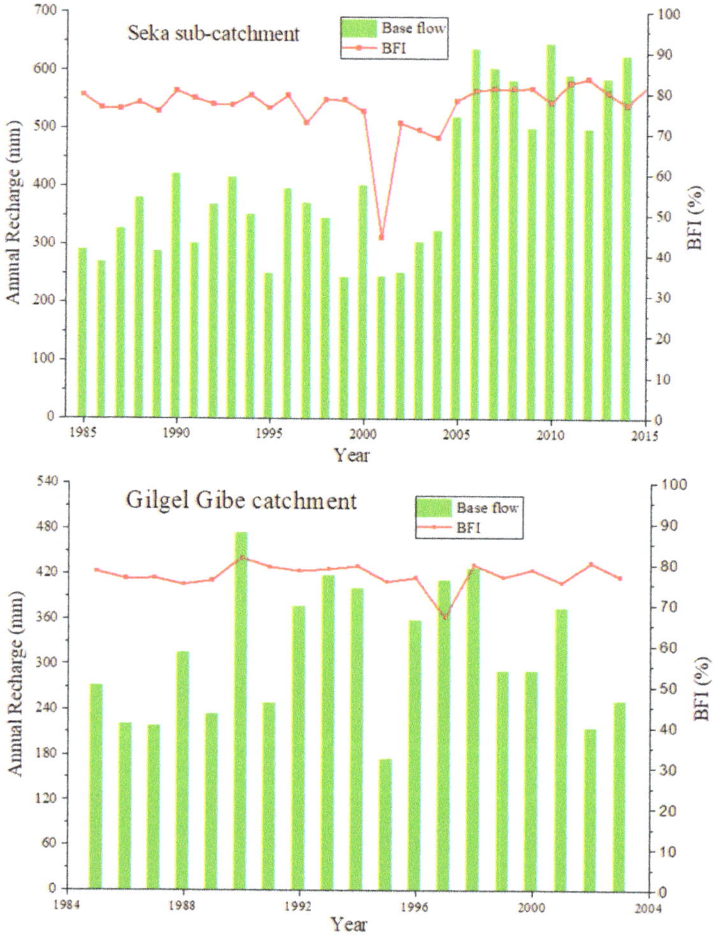

Figure 11. Annual groundwater recharge of the main catchment and sub-catchments with corresponding baseflow index (BFI).

The recharge estimated at Seka, Bulbul, and Bidru Awana was hydrologically and geologically reliable.

3.3. Comparison of SMB and BFS Methods

The groundwater recharge estimated using the SMB method based on the hydrometeorological data were compared with that of the BFS methods (Table 7). The sub-catchments considered for the BFS methods and the Thiessen polygons considered for the SMB method do not coincide with each other in the comparison. Additionally, the analysis periods for the SMB method (1985–2020) and the BFS methods differ for all sub-catchments and the main catchment. Therefore, the average estimate of the SMB method from Thiessen polygons in the sub-catchment or main catchment is used for comparison with the average of estimates of the BFS methods from the corresponding sub-catchment or main catchment.

The annual and seasonal recharge estimated using SMB and BFS methods were comparable, but the estimates using the BFS methods are greater than that of the SMB method for the Bulbul sub-catchment and the whole Gilgel Gibe catchment. For other sub-catchments including Asendabo, Bidru Awana, and Seka, the SMB method estimated higher value than that of the BFS methods. In principle, the SMB method estimates the actual groundwater recharge and the BFS method estimates the net groundwater recharge. However, for the whole catchment, the average annual groundwater recharge estimated using SMB was 313 mm which is almost the same with that estimated using the BFS methods, which is 314 mm. However, the temporal change due to the varying analysis period is not taken into account. Several studies conducted in physiographic settings formed by volcanic eruptions during different geologic times in various parts of Ethiopia revealed that the estimate using the SMB method is lower than the estimates via other methods like BFS, WTF, CMB, and WetSpass [8,13,15,18]. This is due to the fact that the SMB method considers only diffuse recharge from rainfall but not the recharge from preferential flows. The undulating topographic setting of the region, resulting from volcanic eruptions with associated cracks, faults, depressions, and sink holes, makes the preferential flow path recharge significant.

Table 7. Comparison of SMB and BFS methods.

Stream Gauging Stations	Area of Catchment (km^2)	Recharge from BFS Method (mm)	Recharge from SMB Method (mm)	Hydrometeorological Stations
Asendabo	2966	344	368	Seka Chekorsa, Dedo, Jimma, Serbo, Shebe and Yebu
Bidru Awana	41	181	264	Sekoru
Bulbul	526	389	337	Serbo
Gilgel Gibe	5145	314	313	All stations
Seka	280	411	442	Seka Chekorsa

Considering the sub-catchments, the groundwater recharge estimated using BFS for the Seka sub-catchment was 411 mm and the result from the SMB method for the area was 442 mm. The recharge estimated using SMB was greater than that of the BFS method. For the Bidru Awana sub-catchment, the recharge estimated using BFS and SMB were 181 mm and 264 mm, respectively. The result from the SMB method is greater than the result from the BFS method by 31%.

The seasonal recharge (Figure 12) shows that both methods estimated maximum recharge in summer followed by autumn. Based on the SMB method, the groundwater recharge estimated in winter was nil, both at the Seka Chekorsa and Sekoru stations. It is also nil in spring at the Sekoru station, whereas a small recharge value was estimated at the Seka Chekorsa station. However, the BFS methods estimated a significant amount of groundwater recharged in spring and winter in the Seka and Bidru Awana sub-catchments, where the Seka Chekorsa and Sekoru stations are located, respectively. This is attributed to the fact that the recharge that occurred during rainfall will not be discharged at the outlet immediately; it takes groundwater residence time to reach the outlet. BFS methods are not convenient for estimating the groundwater recharge at a daily, monthly, or seasonal time scale but for the annual scale. The baseflow observed during the dry season is the result of recharge during the rainy season. Due to this fact, the comparison should be made on an annual basis for decision-making.

In general, for smaller sub-catchments in our study area, the estimate of the SMB method is larger than that of the BFS method and vice versa except for the Asendabo sub-catchment. The reasons could be the following: (1) The gauging station located at the outlet of the larger sub-catchment has a high possibility of monitoring a higher proportion of groundwater discharge from the aquifer system in the catchment. For the smaller sub-catchments, there could be a possibility for the portion of baseflow to flow through the aquifer to the main outlet without being observed at the gauging stations. (2) The preferential flow path recharge may have a significant contribution to the larger sub-catchment. The recharges from all potential sources, including diffuse recharge and preferential flow path recharge, are captured using BFS methods, whereas the SMB method does not capture the preferential flow path recharge. There is also a high possibility for the surface runoff to terminate into fractures, cracks, and sinkholes, which is not considered in SMB method; (3) the uncertainties introduced during streamflow measurement and assumptions taken may be another reason for their discrepancy. Haile et al. [88] highlighted that the lack of an updated rating curve with changing river morphology in the catchment could be a source of uncertainties in stream flow measurement.

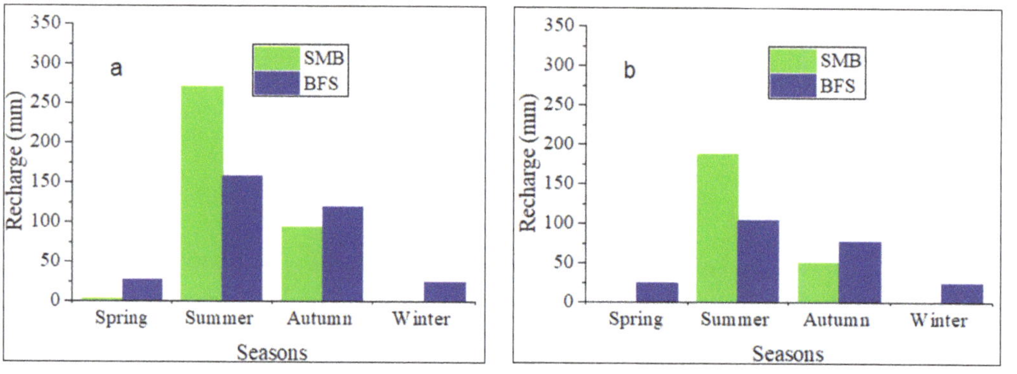

Figure 12. Seasonal recharge estimated using SMB and BFS: (**a**) Seka Chekorsa and (**b**) Bidru Awana.

Regardless of the third reason, previous studies conducted in the catchment and the region revealed that the estimate using the BFS method is greater than that of the SMB method. The study conducted in the upper Gilgel Gibe catchment showed that the BFS (BF + 3) estimate was higher than that of the SMB method by 12% [8]. Another study conducted in the Bulbul sub-catchment showed that the BFS estimate was greater than that of the water balance method by 14% [13].

The SMB method along with two other recharge estimation methods (WetSpass and WTF) was applied in northern Ethiopia, in the Gumera watershed [18]. The hydrogeological characteristics of the catchment are similar to that of the Gilgel Gibe catchment. The recharge estimated using SMB (431 mm) showed that the SMB estimate was the lowest of all. Yenehun et al. [18] highlighted that the SMB method underestimates recharge specifically at mountain fronts due to the presence of groundwater interflow and at flat areas due to the contribution of temporary storage. Another study conducted in Central Ethiopia, in the Akaki River catchment, applied the SMB and CMB methods to estimate recharge. The result showed that the SMB method underestimated recharge by 62% when compared to that of the CMB method [15].

The results from the BFS and SMB methods for the entire Gilgel Gibe catchment are comparable, and the results are reasonably acceptable, despite the fact that the temporal change resulting from varying analysis periods is not taken into account. The BFS method is more reliable for the estimation of the annual recharge of the catchment, whereas the SMB method is reliable for the estimation of diffuse recharge for the specified localities having specific soil and LULC, providing that there is no other source of recharge than direct rainfall.

4. Conclusions

The groundwater recharge of the Gilgel Gibe catchment was estimated using the SMB and the BFS methods. The LULC and soil matrix with hydrometeorological data obtained from CHIRPS, NASA/POWER, and ground-based measurements from 17 stations were used in the SMB method. Six BFS methods (fixed interval, sliding interval, local minimum, BFLOW, Eckhardt, and PART) were applied to six sub-catchments, having measured daily streamflow data.

The catchment annual recharge estimated using SMB was 313 mm. The result shows that there is significant variation in recharge both spatially and temporarily. The up-stream catchment receives a higher annual recharge with a maximum value of 446 mm based on the estimate from the SMB method. The downstream catchment receives a lower annual recharge with a minimum value of 209 mm. The temporal variation is also significant and varies from 115 mm/year to 634 mm/year. The recharge percentage of rainfall of the catchment was estimated to be 20%. BFS methods failed to estimate a realistic value at the Awetu and Kito sub-basins, which might be caused by unreliable stream flow measurement. The main catchment, Gilgel Gibe, and the two major sub-catchments, Asendabo and Bulbul, are estimated to have annual recharge values of 314, 344, and 389 mm, respectively, based on BFS methods. The recharge estimated from the BFS method at Bulbul and Gilgel Gibe was slightly higher than that of the corresponding estimate of the SMB method. Even though errors during stream flow measuring are expected, this could be mainly due to the capability of the method to take into account the recharge from all the potential sources for large catchments.

The results show that estimates from the SMB and BFS methods are comparable. However, for catchments having quality-measured streamflow and no anthropogenic effects, the BFS methods would estimate realistic annual recharge. In volcanic aquifers, where preferential flow path recharge is anticipated, this is the case. On the other hand, the SMB method is an important tool for understanding the diffuse recharge from direct rainfall, but it does not capture the recharge from other sources. The results of this study offer valuable insights for policymakers in terms of formulating strategies for managing groundwater and identifying sub-catchments that require immediate interventions to maintain groundwater sustainability.

Author Contributions: Conceptualization, F.G.T., F.F.F. and K.W.; methodology, F.G.T., M.V.C., A.Y. and K.W.; software, F.G.T., M.V.C., A.Y. and K.W.; validation, F.G.T., F.F.F., A.B.K., W.M.K., B.G.G., S.K.D., B.C.T., A.Y. and K.W.; formal analysis, F.G.T. and K.W.; investigation, F.G.T.; resources, F.G.T. and K.W. data curation, F.G.T., A.B.K., W.M.K. and B.G.G.; writing—original draft preparation, F.G.T.; writing—review and editing, K.W., F.F.F. and A.Y.; visualization, F.G.T., F.F.F. and K.W.; supervision, K.W. and F.F.F.; project administration, K.W.; funding acquisition, F.G.T. and K.W. All authors have read and agreed to the published version of the manuscript.

Funding: F.G.T., W.M.K., B.G.G., S.K.D. and B.C.T. received funding from the NASCERE project for research stays at Ghent University, to do this research. A.B.K. received funding from the Special Research Fund (BOF) from Ghent University.

Data Availability Statement: The data utilized in this research work is available upon request.

Acknowledgments: Authors would like to thank the Network for Advancement of Sustainable Capacity in Education and Research in Ethiopia (NASCERE) project from the Federal Democratic Republic of Ethiopia Ministry of Education and concluded with Jimma University and Ghent University, for providing funds during this study. The authors also acknowledge Ghent University and Jimma University for providing resources and material support for the study.

Conflicts of Interest: The authors declare no conflict of interest.

References

1. Poeter, E.; Ying, F.; John, C.; Warren, W.; Douglas, M.; Poeter, E. *Groundwater in Our Water Cycle*; Groundwater Project: Guelph, ON, Canada, 2020. [CrossRef]
2. Grönwall, J.; Danert, K. Regarding groundwater and drinking water access through a human rights lens: Self-supply as a norm. *Water* **2020**, *12*, 419. [CrossRef]
3. Healy, R.W. *Estimating Groundwater Recharge*; Cambridge University Press: New York, NY, USA, 2010.
4. Giordano, M. Global groundwater? Issues and solutions. *Annu. Rev. Environ. Resour.* **2009**, *34*, 153–178. [CrossRef]
5. Werner, A.D.; Zhang, Q.; Xue, L.; Smerdon, B.D.; Li, X.; Zhu, X.; Yu, L.; Li, L. An initial inventory and indexation of groundwater mega-depletion cases. *Water Resour. Manag.* **2013**, *27*, 507–533. [CrossRef]
6. Arora, N.K.; Mishra, I. Sustainable development goal 6: Global Water Security. *Environ. Sustain.* **2022**, *5*, 271–275. [CrossRef]
7. BGS. Hydrogeology of Ethiopia. 2022. Available online: https://earthwise.bgs.ac.uk/index.php/Hydrogeology_of_Ethiopia (accessed on 19 June 2023).
8. Abebe, W.T. Evaluation of groundwater resource potential by using water balance model: A case of Upper Gilgel Gib Watershed, Ethiopia. *J. Groundw. Sci. Eng.* **2022**, *10*, 209–222. [CrossRef]

9. Takala, W.; Adugna, T.; Tamam, D. The effects of land use land cover change on hydrological process of Gilgel Gibe, Omo Gibe Basin, Ethiopia. *Int. J. Sci. Technol.* **2016**, *7*, 117–128.
10. Fanta, S.S.; Namara, W.G.; Yesuf, M.B. Assessment of water supply and demand in Gilgel Gibe watershed, southwest Ethiopia. *Sustain. Water Resour. Manag.* **2022**, *8*, 104. [CrossRef]
11. Arnold, J.G.; Allen, P.M. Automated methods for estimating baseflow and ground water recharge from streamflow records. *JAWRA J. Am. Water Resour. Assoc.* **1999**, *35*, 411–424. [CrossRef]
12. Mengistu, T.D.; Chang, S.W.; Kim, I.H.; Kim, M.G.; Chung, I.M. Determination of potential aquifer recharge zones using geospatial techniques for proxy data of Gilgel Gibe Catchment, Ethiopia. *Water* **2022**, *14*, 1362. [CrossRef]
13. Shimelis, A.; Megerssa, O.; Fantahun, A. Estimation of groundwater recharge using water balance model coupled with base flow separation in Bulbul River catchment of Gilgel Gibe River Basin, Ethiopia. *Asian J. Appl. Sci. Eng.* **2014**, *3*, 41–49. [CrossRef]
14. Thornwhaite, C.; Mather, J. Instruction and Tables for Computing Evapotranspiration and Water Balance. *Publ. Climatol.* **1957**, *10*, 185–311.
15. Demlie, M. Assessment and estimation of groundwater recharge for a catchment located in highland tropical climate in central Ethiopia using catchment soil–water balance (SWB) and chloride mass balance (CMB) techniques. *Environ. Earth Sci.* **2015**, *74*, 1137–1150. [CrossRef]
16. Nigate, F.; Van Camp, M.; Yenehun, A.; Belay, A.S.; Walraevens, K. Recharge–discharge relations of groundwater in volcanic terrain of semi-humid tropical highlands of Ethiopia: The case of Infranz Springs, in the Upper Blue Nile. *Water* **2020**, *12*, 853. [CrossRef]
17. Walraevens, K.; Vandecasteele, I.; Martens, K.; Nyssen, J.; Moeyersons, J.; Gebreyohannes, T.; De Smedt, F.; Poesen, J.; Deckers, J.; Van Camp, M. Groundwater recharge and flow in a small mountain catchment in northern Ethiopia. *Hydrol. Sci.* **2009**, *54*, 739–753. [CrossRef]
18. Yenehun, A.; Nigate, F.; Belay, A.S.; Desta, M.T.; Van Camp, M.; Walraevens, K. Groundwater recharge and water table response to changing conditions for aquifers at different physiography: The case of a semi-humid river catchment, northwestern highlands of Ethiopia. *Sci. Total Environ.* **2020**, *748*, 142243. [CrossRef]
19. Bayou, W.T.; Wohnlich, S.; Mohammed, M.; Ayenew, T. Application of hydrograph analysis techniques for estimating groundwater contribution in the Sor and Gebba streams of the Baro-Akobo river Basin, southwestern Ethiopia. *Water* **2021**, *13*, 2006. [CrossRef]
20. Ferede, M.; Haile, A.T.; Walker, D.; Gowing, J.; Parkin, G. Multi-method groundwater recharge estimation at Eshito micro-watershed, Rift Valley Basin in Ethiopia. *Hydrol. Sci.* **2020**, *65*, 1596–1605. [CrossRef]
21. Gebert, W.A.; Radloff, M.J.; Considine, E.J.; Kennedy, J.L. Use of Streamflow Data to Estimate Base Flow/Ground-Water Recharge For Wisconsin. *JAWRA J. Am. Water Resour. Assoc.* **2007**, *43*, 220–236. [CrossRef]
22. Lee, C.H.; Chen, W.P.; Lee, R.H. Estimation of groundwater recharge using water balance coupled with base-flow-record estimation and stable-base-flow analysis. *Environ. Geol.* **2006**, *51*, 73–82. [CrossRef]
23. Nathan, R.J.; McMahon, T.A. Evaluation of automated techniques for base flow and recession analyses. *Water Resour. Res.* **1990**, *26*, 1465–1473. [CrossRef]
24. Vu, V.H.; Merkel, B.J. Estimating groundwater recharge for Hanoi, Vietnam. *Sci. Total Environ.* **2019**, *651*, 1047–1057. [CrossRef]
25. Zomlot, Z.; Verbeiren, B.; Huysmans, M.; Batelaan, O. Spatial distribution of groundwater recharge and base flow: Assessment of controlling factors. *J. Hydrol. Reg. Stud.* **2015**, *4*, 349–368. [CrossRef]
26. Furey, P.R.; Gupta, V.K. A physically based filter for separating base flow from streamflow time series. *Water Resour. Res.* **2001**, *37*, 2709–2722. [CrossRef]
27. Tenalem, A. Comparison of Different Base Flow Separation Methods and Drought Vulnerability in a Rift Valley area. *J. Spat. Hydrol.* **2019**, *15*, 1.
28. Danielescu, S.; MacQuarrie, K.T.; Popa, A. SEPHYDRO: A Customizable Online Tool for Hydrograph Separation. *Groundwater* **2018**, *56*, 589–593. [CrossRef]
29. Pettyjohn, W.A.; Henning, R.J. *Preliminary Estimate of Regional Effective Groundwater Recharge Rates in Ohio*; Water Resources Center, Ohio State University: Columbus, OH, USA, 1979.
30. Sloto, R.A.; Crouse, M.Y. HYSEP: A computer program for streamflow hydrograph separation and analysis. *Water Resour. Investig. Rep.* **1996**, *96*, 4040.
31. Lim, K.J.; Engel, B.A.; Tang, Z.; Choi, J.; Kim, K.S.; Muthukrishnan, S.; Tripathy, D. Automated web GIS based hydrograph analysis tool, WHAT. *JAWRA J. Am. Water Resour. Assoc.* **2005**, *41*, 1407–1416. [CrossRef]
32. Chapman, T.G. Comment on "Evaluation of automated techniques for base flow and recession analyses" by RJ Nathan and TA McMahon. *Water Resour. Res.* **1991**, *27*, 1783–1784. [CrossRef]
33. Boughton, W. A hydrograph-based model for estimating the water yield of ungauged catchments. In Proceedings of the Hydrology and Water Resources Symposium, Newcastle, Institution of Engineers, Newcastle, Australia, 30 June–2 July 1993.
34. Chapman, T.G.; Maxwell, A.I. Baseflow separation-comparison of numerical methods with tracer experiments. In Proceedings of the Hydrology and Water Resources Symposium, Water and the Environment, Institution of Engineers, Barton, Australia, 21–24 May 1996.
35. Szilagyi, J.; Parlange, M.B. Parlange, Baseflow separation based on analytical solutions of the Boussinesq equation. *J. Hydrol.* **1998**, *204*, 251–260. [CrossRef]
36. Eckhardt, K. How to construct recursive digital filters for baseflow separation. *Hydrol. Process.* **2005**, *19*, 507–515. [CrossRef]

37. Tularam, G.A.; Ilahee, M. Exponential smoothing method of base flow separation and its impact on continuous loss estimates. *Am. J. Environ. Sci.* **2008**, *4*, 136. [CrossRef]
38. Lyne, V.; Hollick, M. Stochastic time-variable rainfall-runoff modelling. In *Institute of Engineers Australia National Conference*; Institute of Engineers Australia Barton: Barton, Australia, 1979.
39. Arnold, J.G.; Allen, P.M.; Muttiah, R.; Bernhardt, G. Automated base flow separation and recession analysis techniques. *Groundwater* **1995**, *33*, 1010–1018. [CrossRef]
40. Alemu, T. *Geology of the Arjo Area: Memoir 23*; Geological Survey of Ethiopia: Addis Ababa, Ethiopia, 2014.
41. Haro, W.; Besufikad, A.; Bulto, D.; Belete, F.; Edris, M. *Geology, Geochemistry, and Gravity Survey of Jima Area*; Geological Survey of Ethiopia: Addis Ababa, Ethiopia, 2012.
42. Karra, K.; Kontgis, C.; Statman-Weil, Z.; Mazzariello, J.C.; Mathis, M.; Brumby, S.P. Global land use/land cover with Sentinel 2 and deep learning. In Proceedings of the 2021 IEEE International Geoscience and Remote Sensing Symposium IGARSS, Brussels, Belgium, 11–16 July 2021.
43. Yetemegne, A.K. An overview of Omo Gibe River Basin Potential. In Proceedings of the Second National Consultative Workshop on Integrated Watershed Management on Omo-Gibe Basin, Addis Ababa, Ethiopia, January 2016.
44. Jillo, A.Y.; Demissie, S.S.; Viglione, A.; Asfaw, D.H.; Sivapalan, M. Characterization of regional variability of seasonal water balance within Omo-Ghibe River Basin, Ethiopia. *Hydrol. Sci. J.* **2017**, *62*, 1200–1215. [CrossRef]
45. Elias, E. *Soils of the Ethiopian Highlands: Geomorphology and Properties*; CASCAPE Project; ALTERA, Wageningen University and Research Centre: Wageningen, The Netherlands, 2016.
46. Regassa, A. Characterization of Agricultural Soils in CASCAPE Intervention Woredas in Western Oromia Region. *Final. Rep.* **2015**, *18*, 119–120.
47. Woodroofe, R. Omo-gibe River basin integrated development master plan study final Report Vol. In *VI Water Resources Surveys and Inventories*; Ministry of Water Resources: Addis Ababa, Ethiopia, 1996.
48. Zenebe, B.; Agonafir, M.; Taye, M.; Meskerem, T.; Endeis, M.; Bekele, M.; Burusa, G. *Geology Geochemistry and Gravity Survey of the Hosaena Area*; GSE: Addis Ababa, Ethiopia, 2012.
49. Davidson, A. *The Omo River Project, Reconnaissance Geology and Geochemistry of Parts of Ilubabor, Kefa, Gemu Gofa and Sidamo, Ethiopia*; Ethiopian Institute of Geological Surveys: Addis Ababa, Ethiopia, 1983.
50. Merla, G.; Abbate, E.; Azzaroli, A.; Bruni, P.; Canuti, P.; Fazzuoli, M.; Tacconi, P. *A Geological Map of Ethiopia & Somalia (1973) 1: 2,000,000: And Comment with a Map of Major Landforms*; University of Florence: Florence, Italy, 1979.
51. Tefera, M.; Chernet, T.; Haro, W. *Explanation of the Geological Map of Ethiopia*; Geological Survey of Ethiopia: Addis Ababa, Ethiopia, 1996.
52. Regassa, A.; Van Daele, K.; De Paepe, P.; Dumon, M.; Deckers, J.; Asrat, A.; Van Ranst, E. Characterizing weathering intensity and trends of geological materials in the Gilgel Gibe catchment, southwestern Ethiopia. *J. Afr. Earth Sci.* **2014**, *99*, 568–580. [CrossRef]
53. Tadesse, N. Hydrogeology of Jimma Area. Master's Thesis, Addis Abeba University, Addis Abeba, Ethiopia, 1994.
54. Van Daele, K. *Characterization of Geological Materials, Their Weathering Products and Their Relationship with Soils in the Gilgel Gibe Catchment, SW Ethiopia*; Ghent University: Ghent, Belgium, 2011; p. 14.
55. Agezew, T.; Sima, J. *Hydrogeological and Hydrochemical Maps of Jima NB 37–1 Sheet*; AQUATEST as, Prague; Czech Geological Survey: Brno, Czechia, 2015.
56. Koshuma, A.E.; Debebe, Y.E.; Dasho, D.K.; Lohani, T.K. Application of different modelling methods to arbitrate various hydrological attributes using CMORPH and TRMM satellite data in upper Omo-Gibe Basin of Ethiopia. *Math. Probl. Eng.* **2021**, *2021*, 4143958. [CrossRef]
57. Bayissa, Y.; Tadesse, T.; Demisse, G.; Shiferaw, A. Evaluation of satellite-based rainfall estimates and application to monitor meteorological drought for the Upper Blue Nile Basin, Ethiopia. *Remote Sens.* **2017**, *9*, 669. [CrossRef]
58. Sinta, N.S.; Mohammed, A.; Ahmed, Z.; Dambul, R. Evaluation of satellite precipitation estimates over Omo–Gibe River Basin in Ethiopia. *Earth Syst. Environ.* **2022**, *6*, 263–280. [CrossRef]
59. Funk, C.; Peterson, P.; Landsfeld, M.; Pedreros, D.; Verdin, J.; Shukla, S.; Michaelsen, J. The climate hazards infrared precipitation with stations—A new environmental record for monitoring extremes. *Sci. Data* **2015**, *2*, 1–21. [CrossRef]
60. Mocko, D.; Takacs, L.; Bosilovich, M.G.; Robertson, F.R.; Molod, A. Atmospheric water balance and variability in the MERRA-2 reanalysis. *J. Clim.* **2017**, *30*, 1177–1196. [CrossRef]
61. Molod, A.; Takacs, L.; Suarez, M.; Bacmeister, J. Development of the GEOS-5 atmospheric general circulation model: Evolution from MERRA to MERRA2. *Geosci. Model Dev.* **2015**, *8*, 1339–1356. [CrossRef]
62. Monteiro, L.A.; Sentelhas, P.C.; Pedra, G.U. Assessment of NASA/POWER satellite-based weather system for Brazilian conditions and its impact on sugarcane yield simulation. *Int. J. Climatol.* **2018**, *38*, 1571–1581. [CrossRef]
63. Araujo, F.C.; Mello, E.L.D.; Gollin, G.M.; Quadros, L.E.D.; Gomes, B.M. Streamflow regionalization in Piquiri river basin. *Eng. Agrícola* **2018**, *38*, 22–31. [CrossRef]
64. Thornthwaite, C.W. An approach toward a rational classification of climate. *Geogr. Rev.* **1948**, *38*, 55–94. [CrossRef]
65. Thornthwaite, C.W.; Mather, J.R. The water balance. *Climatology* **1955**, *8*, 104.
66. Rushton, K.R.; Ward, C. The estimation of groundwater recharge. *J. Hydrol.* **1979**, *41*, 345–361. [CrossRef]
67. Qablawi, B. *A Comparison of Four Methods to Estimate Groundwater Recharge for Northeastern South Dakota*; South Dakota State University: Brookings, SD, USA, 2016.

68. USDA-SCS. *Hydrology National Engineering Handbook*; U.S Department of Agriculture: Washington, DC, USA, 1972.
69. Mishra, S.K.; Singh, V.P. *Soil Conservation Service Curve Number (SCS-CN) Methodology*, 1st ed.; Springer Science & Business Media: Dordrecht, The Netherlands, 2003; ISBN 978-90-481-6225-32003.
70. Singhal, R.; Kumar, S. Surface runoff estimation by SCS-Curve number method using ArcGIS for Modikuntavagu watershed, Khammam district. In Proceedings of the 19th Esri India User Conference, Delhi, Telangana State, India, 11–16 September 2018.
71. Ethiopian Roads Authority (ERA). *Drainage Design Manual-2013*; ERA: Addis Ababa, Ethiopia, 2013.
72. Huang, M.; Gallichand, J.; Wang, Z.; Goulet, M. A modification to the Soil Conservation Service curve number method for steep slopes in the Loess Plateau of China. *Hydrol. Process.* **2006**, *20*, 579–589. [CrossRef]
73. Sobhani, G. *A Review of Selected Small Watershed Design Methods for Possible Adoption to Iranian Conditions*; Utah State University: Logan, UT, USA, 1976.
74. Monteith, J.L. Evaporation and environment. In *Proceedings of the Symposia of the Society for Experimental Biology*; Cambridge University Press (CUP): Cambridge, UK, 1965.
75. Allen, R.G.; Pereira, L.S.; Raes, D.; Smith, M. *FAO Irrigation and Drainage Paper No. 56*; Food and Agriculture Organization of the United Nations: Rome, Italy, 1998.
76. Carter, M.R.; Gregorich, E.G. *Soil Sampling and Methods of Analysis*, 2nd ed.; CRC Press: Boca Raton, FL, USA, 2007.
77. Assen, M.; Tegene, B. Characteristics and classification of the soils of the plateau of SIMEN Mountains National Park (SMNP), Ethiopia. *SINET Ethiop. J. Sci.* **2008**, *31*, 89–102. [CrossRef]
78. Elias, E. Characteristics of Nitisol profiles as affected by land use type and slope class in some Ethiopian highlands. *Environ. Syst. Res.* **2017**, *6*, 1–15. [CrossRef]
79. Kufa, T. Characterizing major soil physical properties of coffee forest ecosystems in Ethiopia. *Int. J. Soil Sci. Agron.* **2013**, *1*, 1–9.
80. Yitbarek, T.; Beyene, S.; Kibret, K. Characterization and classification of soils of Abobo Area, Western Ethiopia. *Appl. Environ. Soil Sci.* **2016**, *2016*, 4708235. [CrossRef]
81. Broothaerts, N.; Kissi, E.; Poesen, J.; Van Rompaey, A.; Getahun, K.; Van Ranst, E.; Diels, J. Spatial patterns, causes and consequences of landslides in the Gilgel Gibe catchment, SW Ethiopia. *Catena* **2012**, *97*, 127–136. [CrossRef]
82. Sys Ir, C.; Van Ranst, E.; Debaveya Ir, J.; Beernaert, F. *Land Evaluation Part-III: Crop Requirements*; Agricultural Publications: Brussels, Belgium, 1993; p. 199.
83. Foxx, T.S.; Tierney, G.D.; Williams, J.M. *Rooting Depths of Plants Relative to Biological and Environmental Factors*; Los Alamos National Lab: New Mexico, USA, 1984.
84. Brodie, R.; Hostetler, S. A review of techniques for analysing baseflow from stream hydrographs. In Proceedings of the NZHS-IAH-NZSSS 2005 Conference, Auckland, New Zealand, 28 November–2 December 2005.
85. Jean Olivier, K.K.; Brou, D.; Jules, M.O.M.; Georges, E.S.; Frédéric, P.; Didier, G. Estimation of Groundwater Recharge in the Lobo Catchment (Central-Western Region of Côte d'Ivoire). *Hydrology* **2022**, *9*, 23. [CrossRef]
86. Rutledge, A.T. *Program User Guide for PART*; US Geological Survey: Reston, VA, USA, 2007.
87. Rutledge, A.T. *Computer Programs for Describing the Recession of Ground-Water Discharge and for Estimating Mean Groundwater Recharge and Discharge from Streamflow Records-Update*; US Department of the Interior, US Geological Survey: Washington, DC, USA, 1998; pp. 98–4148.
88. Haile, A.T.; Asfaw, W.; Rientjes, T.; Worako, A.W. Deterioration of streamflow monitoring in Omo-Gibe basin in Ethiopia. *Hydrol. Sci.* **2022**, *67*, 1040–1053. [CrossRef]

hydrology

MDPI

Article

Optimizing the Master Recession Curve for Watershed Characterization and Drought Preparedness in Eastern Cape, South Africa

Solomon Temidayo Owolabi * and Johanes A. Belle

Disaster Management Training and Education Centre, Faculty of Natural and Agriculture Sciences, University of the Free State, Bloemfontein 9301, Free State, South Africa
* Correspondence: solomonowolabi11@gmail.com; Tel.: +27-788357662

Abstract: Regions grappling with water scarcity are compelled to fortify their hydrological analytical protocols for efficacious drought disaster preparedness, considering the escalating influence of climate change on river periodicity and the sustainable management of water resources. Hence, this study presents a novel optimization and standardization approach for master recession curve (MRC) parameterization to improve the existing MRC computation for environmental flow (EF) parameterization. The study framework is based on constructing MRC using the RECESS computational tool. The concept involved normalizing quadratic improvement in the digitally filtered, smoothed, and automatically extracted MRC parameters from 24 long-term winter streamflows (2001–2020) in South Africa. The optimum recession length suitable for MRC computation obtained was ten days based on the significant proportion of the variance in streamflow as a function of flow timing ($R^2 > 0.935$), EF consistency in most watersheds (p-value < 0.00), optimum standard error, and the appreciable years of significant discharge. The study obtained the MRC index, EF threshold, and the probable diminution period of 3.81–73.2, 0.001–20.19 m^3/s, and 3.78 to 334 days based on the periods of significant discharge ranging between 4 and 20 years, respectively. The concurrent agreement of rainfall trend and baseflow (p-value < 0.05) with MRC parameters validate their performance as tools for EF conservation. The intra-variation in MRC across the 24 stations alluded to the overriding influence of river aquifer connectivity on watershed viability. The study provides profound insight into perennial and ephemeral rivers' viability/vulnerability, indispensable for watershed prioritization, policy formulation, early warning systems, and drought preparedness.

Keywords: streamflow recession analysis; low flow; SPEI; hydrographic analysis; environmental flow; watershed management; catchment hydrology; South Africa

1. Introduction

Apprehension of a hydrologic system's timing and spatial variability is crucial for adequate watershed management and environmental sustainability management, considering the increasing severity of drought in higher latitudes due to climate change impacts. The recent United Nations and the Intergovernmental Panel on Climate Change (IPCC) documentation accentuates the significance of sustainable water management in climate change [1]. Numerous pieces of the literature allude to the devastating impact of climate change on the limited freshwater system, the sustainable use of the resource, and environmental water allocation [2–5]. Due to the progression of extreme hydrological events such as drought and flood in semi-arid environments, river flow's quantity, quality, and timing are continually being altered and compromised. Consequently, the events severely impact biodiversity, ecosystem services, and the overall sustainability equilibrium among users [3]. With progressive pressure on water resources, the appraisal of the EF threshold is therefore essential to define and assign the minimum flow thresholds required to safeguard and sustain the ecological communities and their functions. This, in return, will aid in prioritizing river networks requiring conservation, provide an early warning system against drought risk, and enable the development of proactive measures of resource sustainability [6].

Environmental flow assessment (EFA) involves the evaluation of the interdependence between the drivers of hydrological processes and their ecological communities within a hydrological catchment [7]. Due to the environmental functions of hydrological resources, EFA is essential for equitable management among competing uses [8]. Notably, the methods of EFA include hydrological, habitat modeling, hydrology and hydraulics, building-block, holistic, and scenario-based evaluation methods [7,9]. Hydrological EFA entails the assessment of how natural flow regime, local flow variability, and anthropogenic activities on the flow impact ecological requirements of the flow such as flow duration curve, master recession curve analysis, range of variability approach, and hydrological indices assessment [10]. Habitat modeling includes methods such as the physical habitat simulation model, instream flow incremental methodology, and fuzzy logic model, in which the flow regime is simulated for its various impacts on habitat availability for target species [11]. Hydrology and hydraulics methods include the wetted perimeter method, R2Cross method, hydraulic habitat methods, hydraulic modeling (1-dimension HEC-RAS and 2-dimension River2D), and hydraulic rating curves (stage-discharge relationships and depth-velocity curves) based on the influence of hydraulic condition on the ecosystem [12]. Building block methods in EFA comprise varying flow level designations, such as low flows, maintenance flows, freshets, small floods, and large floods, for maintaining ecological integrity [13]. The holistic approach in EFA is the integrated approach encompassing the hydrological, hydraulic, biological, and socio-economic aspects of EFA, which include the Instream Flow Incremental Methodologies, Downstream Response to Imposed Flow Transformation, and Ecological Limits Of Hydrological Abstractions [14]. However, none of the EFA approaches target groundwater contributions during recession periods, offering a long-term assessment of water availability and ecosystem resilience to changing flow regimes compared to master recession curve (MRC) analysis [15].

Through mathematical simulation of hydrological processes, MRC enables the computation of the optimum EF threshold essential for nutrient mobilization, sediment transport, habitat conservation, and species of fauna and flora dependent on freshwater can be sustained. The simulation projects the vital properties of the streamflow, such as the flow periodicity, sensitivity, baseflow estimates, recession tendency, and elasticity properties of the watershed in response to the changing flow regime [6]. In doing so, MRC provides a scientific basis for sustainable water allocation and management practices. As a vital watershed and river management tool, MRC can be adapted to strengthen planning frameworks for dam operations, river health monitoring, reservoir exploitation, groundwater pumping rates, and drought risk contingency planning [16]. Developing the EF standard through MRC enhances compliance with the regulatory framework for developing effective water policies. It facilitates the adaptive management approaches for coordinating streamflow response and river level vulnerability to extremities of climate change [17]. With this, human interaction with freshwater sources can be balanced with ecological imperatives to strengthen equitable and sustainable water use for present and future generations [18]. Hence, understanding the streamflow recession threshold and MRC is essential for drought disaster risk reduction, watershed recovery in vulnerable catchments, and developing a drought response strategy. With this, MRC tends to provide a more veritable approximation of the meteorological and land surface complexities [16,18].

Several studies have been conducted to simplify the hydrological processes for gainful insight into EFA parameters for water balance distribution, storage calibration, outflow predictions, and river health integrity studies [16–18]. However, there is no clarity on the specific minimum recession duration for recession curve extraction, considering the call for improvement according to several researchers [19,20]. Earlier, estimations were defined with various hypothetical reasons as Linsley et al. [19] quantified their using the drainage area, Brutsaert and Nieber [21] and Fiorotto and Caroni [20] adopted seven days, and Vogel and Kroll [22] used ten days. Despite the research endeavor on MRC analysis, the application is still unpopular as an EFA tool among environmental stakeholders. The application of MRC parameters as veritable drought risk assessors remains uncertain, while the hypothetical stochasticity of hydrological computation possibly shades its computational simplicity. By and large, MRC is indistinctly recognized as a supportive tool for baseflow estimation.

Aside from the non-linear factors hampering EF's stationarity, the extraction procedures for recession parameter extraction are yet to be standardized in the previous study. That is, the several biases associated with calculating the EF threshold and several uncertainties that culminate into error in the threshold obtained are yet to be logically addressed. Errors in EF computations are primarily due to climate, anthropogenic activities, and the calculation method. We considered it more logical to address the uncertainty in EFA by limiting the river flow assessment to the period when precipitation and anthropogenic influences are at the barest minimum, at the period of groundwater discharge due to prolonged dryness [16]. Contrarily, the improved dynamic calculation method for EF premises its P-III curve and its average monthly flow, a vital aspect of its metrics on the computation of the annual runoff volume and monthly flow for a long series of years [23]. Similarly, the Northern Great Plains Resource Program and the improved monthly frequency calculation compute the environmental flow by calculating the distance from the flat percentages based on the wet and dry years with varying guarantees [24]. The improved range of variability approach is an upgraded version of Indicators of Hydrological Change methods based on the quantification of EF during the primary flood season [25]. Tennant method computes the relationship between runoff and the river ecosystem by separating the runoff data of the river into the dry and wet periods [26]. In the same vein, the Gram–Charlier A Series kurtosis function, derived from the Gram-Charlier series, quantifies the peak functions to reference EF thresholds using the average monthly runoff of a long-term streamflow time series [23,27].

Hence, no study has clarified the threshold indicating the river level at which no further anthropogenic abstraction is advisable. As a rationale unique to this study, we assumed that a standardized MRC approach should be able to provide the probable time of diminution, thus strengthening adequate river health monitoring in terms of provision of the reference time when river flow transit during its low flow from resilient to vulnerable status. None of the previous studies on environmental flow assessment has improved any MRC model to characterize the recession stage of a river in two phases: the resilient and the vulnerable phases. Also, none of the previous studies has factored the variability in parameterization of EFA models across watershed sizes and the influence of leaky or confined. Hence, this study's novelty lies in optimizing recession length (days) and standardizing recession indices across catchment sizes [20,22]. This study synthesized MRC parameters to deduce the probable diminution periods and EF thresholds for effective monitoring, prioritization, and drought preparedness.

Consequently, the MRC coefficient computation was based on the winter streamflow time series. Moreover, the study is carried out in a region within a bimodal climate zone of a semi-arid environment in South Africa, characterized by low mean annual precipitation, high evaporation rates, geohydrological extremes, intra-seasonal and inter-annual fluctuation influenced by El Niño-Southern Oscillation [5,6,8,28]. We characterize the optimum recession length regarding watershed size, EF threshold status quo regarding low-flow (non) conservativeness, and the probable diminution period regarding river vulnerability to drought. The study was based on integrating the smoothed minima method, the linearized Dupuit–Boussinesq aquifer model, and the quadrated matching strip method. We believe this approach is suitable for monitoring and evaluating river and watershed suitability for abstraction and ensuring an effective early warning system against drought impact.

The study aims to investigate the suitability of MRC coefficients as an EFA aid and drought risk marker, considering its potential for watershed prioritization. To this end, the study addresses the following questions: (1) what is the optimum duration required to define a recession segment for an equitable MRC parameterization, considering the sensitivity of the duration length to the catchment scale? (2) To what extent can MRC coefficients aid watershed prioritization against hydrological drought risk?

2. Study Area

The study area includes four water management areas in the South of South Africa comprising the Upper Orange (UO) across the Lesotho boundary, Mzimvubu-to-Keiskamma (MK), and Fish-to-Tsitsikamma (FT) management areas (Figure 1).

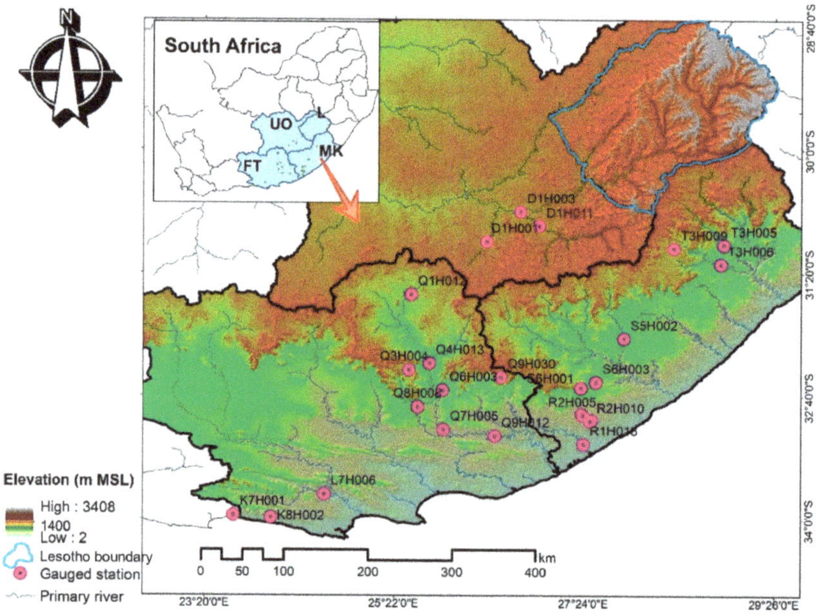

Figure 1. Map showing the elevation, drainage, and location of the selected twenty-five stations at the Fish-Tsitsikamma (FT), Upper Orange (UO), and Mzimvubu-to-Keiskamma (MK), and the boundary of Lesotho at the South of South Africa.

Drawing from Köppen's classification, FT and UO climate is bimodal, exhibiting both the monsoon (westerly climate) and the subtropical (easterly climate), while FT extending to the north and UO extension at Lesotho exhibits the subtropical climate [28]. The average winter and summer temperatures in the area range from −3 °C to 5 °C and 17 °C and 24 °C while the average mean annual rainfall ranges from 280 to 330 mm/yr and 480–560 mm/yr at the northwest (UO) and northeast to 760–1190 mm/yr southwardly (FT) and eastwardly (MK). The mean annual evaporation ranges from 1800 to 2000 mm/yr at the UO (northwest) to 1200–1400 mm/yr eastwardly and southwardly at the coast. The terrain landform is dominated by the plain at the northwest and graduates into the dissected plain and the high-gradient escarpment zone at the northeast and into the high-gradient valley and coastal zone southwardly. The elevation ranges from 2 to 3408 m Mean Sea Level. The geological cover grades from the mountainous northeast from the younger Jurassic Drakensberg of the Stormberg series through the Permian Beaufort and Ecca groups (Southeastern coast) to the older late Carboniferous Dwyka Group at the southern coast [29]. At the southwestern coast, the Karoo Supergroup, initiated by Dwyka groups, tapers underneath the Late Pleistocene Bredasdorp series and the Algoa group of the Cape Supergroup at the Southwest.

Aside from the Basalt of the Drakensberg Formation, the terrain is litho-stratigraphically dominated by mudstone and sandstones of the Beaufort Group [30]. The sedimentary clast of the Permian Ecca Group to the Jurassic Stormberg series is densely intruded by sheets, dyke, and sills of Jurassic Dolerite [29]. The hydrogeological domain is dominated by the fissured aquifer held within an unconfined intergranular hydro-stratigraphic domain and subordinately by a fractured aquifer framed by fractured Jurassic dolerite [30]. The land cover comprises mainly grassland and sparse plantation forest in the northeast, low shrubland in the northwest, and woodland by the coastline. The details of the watershed and station of focus, as shown in Figure 1, are presented in Table 1.

Table 1. Description of the selected streamflow stations in the South of South Africa and designation into classes based on their watershed sizes. 'D' stations belong to the Upper Orange WMA, 'K', 'L', and 'Q' stations belong to Fish-Tsitsikamma WMA, while 'R', 'S', and 'T' stations belong to Mzimvubu-to-Keiskamma WMA.

S/N	Borehole ID	Gauged Station	Latitude	Longitude	Size ($km^{2.2}$)	Streamflow (m^3/s)		
						Min	Mean	Max
Class 1-a	K8H002	Elands	−33.9814	24.05013	35	0.03	0.48	43.47
b.	K7H001	Bloukran	−33.9556	23.63861	57	0.01	1.47	173.00
c.	S6H001	Kubusie	−32.5794	27.36591	90	0.04	0.53	33.35
d.	R2H006	Mgqakwebe	−32.8584	27.37072	119	0.00	0.27	9.73
Class 2-a	R2H011	Yellowwoods	−32.9249	27.4791	197	0.00	0.47	9.98
b.	S6H003	Toise	−32.5163	27.52377	215	0.00	0.45	26.78
c.	Q9H030	Koonap	−32.4656	26.50741	250	0.00	0.70	9.98
d.	T3H009	Mooi	−31.0717	28.35361	307	0.00	3.43	241.93
e.	R2H005	Buffalo @ Tshoxa	−32.875	27.383	411	0.00	0.84	190.73
Class 3-a	R2H010	Buffalo @ Zwelitsha	−32.941	27.461	668	0.06	0.96	12.82
b.	Q6H003	Bavians	−32.6059	25.88461	814	0.01	0.18	75.49
c.	Q3H004	Pauls	−32.3858	25.52055	872	0.00	0.42	54.35
d.	Q8H008	Little Fish	−32.785	25.61483	1512	0.04	0.54	212.48
e.	Q1H012	Teebus	−31.57	25.54388	1652	0.26	22.23	112.51
Class 4-a	S5H002	Tsomo	−32.0443	27.82238	2359	0.00	4.38	212.27
b.	D1H001	Wonderboom	−31.001	26.353	2388	0.00	0.92	216.73
c.	R1H015	Keiskammahoek	−33.1854	27.39075	2530	0.00	3.35	722.27
d.	T3H005	Tina	−31.0318	28.8845	2597	0.12	16.48	540.98
e.	T3H006	Tsitsa	−31.2381	28.85222	4285	0.15	22.27	879.49
f.	D1H011	Kraai	−30.831	26.921	8664	0.00	23.56	1168.78
Class 5-a	Q7H005	Great Fish @S. Vleij	−33.028	25.893	19130	0.49	7.33	308.46
b.	Q9H012	Great Fish @B. Legte	−33.0983	26.44475	23054	0.80	7.18	570.08
c.	L7H006	Groot	−33.731	24.6179	11410	0.00	0.72	9.99
d.	D1H003	Orange River	−30.678	26.716	36975	1.93	147.50	3025.74

3. Materials and Methods

The study framework is based on constructing a master recession curve using the RECESS computational tool. The study engaged twenty-five watersheds with five class sizes to ascertain the statutory reliability of MRC coefficients across varying watershed sizes. The daily streamflow from 1 January 2001 to 31 December 2020, as monthly rainfall and monthly average temperature datasets from January 1981 to December 2020, for the twenty-five stations were downloaded from the hydrology and river portal of the Department of Water and Sanitation of South Africa (https://www.dws.gov.za/hydrology/Verified/hymain.aspx) and accessed on 26 June 2023. A thorough check was performed while downloading and sorting data to ensure that the selected stations met the conditions regarding the missing data of less than 2% in a year. The missing records were addressed using extrapolation based on the average of the preceding and succeeding values in the time series and comparing it with the year record with a similar trend [5,6,28].

3.1. Quantification of Master Recession Curves Parameters

The master recession curve (MRC) parameters were obtained through the following steps:

i. The monthly streamflow time series is discretized into monthly and yearly intervals and grouped into five classes depending on the watershed size.
ii. Due to the concomitancy of hydrograph recession with prolonged low-flow and MRC premise with storage–discharge relationship, the winter flow session (June, July, and

August) is designated for the computation. A single period of recession following a peak streamflow per year, preferably in July, with values nearing zero, is chosen.

iii. The recession limb is filtered using a two-parameter digital filter proposed by Eckhardt [31]. The hydrograph recession analyst is appropriated by Equation (1):

$$B_{i+1} = \frac{(1 - BFI_{max})\alpha B_i + (1 - \alpha) BFI_{max} Q_{i+1}}{1 - \alpha BFI_{max}} \tag{1}$$

where B_{i+1} denotes baseflow at time step $i + 1$, BFI_{max} denotes maximum baseflow index, B_i denotes the antecedent baseflow, Q_{i+1} denotes the streamflow at time step $i + 1$, while α, the filter parameter, is obtained by the expression:

$$\alpha = e^{-\frac{\Delta t}{k}} \tag{2}$$

where k denotes the recession constant and Δt the change in period. The digital filter is based on the assumption that.

$$\left\{ if \begin{array}{c} B_{i+1} \le Q_{i+1}, \forall\, Q_T = B_{i+1} + Q_{i+1} \\ B_{i+1} \ge Q_{i+1}, \forall\, Q_T = B_{i+1} \end{array} \right\} \tag{3}$$

where Q_T is the total streamflow and $\forall$ denotes an equivalent position (a universal quantification) for Q_T concerning the conditional statement.

iv. The time-based recession components in the flow are separated from storage components using the smoothed minima method, defined by the linearized Dupuit–Boussinesq aquifer model, Equation (4) [25]:

$$Q_{i+1} = Q_i e^{-\frac{t}{\tau}} \tag{4}$$

where Q_i denotes the antecedent seepage based on the assumption that the direct runoff is zero, while $\tau = \frac{1}{f_c}$, the mean residence time of groundwater, and f_c is the cut-off frequency. The exponential aspect of the expression denotes the recession constant, K^t whereby; $K^t = e^{-t \times f_c}$. According to Linsley et al. [16],

$$Q_{i+1} = Q_i K^t \tag{5}$$

By differentiating with the time function, Coutagne [32] deduced the recession curve as

$$dQ/dt = -aQ^b \tag{6}$$

The expression is linearized by obtaining its logarithmic expression:

$$log\left(\frac{-dQ}{dt}\right) = log(a) + b \times log(Q) \tag{7}$$

where a and b are the recession indices, deduced as an intercept and slope by linear regression plot of $log(-dQ\,/\,dt)$ against $log(Q)$. It is expected that non-linearity will be repressed should the assessment be focused on the winter flow when (i) precipitation interruption is at the barest minimum, (ii) direct runoff is virtually zero, and (iii) baseflow is largely due to groundwater seepage.

v. The recession curve is extracted from the receding part of the hydrograph, given that $\frac{dQ_t}{dt} > \frac{dQ_{t+1}}{dt} \ge 0$; however, scenarios of minimum flow recession length (k_i) were generated using 2, 5, 7, 10, 15, 20, 25, 30, and 35 to assess the optimum zero-inflow days for MRC assessment. Hence, data deletion is performed on the recession segment to maintain the single recession. The study prioritized data selection in July, being the driest period in the transitional zone [15] and the recession limb with more sufficient data coordinates $\left(\frac{\frac{-dQ}{dt}}{Q}\right)$ higher than 1.

vi. After extracting individual winter recession curves from the annual dataset, the Matching Strip method generated a single curve representing the MRC. The matching strip line is achieved by plotting and superimposing all the individual recession curves to obtain a horizontal arrangement of overlapping recession segments and a best-fit slope regarded as K [33].

vii. The K parameters (minimum, median, and maximum value) were deduced from the mean best-fit logarithm slope. The coefficients of the quadratic curve were drawn. The exercise is repeated for the twenty-five catchments. The baseflow separation and the entire MRC assessment were performed using the RECESS algorithm in the USGS Groundwater Toolbox 1.3.1 [34]. The resulting $log(Q)$ and coefficients -A, -B, and -C were classified and projected in Microsoft Excel v.365.

viii. To validate the optimum recession length, the quadratic value of flow timing t_{k_i} at k_i, were computed from coefficients using Equation (8) and plotted against the logarithm of flow.

$$t_{k_i} = A \times mean\ log(Q)^2 + B \times mean\ log(Q) + C \tag{8}$$

$mean\ log(Q)$ values were generated using the minimum and maximum values of $log(Q)$ generated. The minimum recession length scenario optimization was performed using five streamflows representing different catchment sizes, and the coefficient of determination was extracted.

ix. Using the quadratic coefficients of Equation (8), when $mean\ log(Q)$ is zero, the remaining days to flow inelasticity is derived, such that

$$t_{k_i} = T_{k_i} = C\ since\ \{mean\ log(Q) = 0\} \tag{9}$$

x. The $mean\ log(Q)$ when T_{k_i} is zero is deduced using the quadratic formula

$$q_{\pm} = mean\ log(Q)_{\pm T} = \frac{-B \pm \sqrt{B^2 - 4 \times A \times C}}{2 \times A}\ since\ \{T_{k_i} = 0\} \tag{10a}$$

$$q_{+} = \frac{-B}{2 \times A} + \frac{\sqrt{B^2 - 4 \times A \times C}}{2 \times A}\ since\ \{T_{k_i} = 0\} \tag{10b}$$

$$q_{-} = \frac{-B}{2 \times A} - \frac{\sqrt{B^2 - 4 \times A \times C}}{2 \times A}\ since\ \{T_{k_i} = 0\} \tag{10c}$$

where q_+ and q_- are the flow threshold at the resilient and vulnerability margin measure in m^3/s (though obtained in ft^3/s but converted).

xi. The mean logarithmic streamflow is corroborated to validate the performance of the MRC parameters by calculating the percentage of daily streamflow that meets the threshold, $q_{\pm T}$, accordingly.

$$q_{\pm T} = \left(1 - \frac{Counts\ of\ daily\ flow < q_{\pm}}{Total\ number\ of\ daily\ flow}\right) \times 100 \tag{11}$$

Using the assessment, the rivers are further classified as highly exhausted ($q_{\pm T} \leq 25\%$), moderately exhausted (26% < $q_{\pm T} \leq 50\%$), conserved (51% < $q_{\pm T} \leq 75\%$), and well conserved (76% < $q_{\pm T} \leq 100\%$).

xii. The study proposed a probable diminution period, P_k, based on the effective time log cycle, T_{k_i} [25], defined by the probable deduction of k_i when $mean\ log(Q)_{\pm T}$ is zero.

$$P_k = 1 - \frac{T_{k_i}}{t_{k_i}}\ if\ \{T_{k_i} > t_{k_i},\ \forall\ P_k = 0.01\} \tag{12}$$

where $P_k = 0.01$ is classified as a failed and non-resilient river, $0.01 < P_k < 0.50$ classifies as a highly vulnerable river, $0.5 \leq P_k < 0.75$ represents the lowly resilient, while $0.75 \leq P_k \leq 1$ represents the moderately resilient river. The quotient of drought vulnerability is therefore obtained using 1-P_k.

3.2. Validation of MRC Parameters

Rainfall and temperature datasets were assessed for their trend using Mann–Kendall analysis. The approach for Mann–Kendall analysis of meteorological data was detailed in the authors' previous studies [28,35]. An approach to extracting drought frequency per station was developed and harnessed to validate the assessment. To achieve this, the study adopted the meteorological drought to drive the expected influence of prolonged dryness (zero rainfall) on winter flow. Hence, the recommendation by the World Meteorological Organization, the standard precipitation and evaporation index (SPEI), was used. The computation was achieved through the following steps:

i. The potential evapotranspiration was computed based on the Thornthwaite approach using the monthly rainfall and monthly average temperature dataset. This is based on the monthly heat index, h, annual heat index, H, the mean monthly temperature, m, and factor, φ:

$$h = \left(\frac{m}{5}\right)^{1.514} \quad (13)$$

$$H = \sum_{i=1}^{12} h \quad (14)$$

$$\varphi = \left(6.75 \times 10^{-7}\right) * H^3 - \left(7.71 \times 10^{-5}\right) \times H^2 + \left(1.792 \times 10^{-2}\right) \times H + 0.49239 \quad (15)$$

$$PET_{non\ corrected} = 16 \times \left(\frac{10 \times m}{H}\right)^{\varphi} \quad (16)$$

The potential evapotranspiration is corrected using the respective latitude of the catchments.

ii. The water balance was computed by taking the difference between the monthly precipitation and potential evapotranspiration to calculate the Standard precipitation and evaporation index. The assumed stormflow is computed by fitting the presumed probability density function based on Equation (17):

$$SPEI = z_i = \frac{P_{ij} - P_{ik}}{\sigma} \times 100 \quad (17)$$

where P_{ij} denotes the winter rainfall at the ith time and j^{th} year record, P_{ik} denotes the long-term seasonal mean rainfall, and σ is the standard deviation of the long-term seasonal rain. The potential evapotranspiration and meteorological drought were computed in RStudio using the SPEI package v.1.8 [36,37].

iii. The twenty-five stations' winter SPEI values (June–July–August) were extracted, classified, and plotted in Microsoft Excel v.365. The winter SPEI values for 1981 to 2020 were counted for the frequency of drought considered to be severe and more $f_{z_w<-1}$ and expressed as a percentage.

$$F_{z_w<-1} = \frac{f_{z_w<-1}}{f_0} \times 100 \quad (18)$$

where f_0 denotes the number of winter months within the time series, which is 120 for this study.

Following the computation of the MRC parameters, the catchment size would be expected to have profound influences on the underlying aquifer storage capacity, S, in a sedimentary terrain such as the study terrain, where S at a time, t, and the function is in agreement with the equation of continuity:

$$\frac{dS}{dt} = I - Q \tag{19}$$

such that when catchment inflow due to total precipitation, I_p, is zero, the receding streamflow is the total outflow, Q, and provides information about S, accordingly (except in cases when the free surface is poorly connected to the aquifer):

$$\frac{dS}{dt} = -aQ^b \text{ when } I_p = 0 \tag{20}$$

Non-parametric correlation matrices, namely the Kendall tau and Spearman rho, were used to investigate the depth of the relationship and corroborate the MRC parameters against the rainfall trend computed using the Mann–Kendall test, the meteorological drought parameter, and baseflow.

4. Results

4.1. Optimization of Minimum Recession Length for MRC Assessment

Following the classification of the watersheds into five classes due to the distinct variation in watershed sizes, the scenarios assessment of the optimum flow recession length was performed using a representative from each class for 2, 3, 5, 6, 7, 10, 15, 20, 22, 25, 30, 33, and 35 days of recession. Each scenario assessment generated a fitted quadratic equation; the parameters are presented in Table 2. K_i is the scenario examined in this section, and it indicates the duration over which streamflow decreases after a peak flow event, thus typifying the variation in catchment responses to rainfall [38]. The min logQ and max logQ represent the low and high flow conditions during a recession. The recession constants Kmin, Kmed, and Kmax indicate the slowest, typical, and fastest streamflow decline rates, respectively. The recession constants provide insight into the rate at which streamflow recedes, and these are critical for water resource management and predicting future streamflow behavior. Generally, the smaller catchments, like the Bloukrans (6.15–32.65) and Tshoxa (18.23–24.39) catchments, showed lower maximum recession constants, suggesting longer water retention capacity compared to the larger catchments like Kraai (38.55–75.62) and Orange (35.04–89.66). The larger catchments' higher recession constants indicate faster streamflow declines and greater capacity to handle runoff. The variation in recession constants corresponds to the streamflow within the river catchment, such that, as the capacity for higher flow increases, the catchment's ability to maintain flow during the recession also increases.

Coefficients A, B, and C represent the intercept, slope, and curvature characteristics derived from the fitted quadratic equation to the MRC. The coefficients are essential for modeling how streamflow decreases over time after a flood event. The coefficients A, B, and C vary significantly across stations, reflecting different hydrological responses based on catchment size and characteristics. Coefficients A, B, and C corresponded to the graduation values relative to catchment runoff capacity. Coefficient A depicts the initial flow condition's influence on the catchment's hydrological property, and B shows that the steepness in streamflow decline corresponds to the catchment capacity. At the same time, C exhibits the influence of catchment flow capacity on the significant changes in flow rates over time. The years of significant discharges, n, indicate how often the stations experience meaningful flow events relative to the recession constants, which is crucial for understanding long-term hydrological trends. There is significant variability in the number of years with considerable discharge due to the substantial variability in catchment hydrological properties. However, the larger catchments exhibited a high significant discharge years ratio, indicating a consistent hydrological condition and a stable hydrological regime compared to the smaller catchments. Drawing from Pauls and Bloukrans, which exhibited the poorest significant discharge years, the optimum recession length with a substantial discharge year ratio (25%) is 10 days.

Table 2. The summary of deductions following the optimization of the minimum recession length required for master recession curve assessment.

Station	k_i	Index	Min logQ	Max logQ	Coeff. A	Coeff. B	Coeff. C	n	Kmin	Kmed	Kmax
Bloukrans	2	6.15	0.15	0.74	24.75	−32.23	10.31	12	4	6.2	55
	5	10.16	0.12	0.42	75.10	−60.71	12.25	10	5.1	10.2	64
	10	15.47	0.17	0.39	78.26	−61.50	12.08	8	5.1	15.5	65.8
	15	25.69	0.12	0.30	87.37	−68.73	12.72	8	7.1	25.7	70.2
	20	32.65	0.14	0.22	−68.24	−6.06	4.60	5	6.4	32.7	55.4
	25	6.20	0.27	0.31	496.14	−309.06	48.13	3	4.1	6.2	41
Tshoxa	2	18.23	0.29	1.49	5.59	−32.31	35.79	17	8.6	18.2	51.6
	5	18.22	0.29	1.87	8.26	−35.90	38.22	18	6.2	18.2	50.6
	10	18.23	0.33	1.37	6.36	−32.23	32.18	13	6.7	18.2	49.2
	15	14.41	0.78	1.07	−17.13	14.79	3.77	4	8.1	14.4	31
	20	24.39	0.71	1.20	9.56	−42.71	37.56	2	19.7	24.4	29.1
Pauls	2	20.43	0.09	1.18	19.07	−55.48	38.86	8	11.4	20.4	109.8
	5	27.62	0.14	1.18	20.52	−58.96	40.95	7	13.5	27.6	109.8
	10	17.63	0.14	1.18	5.19	−28.43	26.25	5	11.4	17.6	33.8
	15	17.63	1.07	1.18	−23.61	36.76	−10.59	3	11.4	17.6	18.7
	20	14.49	1.07	1.18	−29.00	50.54	−19.34	2	11.4	14.5	17.6
Kraai	2	64.17	0.76	2.25	30.26	−148.04	179.86	20	12.4	64.2	179.9
	5	75.62	0.63	2.25	32.44	−157.00	189.02	20	12.4	75.6	176.6
	10	75.47	0.76	2.25	35.51	−164.59	190.55	19	12.4	75.5	176.6
	15	68.47	0.59	2.26	23.96	−126.37	163.19	19	15.3	68.5	154.4
	20	45.70	0.59	2.26	13.25	−86.47	127.65	14	21.8	45.7	77.1
	25	68.47	0.70	1.93	23.76	−123.33	149.47	9	29.7	68.5	96.9
	30	38.55	0.89	1.93	20.10	−108.32	134.12	6	29.7	38.6	75.8
	35	43.68	0.89	1.65	21.95	−108.37	119.09	3	36.9	43.7	75.8
Orange	2	76.06	1.31	2.85	46.19	−262.03	371.60	20	12.5	76.1	222
	5	77.54	1.44	2.85	43.50	−247.55	352.16	20	12.5	77.5	155.2
	10	73.16	1.44	2.85	37.46	−223.90	333.90	19	12.5	73.2	166.4
	15	63.65	1.44	2.79	38.83	−227.41	332.20	18	19.9	63.3	163.5
	20	35.04	1.51	2.85	35.09	−207.68	306.91	15	19	35	152
	25	30.94	1.61	2.61	47.85	−258.11	347.64	6	18.2	30.9	134.4
	30	38.50	1.62	2.61	61.01	−327.54	439.25	4	29.1	38.5	141.9
	35	89.66	1.61	2.18	77.88	−385.04	469.34	2	45.3	89.77	134

Note: k_i is the assessment recession length, and Min (Max) logQ is the minimum (maximum) streamflow expressed in logarithm, Coeff. (A, B, and C) is the derived coefficient of the curve based on the fitted quadratic equation, n is the number of years with significant discharge greater than zero, while Kmin, Kmax, and Kmed represent the minimum, maximum, and medium value of MRC constants.

The fitted quadratic expression generated from coefficients A, B, and C was used to quantify the quadratic value of flow timing, t_{k_i}, according to Equation (8) using the recession days above. Figure 2 and Table 3 provide each scenario's coefficient of determination and multilinear regression (MLR) based on the relationship between flow timing and the streamflow log. The result showed that a p-value < 0.000 is statistically significant for the relationship between streamflow and timing across all the stations except in Bloukrans and Pauls stations. For Pauls station, the coefficient of determination is maximum (99.5%) at $k_i = 10$, but dropping to 47.9% at $k_i = 15$. Conversely, moderate and robust fit were attained at $k_i = 2$ and 20 in Bloukrans. Tshoxa, Kraai, and Orange stations are characterized by excellent fit across all the scenarios, thus highlighting the significance of recession length for effective water management. Generally, a vivid trend depicts the influence of the increase in recession length on the model's performance, as evident in the Tshoxa and

Kraai stations. Based on the coefficient of determination, the most consistent statistical significance (p-value = <0.000), recession length's effectiveness across hydrological dynamics of all catchment sizes, flow timing modeling, and streamflow behavioral prediction, the optimum recession length is 20 days.

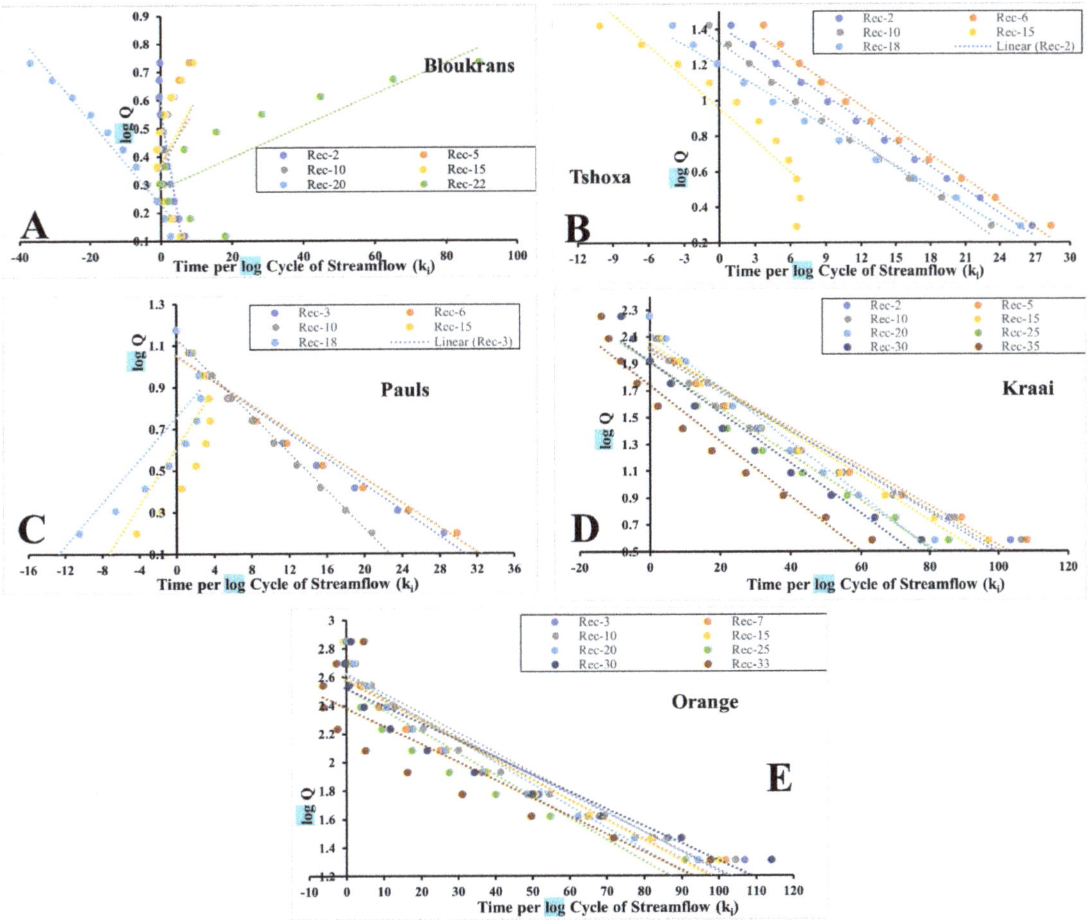

Figure 2. The plot of recession length scenarios for MRC parameters for the five classified stations on (**A**). Bloukrans, (**B**). Tshoxa, (**C**). Pauls, (**D**). Kraai, and (**E**). Orange rivers. It showed the relationship between the plot of flow timing and streamflow for different scenarios to obtain the R^2 and approve the most optimum recession length.

Compared to k_2 and k_{20}, k_{10} showed the most decisive impact, drawing from its R^2 and standard error. As a result, the adoption of 10 days is validated as the most suitable recession length, and this was used to generate the MRC index for the remaining stations. While k_{10} is adopted as the optimum recession length, k_2 and k_{20} assume the benchmark for river sensitivity and prolonged dryness. The study noted the haphazard progradation in the MRC index of Bloukrans, contradicting the general retrogradation in the MRC index as the recession length increases, thus depicting the unpredictability of flow in the station.

Table 3. Multilinear regression of flow timing and streamflow log.

Station	k_i	R	R^2	Adjusted R^2	Standard Error	*p*-Value
Bloukrans	2	0.933	0.871	0.856	0.077	<0.000
	5	0.266	0.071	−0.032	0.208	0.428
	10	0.378	0.143	0.047	0.199	0.252
	15	0.374	0.140	0.044	0.200	0.257
	20	0.984	0.968	0.964	0.039	<0.000
	22	0.805	0.647	0.608	0.128	0.003
Tshoxa	2	0.997	0.994	0.993	0.031	<0.000
	6	0.992	0.985	0.983	0.047	<0.000
	10	0.995	0.991	0.990	0.037	<0.000
	15	0.939	0.881	0.868	0.133	<0.000
	18	0.993	0.986	0.985	0.045	<0.000
Pauls	3	0.983	0.967	0.964	0.069	<0.000
	6	0.983	0.966	0.962	0.070	<0.000
	10	0.997	0.995	0.994	0.027	<0.000
	15	0.692	0.479	0.421	0.273	0.018
	18	0.844	0.712	0.680	0.203	0.001
Kraai	2	0.975	0.951	0.945	0.130	<0.000
	5	0.974	0.948	0.942	0.133	<0.000
	10	0.968	0.936	0.929	0.148	<0.000
	15	0.982	0.964	0.960	0.110	<0.000
	20	0.992	0.984	0.982	0.074	<0.000
	25	0.981	0.962	0.958	0.114	<0.000
	30	0.984	0.967	0.964	0.106	<0.000
	35	0.976	0.953	0.947	0.127	<0.000
Orange	3	0.962	0.925	0.916	0.148	<0.000
	7	0.962	0.926	0.918	0.146	<0.000
	10	0.973	0.947	0.941	0.124	<0.000
	15	0.969	0.939	0.932	0.133	<0.000
	20	0.971	0.943	0.937	0.128	<0.000
	25	0.944	0.891	0.879	0.178	<0.000
	30	0.942	0.887	0.874	0.181	<0.000
	33	0.875	0.766	0.740	0.260	<0.000

4.2. Parameterization of Master Recession Curve Based on k_{10}

The distribution of the MRC index across the selected river stations is presented in Table 4 and Figure 3. While the winter discharge ranges from 0.015 (Groots) to 2.853 (Orange), the MRC index ranges from 3.81 (Teebus) to 75.5 (Kraai). A class-4 catchment (Tsitsa; n = 20) exhibited the highest significant periodic discharge ratio, n, while the medium and the least n are exhibited by catchments in class-3 (Botman and Zwelitsha; n = 4 and 9) stations. The deduced coefficients of MRC were used to generate the quadratic plots of MRC, as illustrated in Figure 3.

Table 4. Summary of master recession curve index assessment at optimum recession length (k_{10}).

S/N	Rivers	Kmin	Kmed	Kmax	Min logQ	Max logQ	Coeff. A	Coeff. B	Coeff. C	n
1	Elands	2.7	6.07	13.6	0.09	0.57	7.64	−11.58	4.11	15
2	Bloukrans	3.3	5.22	55	0.15	0.62	34.25	−38.67	10.82	12
3	Kubusie	3.2	6.07	13.6	0.17	0.57	8.06	−12.10	4.27	13
4	Mgqakwebe	6.4	7.83	15.2	0.05	0.41	9.10	−13.99	4.23	5
5	Yellowwoods	3.5	4.06	8	0.28	0.75	0.60	3.78	3.78	6
6	Toise	4.9	6.76	10.7	0.08	0.98	0.67	−7.67	6.88	6
7	Koonap	6.5	16.88	21.2	0.07	1.12	−3.30	−12.33	17.86	8
8	Mooi	7.7	30.65	171.2	0.02	0.86	34.70	−63.82	29.22	15
9	Tshoxa	6.7	18.2	49.2	0.33	1.37	6.36	−32.23	32.18	13

Table 4. *Cont.*

S/N	Rivers	Kmin	Kmed	Kmax	Min logQ	Max logQ	Coeff. A	Coeff. B	Coeff. C	n
10	Buffalo-Zwe	5.7	11.58	31.4	0.16	0.76	7.16	−18.88	10.22	9
11	Bav-Botman	2.9	4.27	13.6	0.22	1.22	3.02	−10.10	7.82	4
12	Pauls	11.4	17.6	33.8	0.14	1.18	5.19	−28.43	26.25	5
13	Lilfish	5	27.47	115.4	0.37	1.45	21.76	−61.85	43.93	5
14	Teebus	2.9	3.81	60.8	0.23	1.29	26.12	−57.24	30.43	8
15	Tsomo	8.7	12.31	72.8	0.30	1.53	10.08	−37.50	33.80	8
16	Wonderboom	6.7	14.55	72.7	0.07	1.63	6.71	−33.10	36.16	5
17	Keiskammahoek	10.4	19.4	54.8	0.23	1.15	10.95	−42.27	34.04	11
18	Tina	8	60.16	161.8	0.45	1.70	27.91	−115.50	115.64	19
19	Tsitsa	13.6	57.92	173.1	0.66	1.61	23.97	−117.53	127.19	20
20	Kraai	12.4	75.5	176.6	0.76	2.25	35.51	−164.59	190.55	19
21	GF-Vleij	5.2	14.68	41.5	1.13	2.09	−5.51	0.08	23.95	10
22	GF-Brandt	6.7	32.06	59.3	1.24	2.31	12.39	−69.04	93.36	8
23	Groots	12.6	43.31	98.9	0.02	1.08	22.75	−59.99	38.21	11
24	Orange	12.5	73.2	166.4	1.44	2.85	37.46	−223.90	333.90	19

Note: Kmin, Kmed, and Kmax denote the minimum, medium, and maximum values of the MRC constant per station.

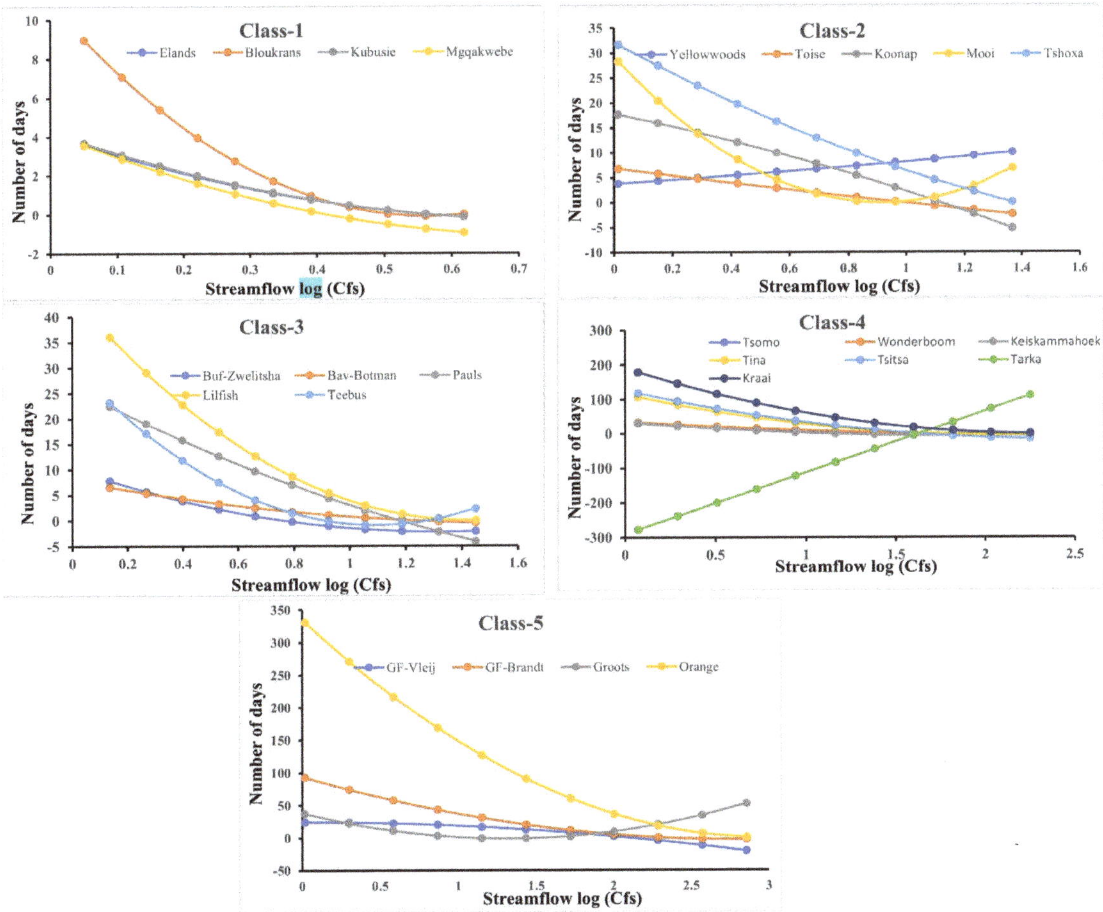

Figure 3. Plot of logarithmic streamflow against K for the Class 1 to 5 river with the number of days at 0 cfs of the streamflow log depicting the diminution days when flow recedes.

Generally, a noticeable increase in recession constants (Kmin, Kmed, and Kmax) indicates that larger catchments tend to have more pronounced declines in streamflow during recession periods. The coefficients A and C generally showed increasing trends with larger catchments, indicating that larger systems can manage higher flows effectively while exhibiting more significant flow timing variability. The number of years with significant discharge, n, varies among the smaller stations but tends to be higher in larger catchments. The quadratic plot provides the probable depletion rate of flow such that the right (left) side of the symmetry line $\left(\frac{-b}{2a}\right)$ represent the resilient (vulnerable) property of the river on account of recharge (discharge) (as vividly depicted by the loops on Mooi (Figure 3; Class 2), Teebus (Figure 3; Class 3), Tsomo, and Keiskammahoek (Figure 3; Class 4) plots.

The streamflow recession and drought vulnerability begins when streamflow recharge reduces to zero at $\frac{-b}{2a}$ and switch into a discharge state until it approaches 0 cfs of the streamflow log. The axis of symmetry (the quadratic loop) denotes the actual threshold of departure from resilience to vulnerability. Hence, the study showed that Yellowwoods has the shortest (about four days) probable diminution period while Orange River has the highest (about 334 days) probable diminution period, respectively. The selected stations are characterized by six non-resilient rivers, three highly vulnerable rivers, one moderately vulnerable river, nine relatively resilient rivers, and five resilient rivers Table 5. While the most susceptible rivers are predominantly among river basins less than 850 km^2, the most resilient are associated with river basins larger than 2500 km^2, and 58% of the rivers are considered resilient based on the probable diminution period. The resulting mean EF thresholds, Q_+ and Q_-, were quantified to range from 3.9×10^{-7} m^3/s to 573.96 m^3/s and 2.3×10^{-7} m^3/s to 20.19 m^3/s, respectively (Table 5). Verification of the Q_+ (q_{+T}%) proved undefined in 5 stations and showed that only 5 out of 19 stations have viable winter flow above the threshold. Meanwhile, Q_- verification (q_{-T}%) depicted only Yellowwoods as undefined and 13 stations as having viable winter flow above the threshold. Hence, based on q_{+T}%, the study showed that 5 of the stations are highly exhausted ($q_{+T}\% < 25\%$), five are moderately exhausted ($25 \leq q_{+T}\% < 50\%$), five are conserved ($50 \leq q_{+T}\% < 75\%$), and eight are well-conserved ($75 \leq q_{+T}\% < 100\%$). Class 2 and 3 are the most exhausted, while Class 1 is the most conserved, though with the lowest probable diminution period, possibly because they constitute the headwaters within the natural landscape.

Table 5. Summary of the probable diminution period (P_k), the streamflow threshold ($Q_\pm$), and the river conservation percentage (q_{+T}).

S/N	Rivers	P_k	Q_+ (m^3/s)	Q_- (m^3/s)	q_{+T} (%)	q_{-T} (%)
1	Elands	0.01	0.25	0.10	30.47	67.23
2	Bloukrans	0.08	0.12	0.09	91.06	97.34
3	Kubusie	0.01	0.24	0.10	47.39	88.62
4	Mgqakwebe	0.01	0.38	0.07	13.18	56.82
5	Yellowwoods	0.01	0.00	0.00	Undefined	Undefined
6	Toise	0.01	67.66	0.47	Undefined	20.21
7	Koonap	0.44	0.00	0.37	Undefined	31.91
8	Mooi	0.66	0.27	0.21	76.71	84.24
9	Tshoxa	0.69	141.21	0.66	0.01	21.6
10	Buffalo-Zwe	0.02	2.14	0.16	10.08	92.5
11	Bav-Botman	0.01	3.72	0.47	1	6.72
12	Pauls	0.62	573.96	0.42	Undefined	17.67
13	Lilfish	0.77	0.79	0.70	11.66	13.08
14	Teebus	0.67	0.55	0.23	98	100
15	Tsomo	0.70	4.38	0.96	17	43.63
16	Wonderboom	0.72	56.26	1.22	0.2	11.28
17	Keiskammahoek	0.71	14.65	0.40	3.18	72.5
18	Tina	0.91	7.82	1.41	38.83	78.23

Table 5. *Cont.*

S/N	Rivers	P_k	Q_+ (m³/s)	Q_- (m³/s)	q_{+T} (%)	q_{-T} (%)
19	Tsitsa	0.92	55.27	1.16	10.51	93.69
20	Kraai	0.95	6.87	5.03	52.18	56.43
21	GF-Vleij	0.58	0.00	3.49	Undefined	82.12
22	GF-Brandt	0.89	51.29	5.81	0.62	35.59
23	Groots	0.74	1.03	0.34	16.95	41.44
24	Orange	0.97	37.72	20.19	57.25	70.46

4.3. *Validation of Master Recession Curve Parameters*

The extracted average baseflow ranges from 0.102 to 1.064 at the Baviaan-Botmans and Kraai basins, respectively (Table 6 and Figure 4). Generally, the baseflow trend is influenced by the climatic type as baseflow increases from the west, the monsoon climate to the east, and the subtropical climate, concerning the trend of rainfall intensity. The Mann–Kendall trend showed that most stations are characterized by a decrease in rainfall corresponding to the increasing temperature trend. While none of the trend increases in temperature is significant, eight stations, including all the Class-3 stations, exhibited a substantial decrease in rainfall trends (Table 5). In contrast, only Groot's station exhibited a significant increase in rainfall.

Table 6. Summary of the Mann–Kendall test for rainfall and temperature, meteorological drought session, and baseflow extracted to validate MRC assessment.

S/N	Rivers	MK-P	MK-T	Drought Session	Drought%	Baseflow
1	Elands	0.01928	0.014188	15	12.5	0.373
2	Bloukrans	0.027321	0.014188	16	13.3	0.697
3	Kubusie	−0.04452	0.049845	17	14.2	0.603
4	Mgqakwebe	**−0.25415**	0.027153	20	16. 7	0.859
5	Yellowwoods	**−0.11075**	0.049845	23	19.2	0.432
6	Toise	−0.00533	0.050561	23	19.2	0.635
7	Koonap	**−0.08706**	0.047213	20	16. 7	0.478
8	Mooi	0.034427	0.030283	18	15	0.673
9	Tshoxa	−0.03087	0.023909	22	18.3	0.711
10	Buffalo-Zwe	**−0.2203**	0.031873	26	21. 7	0.617
11	Bav-Botman	**−0.2203**	0.031873	18	15	0.102
12	Pauls	**−0.08769**	0.036972	18	15	0.296
13	Lilfish	**−0.0616**	0.035297	17	14.2	0.314
14	Teebus	**−0.06098**	0.025523	20	16. 7	0.545
15	Tsomo	−0.02248	0.024638	18	15	0.71
16	Wonderboom	−0.05004	0.019882	25	20.8	0.408
17	Keiskammahoek	0.015454	0.023909	13	10.8	0.678
18	Tina	−0.03302	0.032165	19	15.8	0.961
19	Tsitsa	−0.04033	0.032165	21	17.5	0.931
20	Kraai	−0.02899	0.016925	26	21.7	1.064
21	GF-Vleij	−0.05055	0.035297	18	15	0.295
22	GF-Brandt	−0.06687	0.035392	20	16.7	0.339
23	Groots	**0.093336**	0.014236	22	18.3	0.499
24	Orange	−0.03027	0.018516	23	19.2	0.553

Note: The significant rainfall trends are bolded and underlined.

The extracted winter SPEI-3, illustrated in Figure 4, showed the variability in drought severity across the watershed. Kraai and Buffalo-Zwelitsha rivers exhibit the highest frequency (21%) of drought severity ($z_i < -1$) while Keiskammahoek River exhibits the lowest drought frequency (11%). The drought frequency varies distinctively across all the station classes with no specific trend. However, the class-1 stations that are attributed with high vulnerability are contrarily associated with low drought frequency, and perhaps this is what accounted for their high conservation based on their q_{+T}%. Similarly, some stations in the large class, Class 4 and 5, also exhibited high drought frequencies associated with high resilience. Nonetheless, Kubusie has the highest drought frequency (6) for the extreme drought severity ($z_i < -2$), followed by Keiskammahoek (4). Aside from Great Fish at Vleij (2), other Class 4 and 5 stations have zero drought frequency for extreme drought severity ($z_i < -2$).

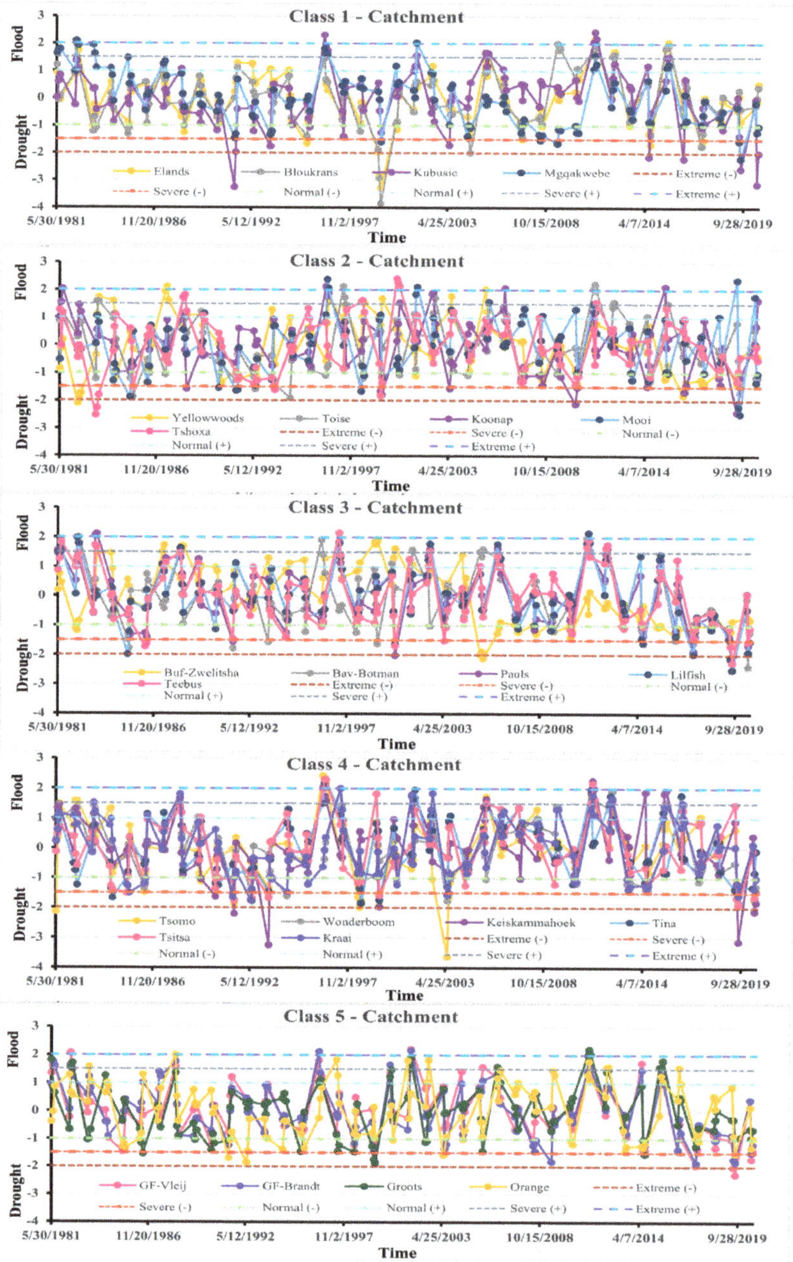

Figure 4. The standard precipitation and evapotranspiration index 3 (SPEI-3) plot of winter rainfall (January 1981 to December 2020) for the selected stations in the South of South Africa with normal (±) depicting the normal rainfall ($-1 \leq x \leq 1$; x is the rainfall value), severe drought ($-2 < x < -1$), extreme drought ($x < -2$), severe flood ($2 > x > 1$), and extreme flood ($x > 2$). The plot showed the intra- and inter-class drought severity variability across the survey stream and confirmed the distinct variability to drought and atmospheric circulation across the entire watershed.

Drawing from the correlation matrices (Table 7), the study confirmed that the catchment size has an enormously substantial influence (p-value < 0.01) on Kmed (29%), Min logQ (22%), Max logQ (52%), B (25%), C (46%), Q_- (41%), and P_k (45%). The study confirmed the significant influence of baseflow on A (12%) and the number of years with significant discharge (n; 19%). It also confirmed the significant impact of rainfall trends on the periods of significant discharge (18%) of the rivers. Meanwhile, meteorological drought exhibits no significant trend with any MRC parameter and showed a 6% (R = 0.24) and 4% (R = 0.20) relationship with watershed size, maximum discharge (MaxQ), streamflow threshold (Q_-), and diminution period (P_k) under the Kendall Tau matrices.

Table 7. Correlation matrices of master recession curve parameters and hydro-meteorological information of the study area based on Kendall Tau and Spearman rho.

		Size	*Kmed*	*MinQ*	*MaxQ*	*A*	*B*	*C*	*n*	Q_-	$q_{-T}\%$	P_k	*MK-P*	B_{i+1}
Kendall Tau	*Size*	1	***0.54***	***0.47***	***0.72***	0.28	***−0.49***	***0.68***	0.16	***0.64***	−0.04	***0.67***	0.11	0.03
	Kmed	***0.54***	1	0.23	***0.44***	***0.4***	***−0.63***	***0.69***	**0.37**	***0.5***	−0.06	***0.71***	0.2	0.23
	MinQ	***0.47***	0.23	1	***0.59***	0.17	−0.27	***0.38***	0.23	***0.53***	0.08	***0.4***	−0.08	0.07
	MaxQ	***0.72***	***0.44***	***0.59***	1	0.18	***−0.42***	***0.66***	0.09	***0.88***	−0.14	***0.65***	0.02	0
	A	0.28	***0.4***	0.17	0.18	1	***−0.72***	***0.46***	***0.4***	0.13	**0.3**	***0.47***	**0.33**	**0.35**
	B	***−0.49***	***−0.63***	−0.27	***−0.42***	***−0.72***	1	***0.75***	**−0.36**	**−0.37**	−0.15	***−0.8***	−0.25	−0.28
	C	***0.68***	***0.69***	***0.38***	***0.66***	***0.46***	***−0.75***	1	0.27	***0.63***	−0.01	***0.97***	0.21	0.19
	n	0.16	**0.37**	0.23	0.09	***0.4***	**−0.36**	0.27	1	0.1	***0.45***	**0.31**	***0.42***	***0.44***
	Q_-	***0.64***	***0.5***	***0.53***	***0.88***	0.13	**−0.37**	***0.63***	0.1	1	−0.21	***0.6***	0.02	0.01
	$q_{-T}\%$	−0.04	−0.06	0.08	−0.14	**0.3**	−0.15	−0.01	***0.45***	−0.21	1	0.01	0.14	**0.3**
	P_k	***0.67***	***0.71***	***0.4***	***0.65***	***0.47***	***−0.8***	***0.97***	**0.31**	***0.6***	0.01	1	0.22	0.2
	MK-P	0.11	0.2	−0.08	0.02	**0.33**	−0.25	0.2	***0.42***	0.02	0.14	0.22	1	0.24
	B_{i+1}	0.03	0.23	0.07	0.00	**0.35**	−0.28	0.19	***0.44***	0.01	**0.3**	0.2	0.24	1
	z_i	0.24	0.18	0.08	0.2	−0.01	−0.12	0.19	0.01	0.2	−0.12	0.2	−0.11	0.12
Spearman rho	*Size*	1	***0.71***	***0.55***	***0.85***	**0.35**	***−0.6***	***0.85***	0.22	***0.8***	−0.04	***0.83***	0.1	−0.02
	Kmed	***0.71***	1	0.38	***0.62***	**0.5**	***−0.8***	***0.85***	**0.51**	***0.67***	−0.07	***0.85***	0.3	0.3
	MinQ	***0.55***	0.38	1	***0.73***	0.22	−0.37	***0.53***	0.33	***0.67***	0.15	***0.54***	−0.11	0.07
	MaxQ	***0.85***	***0.62***	***0.73***	1	0.25	***−0.6***	***0.83***	0.17	***0.96***	−0.19	***0.81***	0	−0.03
	A	0.35	**0.5**	0.22	0.25	1	***−0.9***	***0.63***	***0.58***	0.19	**0.47**	***0.64***	**0.48**	**0.51**
	B	***−0.62***	***−0.77***	−0.37	***−0.56***	***−0.89***	1	***−0.89***	***−0.58***	**−0.5**	−0.22	***−0.9***	−0.35	**−0.43**
	C	***0.85***	***0.85***	***0.53***	***0.83***	***0.63***	***−0.9***	1	0.4	***0.79***	−0.04	***0.99***	0.28	0.26
	n	0.22	**0.51**	0.33	0.17	***0.58***	***−0.5***	0.4	1	0.19	***0.61***	**0.43**	***0.61***	***0.57***
	Q_-	***0.8***	***0.67***	***0.67***	***0.96***	0.19	**−0.5**	***0.79***	0.19	1	−0.31	***0.75***	0.07	−0.01
	$q_{-T}\%$	−0.04	−0.07	0.15	−0.19	**0.47**	−0.22	−0.04	***0.61***	−0.3	1	−0.01	0.24	**0.42**
	P_k	***0.83***	***0.85***	***0.54***	***0.81***	***0.63***	***−0.9***	***0.99***	***0.43***	***0.75***	−0.01	1	0.28	0.27
	MK-P	0.1	0.3	−0.11	0	**0.48**	−0.35	0.28	***0.61***	0.07	0.24	0.28	1	0.38
	B_{i+1}	−0.02	0.3	0.07	−0.03	**0.51**	**−0.42**	0.26	***0.57***	−0	**0.42**	0.27	0.38	1
	z_i	0.31	0.23	0.09	0.29	−0.02	−0.14	0.24	0.01	0.28	−0.16	0.25	−0.2	0.2

Note: Strongly significant correlation at the 0.01 level (2-tailed) is italicized, bolded, and underlined. Significant correlation at the 0.05 level is bolded. Note: Strongly significant correlation at the 0.01 level (2-tailed) is italicized, bolded, and underlined. Significant correlation at the 0.05 level is bolded.

5. Discussion

5.1. Deductions from the Optimization of Recession Length

The National Water Act of South Africa (NWA, Act 36 of 1998) demands the quantification and ascertainment of water reserves, among other competing allocations, such as the basic human needs and reserve for ecological integrity [39]. This is required to ensure the sustainable conservation of water resources amidst the increasing impact of climate extremes such as drought while keeping to its efficient and equitable distribution goals [40,41]. To achieve this, this study began with the recommendation of Fiorotto and Caroni [25], requesting the need to improve and optimize the value of the recession constant used to estimate MRC parameters. The study adopted ten days of recession length based on the central tendencies of MRC indices using flows from five classes of catchment, optimum value of periods with significant discharge (years), consistency of Kmed, and regression analysis of streamflow log.

The finding here agrees with Vogel and Kroll's [27] study, which also considered ten days of recession length as the appropriate optimum length for the approximation of low-flow behavior based on the application of the linear reservoir hypothesis and the Dupuit–Boussinesq aquifer model to 23 river catchments. In addition, EF estimation using a two-day recession length cannot provide a logical guide against drought preparedness. At the same time, the study confirmed that 15 to 20 days of recession length is "overstretching" micro-mini and mini catchments based on the years with significant discharge. The study confirmed the influence of catchment size on the exhaustion of the connecting reservoir, which is evident from the storage-discharge relationship. As the recession length increases, the years with significant discharge reduce for the small catchments compared to the large ones [25]. As a result, the years with significant discharge, n, can provide assumptions about the rainfall-dependent rivers, as significantly validated by baseflow and rainfall trends. Hence, Baviaans River at Botmans can be identified as a rainfall-dependent river inferred by the probable discharge period.

5.2. MRC Parameterization and Implications

The findings on the extent of correlation between baseflow and MRC index, using the generally adopted central tendency, the median curve, Kmed [25], agree with Ebrahim and Villholth's [41] study, which was also carried out in South Africa. However, beyond complementing baseflow as posited by Ebrahim and Villholth [41], the finding here provides more clarity about the extensive function of MRC parameters to aid hydrologic drought risk assessment. While differing from baseflow estimate, which includes distinct information about river seasonality [6], MRC parameters have been synthesized to project other vital information required in aiding environmental flow estimation, particularly against drought risk. For example, in this study, the Great Fish River at Vleij (B_{i+1} = 0.295) and Elands River (B_{i+1} = 0.373) are characterized by low baseflow (average B_{i+1} < 0.5) and could be classified as ephemeral in flow, whereas their period of significant discharge (n) and conservation extent ($q_{-T}\%$) suggest their resilience. Meanwhile, Tsomo (B_{i+1} = 0.700) and Toise (B_{i+1} = 0.635) showed viable average baseflow, which classified them as perennial, whereas n and $q_{-T}\%$ suggest that the rivers are vulnerable, possibly due to the variation in discharge rate to quantity of discharge.

As an extension of the MRC, the probable diminution period, P_k, and the EF threshold, Q_-, were deduced to infer the streamflow's vulnerability or resilience to drought risk, thereby addressing the concerns and reservations of Sharma and Dutta [23]. The concurrent agreement of rainfall trend and baseflow with the years of significant discharge (n) and conservation extent ($q_{-T}\%$) validates their applicability as tools for EF conservation. By benchmarking the EF threshold, the minimum flow required to sustain river flow and the ecological function of the environment can be calibrated and monitored while keeping to the aquatic-riparian habitat sensitivity to such flow. At the same time, considering their variability in drought sensitivity, the probable diminution period can be viewed as a yardstick for prioritizing rivers. With the evidence of an increasing trend of climate change impact and drought, especially in the riparian ecosystem of the Southern part of South Africa [8], the probable diminution period can serve as a viable tool for drought disaster preparedness and equilibrating water use. It can be incorporated into impact assessment requirements while carrying out zone budgeting for water-driven enterprises.

Moreover, the study showed that the stations associated with excessive diminution below the threshold, especially the Class-2 and Class-3 stations, were majorly associated with a significant decline in rainfall trend (Tables 5 and 6). On this note, the finding here refutes the assertion of Bogaart et al. [42], who posited that streamflow recession is independent of climate change. Contrarily, drought significantly impacted the river's diminution rate; hence, its position as a slow-onset effect on the periods of significant discharge was at a 19% rate in the case of the rivers in this study. The variations in drought sensitivity across the water management areas (WMA) with distinct climate types, particularly within the tertiary (Class 3) and quaternary catchments (Class 1 and 2) in Fish-Tsitsikamma WMA (Q9H030, Q6H003, Q3H004, and Q8H008) and Mzimvubu-to-Keiskamma WMA (R2H005, R2H006, R2H010, R2H011, T3H009, and S6H001) significantly revealed the impact of climate change (Tables 4–6). A similar view is shared by Gunawardhana and Al-Rawas [43], who proved the dependency of river recession on rainfall events using the recession rates of rainfall hyetograph, and Parchami et al. [10], using a visual basic program for rivers in a semi-arid environment. This discovery further justifies the assumption of one-dimensional horizontal flow properties of groundwater-surface water flow [41,44], such that the quality of inflow influences storage and discharge as in Equations (19) and (20)

Based on the catchment size, it can be inferred that the resilience and conservation of the Class 4 and 5 rivers is possibly due to the interconnection with intermediate and regional groundwater flow. Toth [44] affirms that local topographical conditions most likely unaffected catchments with larger geometrical dimensions. The study showed that rivers with catchment size above 800 m^2 are resilient, while catchment size above 5000 km^2 would be appropriate for enhancing the inter-basin transfer scheme. The major limitation of this study is the 20 years of study; however, since the study focuses on the current performance of the river vis à vis its aquifer connectivity. Moreover, the time series length was influenced by missing data within some stations and the inadequacy of concurrent monitoring for all the South African rivers. However, future studies may consider undertaking a few stations with more than 50 years of time series. Future study may also investigate the variability of MRC parameters of mini watersheds, with other hydrogeologic parameters, and their influence on the spatial dynamics of vegetation health and riparian function. While considering the socioeconomic and cultural justification for water abstraction, stakeholders should factor in climate change extremes' impact, the river's vulnerability to degradation, and the unrecoverable biodiversity loss to their estimate. Hence, the approach presented in this study can be incorporated into the national action program methodologies for river prioritization for drought preparedness to strengthen the United Nations Convention to Combat Desertification initiatives. Step 4 of the ten-step planning process for drought preparedness requires preparing resource inventory and identifying risky groups, while step 7 advises integrating science [45]. The validation presented in this study fulfills the scientific preparations necessary for drought preparedness.

6. Conclusions

The present study explored the optimum recession length required to standardize the derivation of master recession curve (MRC) parameters and the validatory performance of the MRC parameters, considering their potential as an indicator of drought disaster preparedness. Hence, the study's novelty lies in standardizing the MRC approach for estimating the environmental flow threshold and the probable diminution period based on an optimum recession length. The following conclusions can be drawn from the study:

- The analysis provides a comprehensive view of how different catchments behave under varying hydrological conditions, essential for effective water resource management and planning.
- The correlation between recession constants and streamflow values highlights the hydrological dynamics within different catchments. It emphasizes that larger river systems exhibit larger recession constants and faster declines in streamflow following peak events due to their greater runoff capacity and drainage efficiency.
- The ratio of significant discharge years can serve as a vital aid for classifying river basins concerning streamflow consistency, water retention, and management capabilities.
- The variability assessment between streamflow and its timing at varying length scenarios suggests that the hydrological dynamics of varying catchments can be captured at the optimum recession length of 10 and 20 days. The number of years with significant discharge supports 10 days, while the predictive power of streamflow behavior and the hydrological dynamics of the river supports 20 days.
- MRC coefficients (A, B, and C) characterize environmental flow performance by informing on the initial flow condition, rate of decline, and the overall flow dynamics during the recession phase.
- The probable diminution period enables the assessment of river vulnerability. In contrast, the EF threshold enables the investigation of the river's exhaustion quotient, thus providing a basis for measuring river drought risk level and preparedness.

- Quaternary and tertiary catchments of South Africa are sensitive to severe drought due to low capacity; hence, conserving them for ecological and river health management during winter would be appropriate.

Overall, the proposed approach can provide valuable information for watershed prioritization, river forecast, and drought risk assessment, which could aid planners in allocating water resources appropriately to environmental conditions.

Author Contributions: Conceptualization, S.T.O.; methodology, S.T.O.; software, S.T.O.; validation, S.T.O. and J.A.B.; formal analysis, S.T.O.; investigation, S.T.O.; resources, J.A.B.; data curation, S.T.O. writing—original draft preparation, S.T.O.; writing—review and editing, J.A.B.; visualization, J.A.B.; supervision, J.A.B.; project administration, J.A.B.; funding acquisition, J.A.B. All authors have read and agreed to the published version of the manuscript.

Funding: This research received no external funding, and the APC was funded by the University of the Free State Library.

Data Availability Statement: The streamflow and rainfall data used in the study are obtained from the Department of Water and Sanitation, South Africa, portal, https://www.dws.gov.za/Hydrology/Verified/hymain.aspx and accessed on 26 June 2023.

Acknowledgments: The authors are grateful to the entire DiMTEC team for making the center comfortable for research.

Conflicts of Interest: The authors declare no conflicts of interest.

References

1. Caretta, A.M.M.A.; Arfanuzzaman, R.B.M.; Morgan, S.M.R.; Kumar, M. Water. In *Climate Change 2022: Impacts, Adaptation, and Vulnerability. Contribution of Working Group II to the Sixth Assessment Report of the Intergovernmental Panel on Climate Change*; IPCC: Geneva, Switzerland, 2022. Available online: https://msuweb.montclair.edu/~lebelp/PSC643IntPolEcon/IPCCClimateChange2007.pdf (accessed on 22 January 2024).
2. Orimoloye, I.R.; Mazinyo, S.P.; Kalumba, A.M.; Ekundayo, O.Y.; Nel, W. Implications of climate variability and change on urban and human health: A review. *Cities* **2019**, *91*, 213–223. [CrossRef]
3. Mishra, A.; Alnahit, A.; Campbell, B. Impact of land uses, drought, flood, wildfire, and cascading events on water quality and microbial communities: A review and analysis. *J. Hydrol.* **2021**, *596*, 125707. [CrossRef]
4. Belle, J.; Mapingure, T.; Owolabi, S.T. Factors Influencing Rural Women's Adoption of Climate Change Adaptation Strategies: Evidence from the Chivi District of Zimbabwe. *Climate* **2024**, *12*, 191. [CrossRef]
5. Owolabi, S.T.; Belle, J.A. Investigating extreme hydrological risk impact on water quality; evidence from Buffalo catchment headwater, Eastern Cape, South Africa. *Environ. Sci. Pollut. Res.* **2023**, *30*, 73425–73450. [CrossRef] [PubMed]
6. Owolabi, S.T.; Madi, K.; Kalumba, A.M.; Alemaw, B.F. Assessment of recession flow variability and the surficial lithology impact: A case study of Buffalo River catchment, Eastern Cape, South Africa. *Environ. Earth Sci.* **2020**, *79*, 187. [CrossRef]
7. Poff, N.L.; Tharme, R.E.; Arthington, A.H. Evolution of environmental flows assessment science, principles, and methodologies. In *Water for the Environment*; Academic Press: New york, NY, USA, 2017; pp. 203–236. [CrossRef]
8. Afuye, G.A.; Kalumba, A.M.; Owolabi, S.T.; Thamaga, K.H.; Ndou, N.; Sibandze, P.; Orimoloye, I.R. Analyzing spatiotemporal variations and dynamics of vegetation over Amathole district municipality in South Africa. *Environ. Dev. Sustain.* **2024**, 1–26. [CrossRef]
9. Volchek, A.; Kirvel, I.; Sheshko, N. Environmental flow assessment for the Yaselda River in its Selets reservoir section. *Ecohydrol. Hydrobiol.* **2019**, *19*, 109–118. [CrossRef]
10. Parchami, N.; Mostafazadeh, R.; Ouri, A.E.; Imani, R. Spatio-temporal analysis of river flow master recession curve characteristics over Ardabil province rivers. *Phys. Chem. Earth Parts A/B/C* **2024**, *134*, 103586. [CrossRef]
11. Sedighkia, M.; Datta, B.; Abdoli, A. Utilizing classic evolutionary algorithms to assess the Brown trout (Salmo trutta) habitats by ANFIS-based physical habitat model. *Model. Earth Syst. Environ.* **2022**, *8*, 857–869. [CrossRef]
12. Zhang, X.R.; Zhang, D.R.; Ding, Y. An environmental flow method applied in small and medium-sized mountainous rivers. *Water Sci. Eng.* **2021**, *14*, 323–329. [CrossRef]
13. Arnillas, C.A.; Yang, C.; Zamaria, S.A.; Neumann, A.; Javed, A.; Shimoda, Y.; Feisthauer, N.; Crolla, A.; Dong, F.; Blukacz-Richards, A.; et al. Integrating watershed and ecosystem service models to assess best management practice efficiency: Guidelines for Lake Erie managers and watershed modellers. *Environ. Rev.* **2021**, *29*, 31–63. [CrossRef]
14. Abebe, W.B.; Tilahun, S.A.; Moges, M.M.; Wondie, A.; Dersseh, M.G.; McClain, M.E. Environmental flow assessment and implications on sustainability of aquatic ecosystems in Ethiopia: A literature review on global and national evidences. *Environ. Dev.* **2022**, *44*, 100758. [CrossRef]

15. Yang, W.; Xiao, C.; Liang, X. Extraction method of baseflow recession segments based on second-order derivative of streamflow and comparison with four conventional methods. *Water* **2020**, *12*, 1953. [CrossRef]
16. Ambroise, B. Variable water-saturated areas and streamflow generation in the small Ringelbach catchment (Vosges Mountains, France): The master recession curve as an equilibrium curve for interactions between atmosphere, surface and ground waters. *Hydrol. Process.* **2016**, *30*, 3560–3577. [CrossRef]
17. Sharma, D.; Kadu, A.; Biswal, B. Universal recession constants and their potential to predict recession flow. *J. Hydrol.* **2023**, *626*, 130244. [CrossRef]
18. Carlotto, T.; Chaffe, P.L.B. Master recession curve parameterization tool (MRCPtool): Different approaches to recession curve analysis. *Comput. Geosci.* **2019**, *132*, 1–8. [CrossRef]
19. Linsley Jr, R.K.; Kohler, M.A.; Paulhus, J.L. Hydrology for Engineers. Available online: https://trid.trb.org/View/42130 (accessed on 22 January 2024).
20. Fiorotto, V.; Caroni, E. A new approach to master recession curve analysis. *Hydrol. Sci. J.* **2013**, *58*, 966–975. [CrossRef]
21. Brutsaert, W.; Nieber, J.L. Regionalized drought flow hydrographs from a mature glaciated plateau. *Water Resour. Res.* **1977**, *13*, 637–643. [CrossRef]
22. Vogel, R.M.; Kroll, C.N. Regional geohydrologic-geomorphic relationships for the estimation of low-flow statistics. *Water Resour. Res.* **1992**, *28*, 2451–2458. [CrossRef]
23. Zhang, X.; Yu, J.; Wang, L.; Zhang, R. Determination of River Ecological Flow Thresholds and Development of Early Warning Programs Based on Coupled Multiple Hydrological Methods. *Water* **2024**, *16*, 1986. [CrossRef]
24. Shah, S.J.; van Walraven, C.; Jeon, S.Y.; Boscardin, J.; Hobbs, F.R.; Connolly, S.J.; Ezekowitz, M.D.; Covinsky, K.E.; Fang, M.C.; Singer, D.E. Estimating Vitamin K Antagonist Anticoagulation Benefit in People With Atrial Fibrillation Accounting for Competing Risks: Evidence From 12 Randomized Trials. *Circ. Cardiovasc. Qual. Outcomes* **2024**, *17*, e010269. [CrossRef] [PubMed]
25. Richter, B.D.; Baumgartner, J.V.; Wigington, R.; Braun, D.P. How much water does a river need? *Freshw. Biol.* **1997**, *37*, 231–249. [CrossRef]
26. Tennant, D.L. Instream Flow Regimens for Fish, Wildlife, Recreation and Related Environmental Resources. *Fisheries* **1976**, *1*, 6–10. [CrossRef]
27. Wang, W.; Seta, K.; Ikegaya, N. Modelling Probability Density Functions Based on the Gram–Charlier Series with Higher-Order Statistics: Theoretical Derivation and Application. *J. Wind. Eng. Ind. Aerodyn.* **2022**, *231*, 105227. [CrossRef]
28. Owolabi, S.T.; Belle, J.A.; Mazinyo, S. Quantifying intra-catchment streamflow processes and response to climate change within a climatic transitional zone: A case study of Buffalo catchment, Eastern Cape, South Africa. *Mathematics* **2022**, *10*, 3003. [CrossRef]
29. Baiyegunhi, C.; Liu, K. Sedimentary facies, stratigraphy, and depositional environments of the Ecca Group, Karoo Supergroup in the Eastern Cape Province of South Africa. *Open Geosci.* **2021**, *13*, 748–781. [CrossRef]
30. Owolabi, S.T.; Madi, K.; Kalumba, A.M.; Baiyegunhi, C. A geomagnetic analysis for lineament detection and lithologic characterization impacting groundwater prospecting; a case study of Buffalo catchment, Eastern Cape, South Africa. *Groundw. Sustain. Dev.* **2021**, *12*, 100531. [CrossRef]
31. Eckhardt, K. How to construct recursive digital filters for baseflow separation. *Hydrol. Process. Int. J.* **2005**, *19*, 507–515. [CrossRef]
32. Coutagne, A. Les variations de dèbit en pèriode non influencèe par les prècipitations. Le dèbit d'inflitration (corrèlations fluviales internes)—2me partie. *Meteorol. Et Hydrol. La Houille Blanche* **1948**, *34*, 416–436. [CrossRef]
33. Sujono, J.; Shikasho, S.; Hiramatsu, K. A comparison of techniques for hydrograph recession analysis. *Hydrol. Process.* **2004**, *18*, 403–413. [CrossRef]
34. Barlow, P.M.; Cunningham, W.L.; Zhai, T.; Gray, M. *US Geological Survey Groundwater Toolbox, a Graphical and Mapping Interface for Analysis of Hydrologic Data (Version 1.0): User Guide for Estimation of Base Flow, Runoff, and Groundwater Recharge from Streamflow Data (No. 3-B10)*; US Geological Survey: Reston VA, USA, 2015. Available online: https://pubs.usgs.gov/tm/03/c05/tm3c5.pdf (accessed on 22 January 2024).
35. Owolabi, S.T.; Madi, K.; Kalumba, A.M. Comparative evaluation of spatio-temporal attributes of precipitation and streamflow in Buffalo and Tyume Catchments, Eastern Cape, South Africa. *Environ. Dev. Sustain.* **2021**, *23*, 4236–4251. [CrossRef]
36. Thornthwaite, C.W. An approach toward a rational classification of climate. *Geogr. Rev.* **1948**, *38*, 55–94. [CrossRef]
37. Vicente-Serrano, S.M.; Beguería, S.; López-Moreno, J.I. A multiscalar drought index sensitive to global warming: The standardized precipitation evapotranspiration index. *J. Clim.* **2010**, *23*, 1696–1718. [CrossRef]
38. Rutledge, A.T. *Computer Programs for Describing the Recession of Ground-Water Discharge and for Estimating Mean Ground-Water Recharge and Discharge from Streamflow Records: Update (No. 98)*; US Department of the Interior, US Geological Survey: Reston VA, USA, 1998.
39. Xu, Y.; Colvin, C.; Van Tonder, G.J.; Hughes, S.; Le Maitre, D.; Zhang, J.; Mafanya, T.; Braune, E. Towards the Resource Directed Measures: Groundwater Component (Version 1.1). *Water Res. Comm.* 2002. Available online: https://citeseerx.ist.psu.edu/document?repid=rep1&type=pdf&doi=9e5aa720259fbc6b6918d81ced396a202e015d2d (accessed on 22 January 2024).
40. Kelbe, B.; Germishuyse, T. Groundwater/surface water relationships with specific reference to Maputaland. In *Water Research Commission*; University of Zululand: Richards Bay, South Africa, 2010; Volume 10.
41. Ebrahim, G.Y.; Villholth, K.G. Estimating shallow groundwater availability in small catchments using streamflow recession and instream flow requirements of rivers in South Africa. *J. Hydrol.* **2016**, *541*, 754–765. [CrossRef]

42. Bogaart, P.W.; Van Der Velde, Y.; Lyon, S.W.; Dekker, S.C. Streamflow recession patterns can help unravel the role of climate and humans in landscape co-evolution. *Hydrol. Earth Syst. Sci.* **2016**, *20*, 1413–1432. [CrossRef]
43. Gunawardhana, L.N.; Al-Rawas, G.A. Investigating meteorological effect on river flow recession rate in an arid environment. *Hydrol. Sci. J.* **2020**, *65*, 2249–2255. [CrossRef]
44. Toth, J. A theoretical analysis of groundwater flow in small drainage basins. *J. Geophys. Res.* **1963**, *68*, 4795–4812. [CrossRef]
45. Wilhite, D.A.; Hayes, M.J.; Knutson, C.L. Drought preparedness planning: Building institutional capacity. In *Drought and Water Crises: Science, Technology, and Management Issues*; Taylor Francis Group: Abingdon, UK, 2005; Volume 107, pp. 93–136. Available online: https://tailieu.vn/docview/tailieu/2011/20111218/chimungdauhoi/drought_and_water_cruises_1__9341.pdf (accessed on 22 January 2024).

hydrology

MDPI

Article

Assessing Groundwater Connection/Disconnection to Waterholes Along the Balonne River and in the Barwon–Darling River System in Queensland and New South Wales, Australia, for Waterhole Persistence

Harald Hofmann [1,2,*] and Jonathan Marshall [3,4]

1 School of the Environment, The University of Queensland, Brisbane, QLD 4072, Australia
2 Commonwealth Scientific and Industrial Research Organisation, Brisbane, QLD 4102, Australia
3 Department of the Environment, Tourism, Science and Innovation, Brisbane, QLD 4102, Australia; jonathan.marshall@des.qld.gov.au
4 Australian Rivers Institute, Griffith University, Brisbane, QLD 4111, Australia
* Correspondence: h.hofmann@uq.edu.au

Abstract: Waterholes in semi-arid environment are sections of rivers that fill during high river flows or floods and keep water once flow ceases. They are essential water sources for rive ecosystems. Some waterholes remain even during prolonged droughts. The resilience of ecosystems in these environments depends on the persistence of the waterholes. While most semi-arid, ephemeral river systems are disconnected from regional groundwater and losing in most parts there may be some sections that can be connected to localised groundwater or parafluvial areas. To assess the persistence of waterholes the groundwater contribution to the water balance needs to be addressed. This study assesses groundwater connectivity to waterholes in a part of the Murray-Darling Basin, one of the largest watersheds in the world, using environmental tracers radon and stable isotopes. Approximately 100 samples were collected from 27 waterholes along the Narran, Calgoa, Barwon and Darling rivers, as well as 8 groundwater bore samples. The assessment of groundwater connectivity or the lack of is necessary from water balance modelling and estimation of persistence of these waterholes. As expected, the results indicate consistently low radon concentrations in the waterholes and very small deviation in stable isotopes δ^{18}O and δ^{2}H. In general, most of these waterholes are losing water to groundwater, indicated by low salinity (EC values) and low radon concentrations. While radon concentrations are small in most cases and indicative of little groundwater contributions, some variability can be assigned to bank return and parafluvial flow. It indicates that these contributions may have implications for waterhole persistence in ephemeral streams. The study demonstrates that in some cases local bank return flow or parafluvial flow may contribute to waterhole persistence.

Keywords: waterholes; groundwater/surface water; river bank storage; inland freshwater lens; radon; stable isotopes; Queensland; New South Wales; Darling river; Australia

1. Introduction

Ephemeral rivers and streams are common landscape features in arid and semi-arid environments across the globe. Approximately half of the global streams and rivers are ephemeral or non-perennial, which means that flow ceases for certain periods throughout the year, resulting in highly variable and transient storage in these systems [1–4]. Hence, there is a fine balance in sediment and nutrient transport and recycling as well as water storage across the hydrological compartments in these systems [5]. They consist of the stream channel, the banks, the floodplains and regional aquifers. The latter are of minor importance as groundwater levels are often below the riverbed in arid and semi-arid environments, and therefore, the stream is rather losing water to the regional groundwater than gaining. Local exceptions may occur where, for example, regional artesian aquifers out- or subcrop the riverbed or when flow along faults contribute to the stream discharge [6]. Most of the water storage in ephemeral streams is therefore concentrated in the channel itself, the river banks and the floodplain providing a temporal storage [7]. This is subsequently returned to the stream when river levels decline [8]. Waterholes on the other hand retain residual stream water after flow has ceased, building up series of pools along the stream network [9]. In arid and semi-arid environments, these waterholes are arguably one of the most important water sources as they provide water for riparian vegetation, native fauna as well as being refugia for aquatic ecosystems. Furthermore, they have significant economic and social value for the local population and are culturally important for indigenous populations [10].

The rivers of Australia are among the world's most seasonal, many having more than 80% of their total annual streamflow during the peak flow periods [9]. During prolonged drought periods, rivers dry into a series of waterholes which are an important resource for agriculture, town water supplies, industry and other human consumptive uses. Similarly to other semi-arid locations globally, they also serve an important ecological function by providing refuge habitat for aquatic organisms during prolonged periods without surface flow. It is for these reasons that waterholes have been identified as important environmental assets in a number of locations, including the northern Murray–Darling Basin [11].

A key indicator for waterholes and their ability to serve as ecosystem refugia is their persistence time [12]. The persistence time is the duration waterholes retain water in the absence of flow [13]. It is a function of how much water is held as flow stops, the time since flow stopped, hydrological processes, such as water loss and gain through rainfall, groundwater inputs, seepage loss and evaporation and the morphology of the channel, especially the depth [12]. Quantifying these processes allows determining how long waterholes can persist through hydrological modelling. Anthropogenic activities and water resource development have altered the arid landscape hydrological systems and changed the frequency and magnitude of flow events through weirs and locks, thus impacting how often waterholes are formed and re-filled. Impacts from climate change are another factor altering the hydrology of the systems. As evaporation is one of the largest loss terms, increased evaporation will most likely lead to faster declines of waterholes in the future [10]. A large unknown in waterhole persistence is their connection to groundwater. Many ephemeral streams are losing streams but may be gaining in some parts of the stream network through processes described above [9]. Most importantly bank return flow and fresh water lenses within the floodplains have the potential to contribute to the water budget of waterholes [7,14].

An accurate understanding of groundwater contributions to the riverine water balance are especially necessary within dryland regions, where surface water resources may diminish during low flow to poorly connected (or disconnected) waterholes [15,16]. These are essentially enlarged channel segments capable of holding larger volumes of water; however, their persistence (and therefore vulnerability) during dry periods may greatly vary as a function of size, shading and potential groundwater connectivity [17]. A critical question is therefore the degree to which groundwater interacts with these waterholes, and the mechanisms by which this occurs. Cendon et al. (2010) [18] suggested that some interaction of waterholes with freshwater lenses and riparian groundwater is physically plausible; however, the absence of groundwater bores to determine local or regional groundwater gradients make it difficult to determine the degree of interaction between the waterhole and groundwater. Studies in dryland river systems in Australia have shown a wide variety of potential interactions with groundwater, from strongly losing to gaining conditions from perched aquifer systems [19], losing but connected [18], and even to gaining conditions [20]. These factors are little studied for waterholes.

While hydraulic gradients between riparian groundwater and the surface water body are good indications for exchange, geochemical tracers, such a major ions or isotopic tracers are excellent tools to determine groundwater contributions to streams. The different geochemical tracers have potential advantages and disadvantages, for example, major ion concentrations and stable isotope ratios are relatively easy to measure and are often measured as part of general water quality studies; hence, databases with time series measurements may exist for some regions [21]. The principle of these methods is that there are distinctly different concentrations of a tracer in the groundwater and the surface water which can be used to determine mass fluxes from one to the other. Physico-chemical processes in the waterhole, such as evaporation, mineral precipitation, or biogeochemical processes may also alter the water chemistry, which need to be assessed and corrections need to be applied. However, providing that groundwater and surface water have distinct geochemical concentrations, changes in concentration of that component in the river may be used to define the distribution of gaining and losing river reaches and to quantify groundwater inflows in gaining reaches [22,23].

Radon (^{222}Rn), which is part of the Uranium (^{238}U) to Lead (^{206}Pb) decay series, is commonly used for quantifying groundwater inflows to rivers. Radon has a half-life of 3.8 days and the activities in groundwater reach secular equilibrium with its parent isotope radium (^{226}Ra) over a few weeks [24]. As the concentration of radium in minerals is several orders of magnitude higher than dissolved radium concentrations in surface water, groundwater radon activities are commonly two or three orders of magnitude higher than those of surface water [23–26]. This makes radon a potentially useful tracer in catchments where groundwater and surface water have similar major ion concentrations or stable isotope ratios, which is often the case with regards to riparian groundwater or freshwater lenses as a potential source [27]. Due to the fact that radon activities reach secular equilibrium with the matrix radium within a few week, radon concentrations are generally high in even relatively recently recharged groundwater, whereas rainfall has very small radon activities.

Radon has been used as a tracer for surface water/ groundwater interactions for decades but studies in arid and semi-arid environments are scarce. Cartwright et al. (2014) [28], for example, demonstrated sections of groundwater return flow to the Murray River in Victoria using radon. Bourke et al. (2014) [29] demonstrated the use of radon to assess parafluvial flow in ephemeral streams in Western Australia. Gilfedder et al. (2019) [30] showed on an example from the Avon River in Victoria, Australia, and a modelling approach the usefulness of radon as a mean residence time estimator for bank return flow. Zhou and Cartwright (2021) [31] demonstrated the use of radon to estimate baseflow fluxes to an intermittent stream in Victoria, Australia.

While radon is an important tracer for this purpose, radon concentrations alone are insufficient to separate different groundwater reservoirs. The distinction between groundwater from more regional aquifers compared to return flow from bank storage or shorter-term localised perched aquifer systems around the stream is important to determine in terms of long-term flow sustainability. The radon concentrations in the reservoirs are only dependant on the amount of uranium/thorium containing minerals or radium and the emanation rate and not on the residence time of the water once secular equilibrium is reached. Therefore, radon concentrations in riverbanks, perched aquifer systems, freshwater lenses or regional groundwater will be indistinguishable if radium concentrations in the host material are also the same.

Combining radon with other geochemical tracers such as major ions or stable isotopes, may allow the groundwater contributions from relatively short-term reservoirs (e.g. bank storage) and regional groundwater to be distinguished. Chloride concentrations for example often increase due to mixing with more evaporation enriched soil waters and significant increases can only be achieved over thousands of years. Hence, short or medium term reservoirs should only have small concentrations of chloride in the absence of mixing compared to some older, regional groundwater.

Here, we investigate the potential interaction of the surface water in waterholes with groundwater, in particular the potential role that groundwater discharge may play in waterhole persistence during low flow conditions within the Culgoa, the Narran and the Barwon-Darling rivers. This study will improve our understanding of groundwater interactions within waterholes across the Lower-Balonne and Barwon-Darling and inform modelling the water balances of these waterholes.

1.1. Geographical and Geological Background

The Narran, the Culgoa, the Barwon and the Darling river are part of the greater Barwon-Darling River system, located in the central northern part of the Murray-Darling River Basin in the border region between Queensland and New South Wales, Australia (Figure 1). The Narran and the Culgoa rivers have their headwaters in the Condamine-Balonne River catchments in Queensland and drain to the west. The Darling River is a confluent of most importantly the Barwon River and many smaller tributaries in the eastern and northern part of the Barwon-Darling River system at Brewarrina and also drains to the west.

The river systems are mostly ephemeral, especially in their upper reaches. They dry up during increased periods of low or no flow and result in a series of waterholes during times of drought [13]. The Lower Balonne floodplain system includes the channels, waterholes and floodplains of the Balonne, Culgoa, Birrie, Bokhara and Narran rivers to the confluence with the Barwon River in New South Wales (Figure 1). The floodplain systems have been identified as essential environmental asset [5,32]. The river/floodplain systems have series of in-channel waterholes, which provide important ecosystem services as habitats and act as refugia during drought [13]. Episodical in-channel flow events fill the waterholes, providing enough water to sustain aquatic biota during prolonged no-flow periods. These waterholes are also important resources of water for local agriculture, town water supplies and other industries in the region [33].

The rivers in the area are incised in larger floodplains comprised of Tertiary and Quaternary riverbed sediments. These sediments are underlain by a variety of Palaeozoic and Mesozoic bedrock formations from the New England Orogen, the Lachlan Orogen and the Gunnedah Basin [34]. The bedrock consists of consolidated, partly metamorphosed and fractured marine sand, silt and clay stones and volcanic and igneous intrusions. The Tertiary and Quaternary sediments provide the main aquifer systems in contact with the river systems. The aquifers in the western part of the study area along the Darling River comprise the Cubbaroo Formation (Miociene), the Gunnedah Formation (Pliocene) and the Narrabri Formation (Quaternary). All formations are terrestrial deposits of fluvio-lacustrine origin consisting of well sorted, quartzose and lithic sand and fine gravels, interbedded with predominantly yellow to brown clays [35].

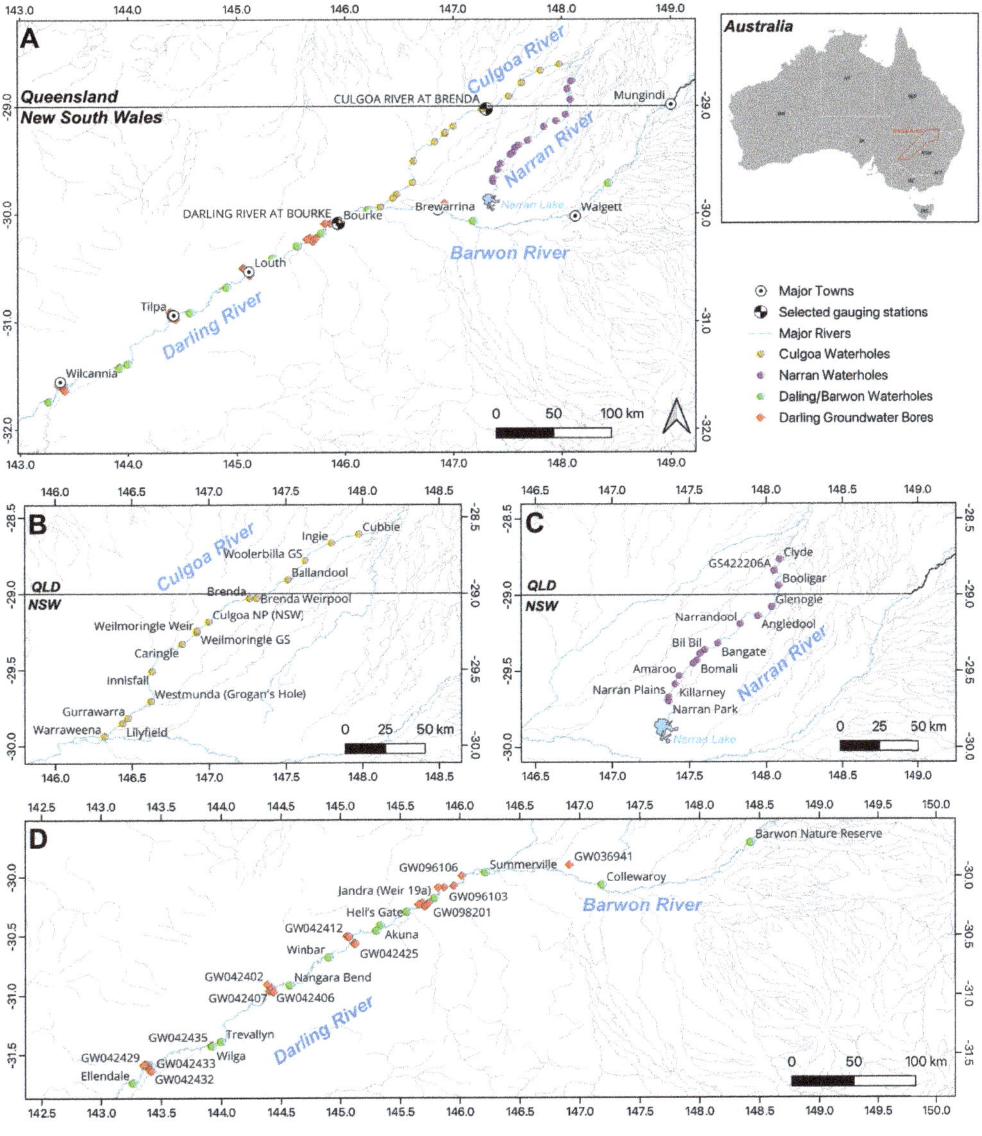

Figure 1. Locations of the waterholes (i.e., those that retained water after 350 days without flow) across the study area. (**A**) Overview of the Culgoa, Narran, Barwon, and Darling Rivers; (**B**) Sampled waterholes along the Culgoa River; (**C**) Sampled waterholes along the Narran River; (**D**) Sampled waterholes and groundwater bores along the Darling and Barwon Rivers. The letter/number sequence GW0XXXXX represents the New South Wales Government Bore number. QLD = Queensland; NSW = New South Wales; ACT = Australian Capital Territory; VIC = Victoria; TAS = Tasmania; NT = Northern Territory; WA = Western Australia; SA = South Australia. Coordinates are in longitude and latitude decimal degrees.

A semi-arid climate dominates the river catchments with long-term average annual rainfall of ~355 mm/year at Bourke, ~399 mm/year at Cobar and ~305 mm/year at Louth [36]. Long-term average rainfall in Bourke indicates generally higher rainfall in the summer months from November to April, which is in line with the general wet season patterns in southeast Queensland and northern New South Wales. Temperatures are also highest during these month and average temperatures range from 18 °C in July to 37 °C in January. The mean annual evaporation of ~2400 mm/year, determined at Cobar, greatly exceeds rainfall for the area.

The area has a low topographic gradient (100/600,000 m) and the rivers developed large meander bends with flood plains more than 12 km wide. The rivers are deeply incised in the flat landscape with approximate bank heights of 6–8 m on the Narran and the Culgoa rivers and more than 10 m in parts of the Darling River [34]. The rivers are highly episodic with large flow increases during the main runoff events and cessation of flow during dry seasons, especially in the Narran and the Culgoa, when waterholes develop. This results in a highly variable flow regime, with significant implications for the way surface and groundwater interacts, as well as ecosystem dependance and adaptation. The Narran River ends in a series of terminal lakes, the Narran Lake (Figure 1A,B) and has an approximate length of 299 km. The Culgoa merges has a length of 489 km and merges with the Darling east of Bourke. The Bowen–Darling River system has a total length of more than 700 km and only the eastern and middle sections were investigated here.

1.2. Waterhole Selection and Extent

The waterholes were selected based on spatial distribution, a mix of natural waterholes and those resulting from weirs (Table 1). Access to waterholes was also an important consideration as well as choosing waterholes that had information from previous studies, e.g. Woods et al. (2012) [37]. Many waterholes have been made artificially deeper by installing weirs, locks and rock walls, which can have heights of up to two metres. Many weirs were installed almost a century ago and have, over time, become a significant part of ecosystem resilience in the region [13].

The size of the waterholes is variable across the river systems but generally increases downstream. Waterholes in the headwater catchments of the Narran and Culgoa rivers are generally smaller than those along the Darling River. The size of the waterholes was assessed by remote sensing in a separate study and details of the methodology can be found in DSITI (2015) [13], Danaher & Collett (2006) [38] and de Vries et al. (2007) [39]. In brief, surface water was identified on Landsat images for the period between 1988–2015 by using a modification of the Landsat-based Water Index threshold. This was combined with hydrological discharge data for gauging stations along the river systems to allow determining a maximum waterhole extent after 30 days of no flow. Waterholes along the Culgoa River range in size from approximately 0.2 km^2 to 1.6 km^2. Those along the Narran River range from 0.4 to approximately 3.1 km^2. In comparison, the waterholes along the Barwon-Darling range in size from 1.4 to 10.8 km^2. Maximum waterhole depth is also variable and ranges from 1.45 to 2.97 m for the Culgoa River and 1.29 to 3.05 m for the Narran River. Waterhole sizes and depths were not determined for the waterholes in the Darling River.

Table 1. Sites in the Culgoa and Narran Rivers. A total of 30 sites, 9 of which had been augmented with weirs, were selected across the two rivers.

Site Name	Latitude	Longitude	Natural/Weir Pool
Culgoa River			
Cubbie	−28.61049	147.97784	Natural
Ingie	−28.6683	147.80261	Natural
Woolerbilla GS	−28.78385	147.63031	Natural
Ballandool	−28.90783	147.51321	Natural
Brenda	−29.02882	147.26564	Natural
Brenda Weir pool	−29.02796	147.31221	Weir pool
Culgoa NP (NSW)	−29.18325	147.00017	Natural
Weilmoringle Weir	−29.25396	146.92188	Weir pool
Weilmoringle GS	−29.24327	146.92389	Natural
Caringle	−29.33033	146.82761	Natural
Innisfail	−29.50803	146.63242	Natural
Westmunda (Grogan's Hole)	−29.70533	146.62436	Natural
Gurrawarra	−29.81595	146.47699	Natural
Lilyfield	−29.85022	146.44121	Natural
Warraweena	−29.93411	146.32320	Natural
Narran River			
Clyde	−28.76764	148.08948	Natural
GS422206A	−28.84097	148.05302	Natural
Booligar	−28.93873	148.08159	Natural
Glenogie	−29.07827	148.03789	Natural
Angledool	−29.13709	147.94771	Natural
Narrandool	−29.19076	147.83507	Natural
Bangate (Sorrento Hole)	−29.31740	147.68872	Natural
Bil Bil	−29.36072	147.60225	Weir pool
Golden Plains	−29.38503	147.57106	Natural
Bomali	−29.42910	147.55040	Natural
Belvedere	−29.44701	147.52798	Natural
Amaroo	−29.53071	147.43658	Natural
Killarney	−29.58680	147.41068	Weir pool
Narran Plains	−29.67077	147.36882	Natural
Narran Park	−29.69681	147.36775	Natural
Darling and Barwon			
Barwon Nature Reserve	−29.71472	148.42895	Weir pool
Collewaroy	−30.06863	147.18817	Weir pool
Summerville	−29.96333	146.21512	Weir pool
Jandra (Weir 19a)	−30.17973	145.78943	Weir pool
Hell's Gate	−30.29673	145.55720	Weir pool
Akuna	−30.40982	145.33432	Weir pool
Weir 20a	−30.45682	145.30090	Weir pool
Winbar	−30.67888	144.90575	Weir pool
Nangara Bend	−30.91258	144.57527	Weir pool
Trevallyn	−31.38992	144.00055	Weir pool
Wilga	−31.42735	143.92540	Weir pool
Ellendale	−31.73642	143.27040	Natural

2. Materials and Methods

This study estimates groundwater inflows to the parts of the Darling and Barwon rivers, the Culgoa River and the Narran River between Dirranbandi and Lightning Ridge in the NE to Tilpa, NSW, in the SW, along a total river reach length of approximately 500 km (Figure 1 and Table 1). The study area is subdivided into upstream sites (the Culgoa River and the Narran River) and downstream sites (the Darling/Barwon River System) sites. Discharge data and semi-continuous electrical conductivity (EC) data in the river were used for two gauging stations, the gauging station at Brenda and the one at Bourke (Figure 1 and Table 1).

The preliminary mapping of the electrical conductivity (EC) in all representative waterholes of the Narran and the Culgoa River was undertaken during bathymetric survey in February/March 2015 [13] and had the purpose of finding potential groundwater discharge regions along the waterholes. A HOBO EC logger (Onset Computer Corporation) was attached to the front of the bathymetry vessel and programmed to record EC-readings every 10 s. A handheld GPS unit recorded the location.

Sampling and Analytical Methods

100 surface water samples from 27 representative waterholes along the Narran and Culgoa Rivers and one groundwater sample from the Narran Park waterhole were taken in two sampling campaigns during February–March 2015 and May–June 2015 (Table 1). 12 surface water samples from 10 representative waterholes along the Darling River and 8 groundwater samples adjacent to these waterholes were taken on a single campaign in September 2015 (Table 1). Surface water samples were taken using a sampling pole from the riverbanks or from a boat on the river by submerging the pole gently under water and filling the sampling cup close to the riverbed. Groundwater samples were obtained from New South Wales Department of Primary Industry (NSW DPI) groundwater monitoring bores by using a SuperTwister groundwater pump after adequate purging of > 5 bore volumes. The groundwater sample on the Narran River was taken from a bank in the waterhole by digging a hole of ~1 m depth and purging it several times. All samples were prepared for major ion, stable isotope ($\delta^2 H$ and $\delta^{18}O$) and radon analysis.

Electrical Conductivity (EC), dissolved oxygen (DO), pH and Temperature were measured in the field using a calibrated Thermo-Fisher AquaRead Meter. Major cation concentrations were analysed filtration through 0.45 µm cellulose nitrate filters and acidification to pH <2 (in the field) before using a PerkinElmer Optima 7300 DV inductively coupled plasma-optical emission spectroscopy (ICP-OES) (Advanced Water Management Centre at the University of Queensland) equipped with WinLab32 for ICP software after digestion using a CEM Mars Xpress Microwave. 9.0ml of sample added to 1.0 mL concentrated HNO_3. Samples were ramped for 10 minutes to 160 °C then held for 10 minutes. Major anion concentrations were analysed using a Thermo-Fisher Dionex ICS2100 Ion Chromatography System at the Advanced Water Management Centre at the University of Queensland. Samples with salinities over 300 mg/L total dissolved solids (TDS) were diluted prior to analysis. The precision of major ion concentrations based on replicate analyses is 2–5%.

Radon activities in groundwater and surface water were determined using a portable radon-in-air monitor (RAD-7, Durridge Co.) following methods described by Burnett [40] and are expressed in Bq/m^3 water. 0.5 L of sample was collected by bottom-filling a glass flask and radon was degassed for 5 min into a closed air loop of known volume. Counting times were 4 × 30 min for surface water and 4 x 15 minutes for groundwater. Typical relative precision is b 3% at 10,000 Bq/m^3 and ~10% at 100 Bq/m^3. Radon emanation rates were measured by sealing a known dry weight of sediment in airtight containers with water for more than 5 weeks by which time the rate of radon production and decay will have reached steady state. Subsequently, 40 ml of pore water was extracted and analysed for radon activities using the same method as above, but with counting times of 12 h.

Oxygen and hydrogen isotope ratios were measured on samples that were filtered and kept in cool in sample containers without headspace. Samples were analysed using an Isoprime dual inlet isotope ratio mass spectrometer (DI-IRMS) coupled to a multiprep bench for online analysis at the Stable Isotope Geochemistry Laboratory at the University of Queensland. $\delta^2 H$ values were analysed after online equilibration at 40 °C with Hokko coils. $\delta^{18}O$ values were analysed as above, but after equilibration with carbon dioxide. $\delta^{2}H$ and $\delta^{18}O$ values (reported in per mil)) were normalised to the standard mean ocean water (VSMOW-SLAP) scale, following a three point normalisation based on four replicate analyses of three laboratory standards per analytical cycle. All laboratory standards were calibrated against International Energy and Atomic Agency (IAEA) (VSMOW, SLAP, GISP) and USGS (USGS45, USGS46) international water standards. Accuracy and precision were better than $\pm$ 2 for $\delta^{2}H$ and $\pm$ 0.1 for $\delta^{18}O$ at 1σ. Both compositions were measured as deviations relative to VSMOW.

3. Results

3.1. Narran and Culgoa Rivers

Quantifying groundwater fluxes to rivers using geochemical tracers requires knowledge of the groundwater composition for the tracers used. While groundwater bores from the local aquifers, river banks and parafluvial zones are used to constrain the 'end member' concentration for the potential groundwater sources, government monitoring bores in the vicinity of the Narran and Culgoa rivers are rare and weren't sampled due to the lack of access. One shallow groundwater/parafluvial zone sample from a sand bank in the Narran Park waterhole was collected. This parafluvial water had an EC value of 2779 µS/cm and a radon concentration of 10600 Bq/m^3. In a regional sense, groundwater in the major aquifers is expected to have at least Total Dissolved Solids (TDS) >500 mg/L (equivalent of approximately 780 µS/cm in EC). Groundwater data from the Australian Bureau of Meteorology [41] indicates a median EC value of 10,000 µS/cm, with a minimum of 184 µS/cm and a maximum of 89,000 µS/cm (n = 3455).

Sampling locations for radon were based on the results from EC mapping. The EC values during the mapping exercise were spread over a large range from waterholes with very low EC values of ~30 µS/cm at Brenda waterhole, to ~600 µS/cm at Cubbie waterhole (Table 2). Variability in waterhole EC-values may indicate potential groundwater discharges to the river due to mixing with the higher saline groundwater under the assumption that no change occurs as a result of nutrients or surface runoff, and tributary contributions. In perennial streams, the EC (TDS proxy) in rivers is generally expected to increase downstream due to increasing baseflow from higher salinity aquifer units. The variability of the EC-values was highest in the Gurrawarra, Golden Plains and Narran Park waterholes and lowest in Caringle, Weilmoringle Weir and Angledool waterholes. This only takes into consideration the upper and the lower quartiles. While higher variability can potentially indicate that the waterholes may have groundwater contributions, robust trends within the waterholes were not observed, and the variability might instead be the result of separated pools or contributions from tributaries rather than groundwater contributions. Furthermore, the variability could not be reproduced by hand measurements at selected points in a later campaign in May/June 2015. The reason for the discrepancy is likely due to the different discharge conditions during the EC-mapping survey at high flow (~2000 ML/Day at Bourke) whereas the spot sampling was conducted at much lower discharges (~100 ML/Day at Bourke) (Figure 2).

ECs in the Culgoa and the Narran Rivers range from 77 to 292 µS/cm during the sampling campaign in March/June 2015 (Table 2). The low flow conditions of ~100 ML/Day allow a better distinction between the waterholes with respect to their EC content. Generally EC-values were higher in the Narran than in the Culgoa with values ranging from 101 to 292 µS/cm (median = 194) in the Narran and 77 to 217 µS/cm (median = 124) in the Culgoa River, respectively. The EC values indicated a small increase along flow paths in the Culgoa, while the EC increase in the Narran River was more pronounced. The EC in the Culgoa River drops from 155 to 96 µS/cm at Innisfalls but increase again towards the confluence with the Darling River (Figure 3).

Table 2. Results from radon analysis and EC measurements for the waterholes of the Culgoa and Narran Rivers. A total of 100 samples were collected and analysed for $\delta^{18}O$ and $\delta^{2}H$. During the bathymetric survey (February/March 2015), 33 samples were collected, for which only EC, $\delta^{18}O$, and $\delta^{2}H$ were determined, while the remaining 67 were also measured for radon.

Sample	Type	River	Date	Radon (Bq/m^3)	EC (µS/cm)	$\delta^{18}O$‰	$\delta^{2}H$‰
Ballandool	sw	Culgoa	12/06/2015	133.33	151.4	−0.07	0.15
Ballandool	sw	Culgoa	10/03/2015	N/A	139.4	−0.19	−2.84
Brenda	sw	Culgoa	10/03/2015	N/A	140.0	−0.31	−3.45
Brenda Waterhole	sw	Culgoa	13/06/2005	174.43	162.8	0.73	3.12
Brenda Waterhole-1	sw	Culgoa	12/06/2015	158.74	109.8	0.52	−0.63
Brenda Waterhole-2	sw	Culgoa	12/06/2015	169.34	135.9	0.31	1.14
Brenda Waterhole-3	sw	Culgoa	12/06/2015	156.33	131.8	0.20	0.32
Brenda Waterhole-4	sw	Culgoa	12/06/2015	159.64	105.9	0.22	−0.05
Brenda Waterhole-5	sw	Culgoa	12/06/2015	166.40	105.2	0.21	−0.22
Brenda Waterhole-6	sw	Culgoa	12/06/2015	205.38	106.5	0.20	0.13
Brenda Waterhole-7	sw	Culgoa	12/06/2015	177.30	96.8	0.19	−0.95
Brenda Waterhole-8	sw	Culgoa	12/06/2015	182.01	125.5	0.05	0.81
Brenda Weirpool	sw	Culgoa	11/02/2015	N/A	137.7	−0.63	−4.19
Caringle	sw	Culgoa	26/02/2015	N/A	126.5	−1.73	−10.16
Cubbie	sw	Culgoa	10/06/2015	158.55	147.6	−0.56	−1.42
Cubbie	sw	Culgoa	18/02/2015	N/A	107.5	−2.24	−10.13
Culgoa	sw	Culgoa	12/03/2015	N/A	150.5	0.03	−3.68
CulgoaNP-1	sw	Culgoa	13/06/2015	128.64	167.8	2.53	8.60
CulgoaNP-2	sw	Culgoa	13/06/2015	169.21	172.5	2.51	7.66
CulgoaNP-3	sw	Culgoa	13/06/2015	200.61	143.7	2.55	7.78
CulgoaNP-4	sw	Culgoa	13/06/2015	172.33	142.3	2.63	9.96
Gurrawarra	sw	Culgoa	15/03/2015	N/A	203.0	1.74	4.30
Ingie	sw	Culgoa	17/02/2015	N/A	108.0	−2.36	−10.12
Ingie-1	sw	Culgoa	26/05/2015	40.13	112.0	−0.43	-2.75
Ingie-2	sw	Culgoa	26/05/2015	42.74	124.0	−0.43	-2.16
Ingie-3	sw	Culgoa	26/05/2015	76.97	123.0	−0.45	−0.55
Ingie-3	sw	Culgoa	26/05/2015	59.96	217.0	−0.43	−3.22
Innisfal	sw	Culgoa	14/03/2015	N/A	125.6	−0.06	−0.47
Innisfall-1	sw	Culgoa	29/05/2015	66.25	96.0	−3.09	−14.84
Innisfall-2	sw	Culgoa	29/05/2015	76.47	84.0	−3.18	−16.28
Innisfall-3	sw	Culgoa	29/05/2015	69.94	85.0	−3.07	−15.32
Innisfall-4	sw	Culgoa	29/05/2015	53.75	87.0	−3.08	−16.12
Lillyfield	sw	Culgoa	15/03/2015	N/A	201.0	1.79	3.81
Lilyfield-1	sw	Culgoa	29/05/2015	53.75	79.0	−2.68	−15.05
Lilyfield-2	sw	Culgoa	29/05/2015	90.83	77.0	N/A	N/A
Lilyfield-3	sw	Culgoa	29/05/2015	71.89	79.0	N/A	N/A
Lilyfield-4	sw	Culgoa	29/05/2015	48.21	78.0	−2.75	−14.93
Warraweena	sw	Culgoa	15/03/2015	N/A	198.8	1.62	3.45
Weilmoringle Weir	sw	Culgoa	12/03/2015	N/A	150.4	−0.31	−3.04
Weilmoringle Weir-1	sw	Culgoa	14/06/2015	124.10	122.9	2.13	4.16
Weilmoringle Weir-2	sw	Culgoa	14/06/2015	165.61	142.8	N/A	N/A
Weilmoringle Weir-3	sw	Culgoa	14/06/2015	119.08	140.8	2.65	8.93
Weilmoringle Weir-4	sw	Culgoa	13/06/2015	234.81	195.2	2.16	6.38
Weilmoringle Station	sw	Culgoa	12/03/2015	N/A	146.7	−0.01	−3.28
Westmunda	sw	Culgoa	13/06/2015	114.72	97.8	−2.26	−12.19
Westmunda	sw	Culgoa	14/03/2015	N/A	143.0	0.19	−1.60
Woolerbilla	sw	Culgoa	17/02/2015	N/A	110.1	−2.38	−10.46

Table 2. *Cont.*

Sample	Type	River	Date	Radon (Bq/m^3)	EC (μS/cm)	$\delta^{18}O$ ‰	$\delta^{2}H$ ‰
Woolerbilla-1	sw	Culgoa	13/06/2015	N/A	161.1	−0.34	−1.43
Woolerbilla-2	sw	Culgoa	13/06/2015	N/A	163.3	−0.29	−0.38
Woolerbilla-3	sw	Culgoa	13/06/2015	N/A	164.3	−0.24	−0.74
Amaroo	sw	Narran	13/05/2015	N/A	156.8	2.04	6.18
Amaroo-1	sw	Narran	28/05/2015	114.86	274.0	4.79	19.32
Amaroo-14	sw	Narran	28/05/2015	84.86	290.0	4.82	18.39
Amaroo-15	sw	Narran	28/05/2015	79.44	289.0	4.81	18.29
Amaroo-17	sw	Narran	28/05/2015	114.44	283.0	4.79	18.86
Amaroo-2	sw	Narran	28/05/2015	78.33	285.0	4.75	18.57
Amaroo-20	sw	Narran	28/05/2015	115.14	292.0	4.79	18.48
Bangate	sw	Narran	21/02/2015	N/A	150.1	−0.52	−0.48
Belvedere	sw	Narran	22/02/2015	N/A	164.8	0.04	1.19
BilBil	sw	Narran	21/02/2015	N/A	158.5	−0.23	0.24
Bomali	sw	Narran	22/02/2015	N/A	170.9	0.03	0.02
Bomali-2	sw	Narran	16/06/2015	225.54	191.8	1.96	8.04
Bomali-3	sw	Narran	16/06/2015	201.95	194.8	4.03	14.02
Booligar	sw	Narran	13/06/2015	116.27	156.7	0.35	1.37
Booligar	sw	Narran	20/02/2015	N/A	119.5	−2.05	−10.99
Clyde	sw	Narran	19/02/2015	N/A	101.6	−2.22	−11.00
Clyde 1	sw	Narran	11/06/2015	176.42	104.6	−0.06	−1.80
Clyde 2	sw	Narran	11/06/2015	170.87	101.2	−0.22	−2.48
Clyde 3	sw	Narran	11/06/2015	155.99	101.1	−0.22	−2.51
Glenogie	sw	Narran	20/02/2015	N/A	124.0	−1.84	−11.00
GoldenPlains	sw	Narran	24/02/2015	N/A	165.9	−0.09	−1.36
GSA22206A	sw	Narran	19/02/2015	N/A	107.6	−2.16	−12.13
GSA22206A-1	sw	Narran	26/05/2015	40.13	122.0	−0.57	−2.53
GSA22206A-2	sw	Narran	26/05/2015	110.97	119.0	−0.49	−2.68
GSA22206A-3	sw	Narran	27/05/2015	135.97	115.0	−0.61	−3.30
GSA22206A-4	sw	Narran	26/05/2015	57.28	114.0	−0.59	−2.11
Kilarney	sw	Narran	13/03/2015	N/A	200.0	1.59	6.84
Killarny-10	sw	Narran	15/06/2015	203.77	165.7	3.55	13.95
Killarny-12	sw	Narran	15/06/2015	203.22	211.0	3.54	14.03
Killarny-13	sw	Narran	15/06/2015	142.18	213.0	3.62	14.93
Killarny-5	sw	Narran	15/06/2015	215.23	211.0	4.40	17.01
Killarny-6	sw	Narran	15/06/2015	269.79	190.2	4.12	16.10
Killarny-7	sw	Narran	15/06/2015	207.73	164.2	3.82	15.14
Killarny-8	sw	Narran	15/06/2015	238.00	213.0	3.55	14.15
Killarny-8	sw	Narran	15/06/2015	142.08	N/A	N/A	N/A
Killarny-9	sw	Narran	15/06/2015	225.29	165.2	3.53	14.11
Narran Park 7	sw	Narran	27/05/2015	N/A	250.0	N/A	N/A
Narran Park 8	sw	Narran	27/05/2015	93.19	253.0	4.59	18.29
Narran Park 9	sw	Narran	27/05/2015	101.53	247.0	4.29	18.50
Narran Park 10	sw	Narran	27/05/2015	79.96	251.0	4.24	16.80
Narran Park 12	sw	Narran	27/05/2015	81.90	248.0	4.21	17.06
Narran Park 13	sw	Narran	27/05/2015	95.28	244.0	4.42	18.44
Narran Park-bank	gw	Narran	27/05/2015	2779.17	10,600.0	4.22	16.53
Narran Plains	sw	Narran	15/06/2015	234.98	233.0	5.14	18.27
Narrandool	sw	Narran	24/02/2015	N/A	141.2	−0.95	−6.04
Narrandool-1	sw	Narran	14/06/2015	187.04	184.6	1.99	6.51
Narrandool-2	sw	Narran	14/06/2015	182.22	125.7	2.12	4.49
Narrandool-3	sw	Narran	14/06/2015	208.59	127.5	N/A	N/A
NarranPark	sw	Narran	23/02/2015	N/A	185.6	0.64	4.97
NarranPlains	sw	Narran	23/02/2015	N/A	184.1	0.57	−5.57
Goodooga Rain	rain	Rain	26/05/2015	N/A	81.0	−2.63	−9.51

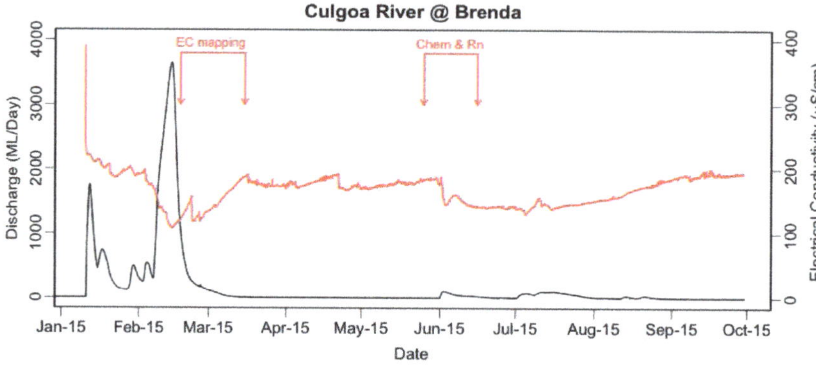

Figure 2. Discharge of the Culgoa River at Brenda.

Radon concentrations in the rivers ranged from 40 to 269 Bq/m^3 (median = 139) and radon concentrations were slightly higher in the Narran River than the Culgoa River, ranging from 40 to 269 Bq/m^3 and 40 to 234 Bq/m^3, respectively. Similar to the EC values, radon concentrations also increased downstream in both rivers (Figure 3B). The radon concentrations also dropped at Innisfalls in the Culgoa River but increased again before the Culgoa merges with the Darling River. There was a general trend that radon concentrations increased when EC also increased (Figure 4A,B).

Stable isotope ratios of $\delta^{18}O$ ranged from −3.18 to 5.14‰ and −16.28 to 19.32‰ for δ^2H. Most isotope values were similar between the Culgoa and the Narran River with the only difference that the most downstream samples on the Narran were strongly enriched with $\delta^{18}O$ values in between 4 and 6‰ and δ^2H values between 12 and 20‰. The stable isotope values all plot along an evaporation trend (Figure 5A,B) with three major groupings. The first group is relatively depleted and plots close to the local meteroric water line, which is representative of rainfall. The second group has a slight evaporation signature with more enriched values, which is emphasised in the Culgoa River (Figure 5A). The third group indicates strong evaporation with highly enriched values.

The waterholes along the Narran River generally showed a trend towards higher enrichment downstream (Figure 3C). The stable isotope values were a lot more variable in the Culgoa and dropped towards much more depleted values at Innisfalls, which is in line with radon and EC values. There is a clear trend between the EC and stable isotope values, e.g., $\delta^{18}O$, in both rivers (Figure 6A,B) and the three groups with increasing enrichment correspond to increasing EC values.

3.2. Darling River

Along the Darling River, 12 surface water samples and 8 groundwater samples were taken at the end September 2015. The sample locations were predetermined from previous work by the NSW DPI Water. The locations are summarised in Table 1. Weir 19A was not part of the original sampling strategy but was included after reviewing a report on the salt interception scheme around Weir 19A (Glen Villa) [42]. Two aquifers were also targeted for sampling: the shallow Narrabri Formation (10 to 25 m depth) and the deeper Gunnedah Formation (30 to 50 m depth) (Table 3). The EC values were variable in the Narrabri formation with fresh groundwater of 541 µS/cm in bores B36842-1 and 377 µS/cm in B36853-1, but brackish to saline water with values of 12,700 µS/cm in B36937-1 and 34570 µS/cm in B56852-1. There was less variability in the groundwater bores in the Gunnedah formation with mostly saline groundwater in bores BB36852-2, B36937-2 and

B36853-2 (EC= 35,089 to 40,370 μS/cm) with the exception of bore B36842-2, which had an EC of 1976 μS/cm (Table 3). Radon concentrations in the groundwater bores were orders of magnitude higher than in the surface water ranging from 74.5 to 13,236.1 Bq/m^3 with a median concentration of 3769.4 Bq/m^3. Bores with relatively fresh groundwater, Bore B36842-1 and B36853-1, also had reasonably high radon concentrations of 5343.8 and 13,236.1 Bq/m^3, respectively.

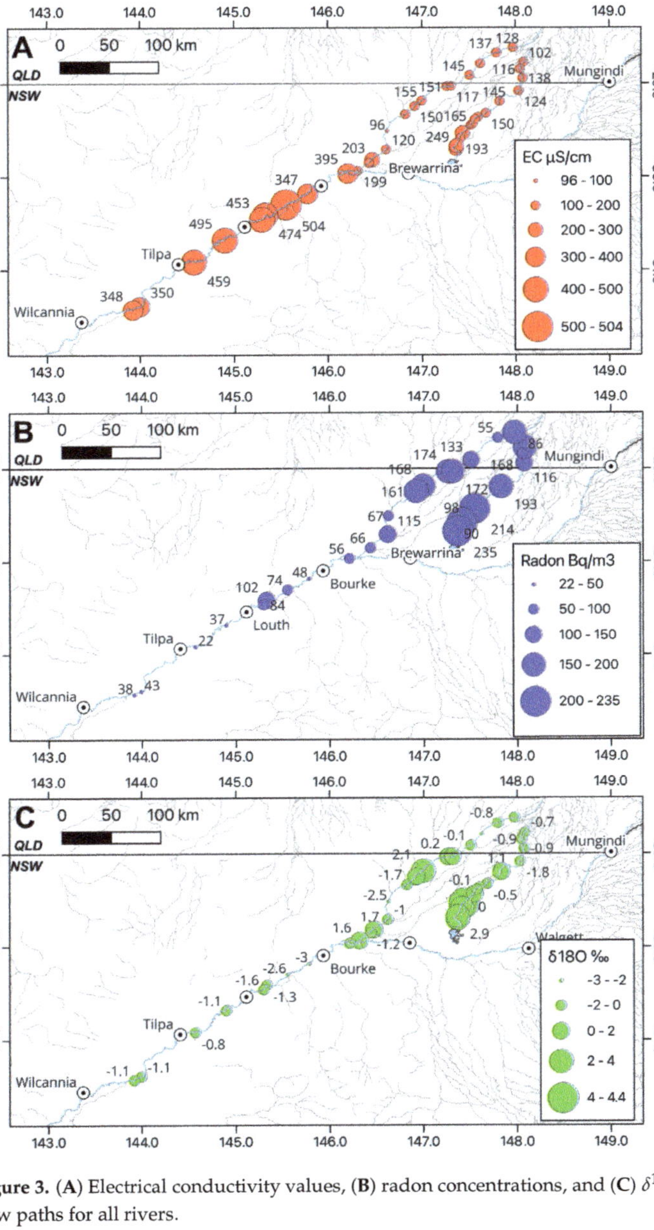

Figure 3. (**A**) Electrical conductivity values, (**B**) radon concentrations, and (**C**) $\delta^{18}O$ values along the flow paths for all rivers.

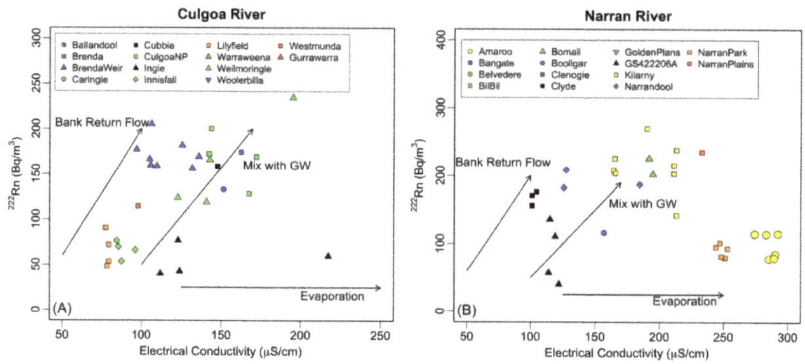

Figure 4. (**A**) Radon concentrations and EC values for water samples from the Narran River. (**B**) Radon concentrations and EC values for water samples from the Culgoa River. The increasing symbol size represents the increasing distance in the flow direction.

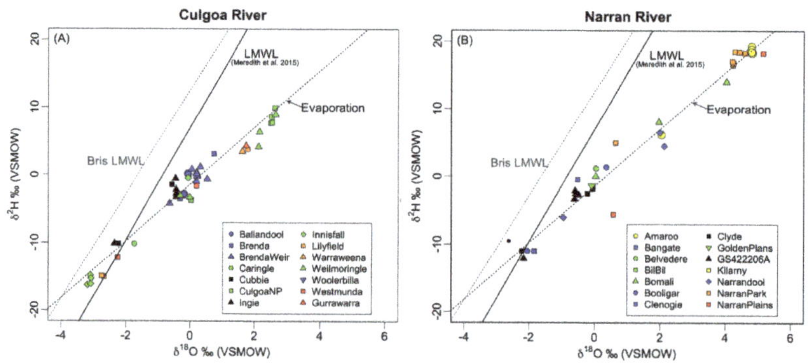

Figure 5. Stable isotopes $\delta^{18}O$ and δ^2H for the water samples from the Narran (**B**) and Culgoa (**A**) Rivers. All samples follow an evaporation trend (black dashed line). The Cobar local meteoric water line (black dashed line) from Meredith [20] and the Brisbane local meteoric water line (grey dashed line) are shown for reference. The symbol size increases with the distance from the most upstream sites.

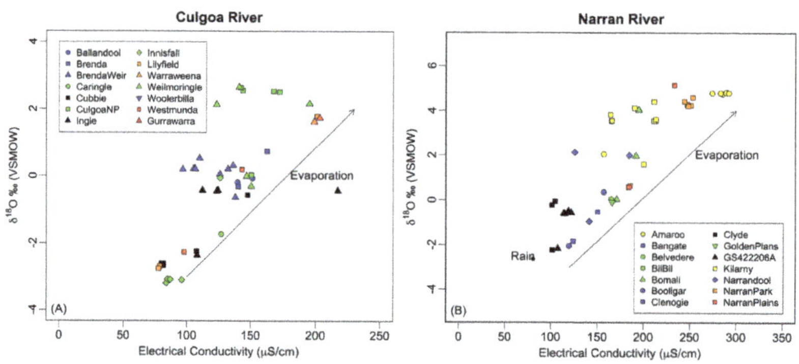

Figure 6. (**A**) $\delta^{18}O$ and EC values for water samples from the Culgoa River. (**B**) $\delta^{18}O$ concentrations and EC values for water samples from the Narran River. The increasing symbol size represents the increasing distance in the flow direction.

Table 3. Data obtained from 12 surface water samples and 8 groundwater samples along the Darling River from Brewarrina to the west of Tilpa.

Sample	Type	Date	EC (μS/cm)	T°C	DO (mg/L)	Radon (Bq/m^3)	$\delta^{18}O$‰	δ^2H‰
Sommerville	sw	28/09/15 12:00	395	20.3	8.9	56.1	−1.67	−13.28
Jandra	sw	28/09/15 14:00	347	22.7	10.12	47.7	−2.95	−19.12
Hells Gate upstream	sw	28/09/15 17:00	499	21.7	10	73.5	−2.60	−18.63
Hells Gate downstream	sw	28/09/15 17:00	504	20.2	9.72	65.9	−2.57	−18.24
Akuna	sw	29/09/15 07:40	474	17.9	6.99	102.2	−1.32	−8.62
Weir 20A	sw	29/09/15 08:30	453	18.3	7	83.8	−1.62	−10.21
Winbar	sw	29/09/15 11:20	495	18.8	8.68	37.1	−1.12	−8.78
Nangara Bend	sw	29/09/15 12:00	459	20.5	8.9	22.1	−0.81	−7.90
Trevallyn	sw	29/09/15 14:00	350	21	8.4	43.4	−1.14	−7.84
Wilga	sw	29/09/15 14:40	348	19.6	8.02	38.3	−1.10	−7.43
Weir 19A downstream	sw	10/01/15 08:00	309	18.1	9.37	23.8	−2.84	−19.64
Weir 19A upstream	sw	10/01/15 08:00	310	17.7	6.84	28.2	−2.95	−19.88
B36842-1	gw	29/09/15 16:00	541	23.3	0.14	5343.8	−3.54	−28.48
B36842-2	gw	29/09/15 16:00	1976	23.2	0.4	74.5	−3.64	−29.24
B36852-1	gw	30/09/15 09:30	34,570	23.9	0.38	6739.6	−4.04	−31.45
B36852-2	gw	30/09/15 09:30	35,120	24	3.7	4687.5	−3.69	−29.24
B36937-1	gw	30/09/15 13:30	12,700	24.4	0.11	2905.2	−4.85	−35.17
B36937-2	gw	30/09/15 13:30	40,370	24.4	0.1	1318.8	−4.16	−32.50
B36853-1	gw	10/01/15 09:15	377	23.8	0.11	13,236.1	−3.45	−25.24
B36853-2	gw	10/01/15 09:15	35,089	23.9	1.37	178.8	−4.12	−31.88

River water EC ranged from 309 to 504 μS/cm and radon concentration were very low, ranging from 22.09 to 102.19 Bq/m^3 at the time of sampling in September 2015 (Figure 7). EC values increased downstream from the confluent of the Darling River with the Culgoa River and reached a peak west of Louth, peaking at Winbar with a value of 495 Bq/m^3, after which they declined again. Radon values are also highest in the area between Weir 19A and Louth (Figure 3A,B). Radon concentrations along the Darling River were generally low, ranging from 22.1 to 102.2 Bq/m^3 with a median value of 45.6 Bq/m^3 (Table 3 and Figure 7). The highest activity also occurred close to Louth, however, further upstream than the highest EC value. At Akuna, radon concentrations peaked at 102.2 Bq/m^3.

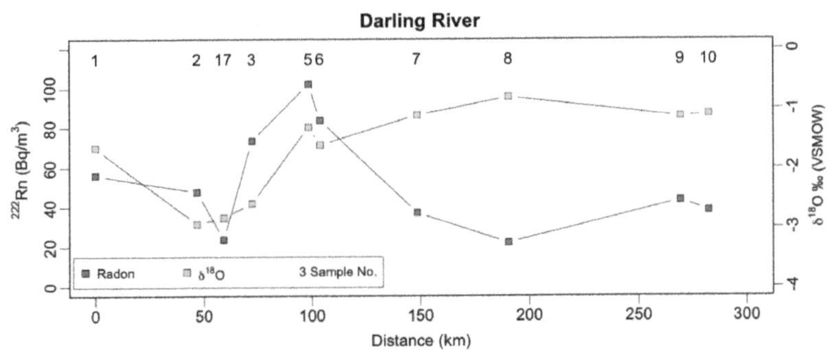

Figure 7. Radon and $\delta^{18}O$ along the flow path on the Barwon and Darling Rivers from Summerville to Wilga. Sample numbers indicate locations.

The stable isotopes values of groundwater from both formations had average $\delta^{18}O$ and δ^2H of −3.93‰ and −30.39‰, respectively (Figure 8), which are close to local and regional long-term rainfall [20]. River $\delta^{18}O$ and δ^2H values ranged from −2.95 to −0.81‰ and −19.88 to −7.84‰, respectively. The lowest, most depleted values occurred in between Brouke and Louth (Figure 3C).

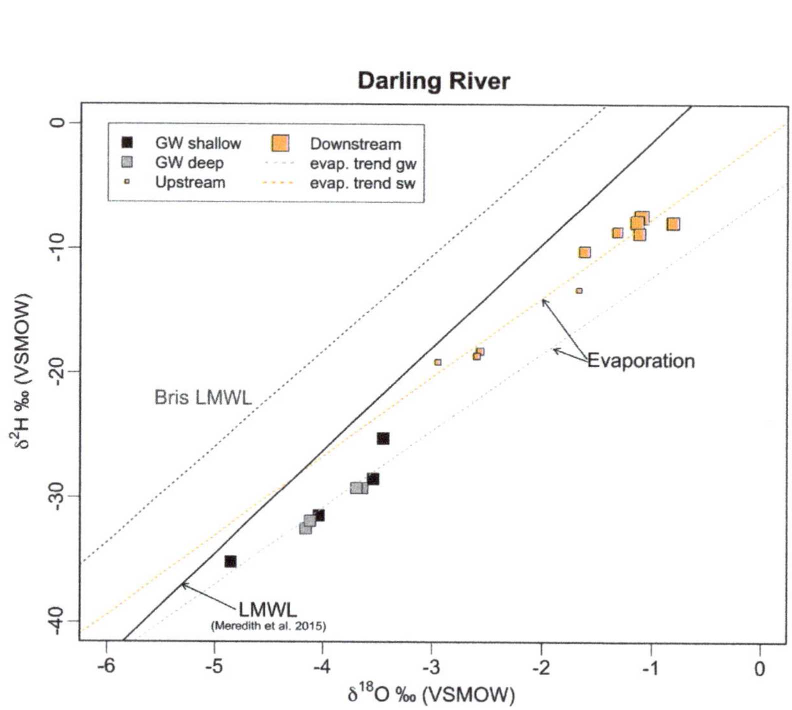

Figure 8. Stable isotope compositions for $\delta^{18}O$ and $\delta^{2}H$ for the samples collected from the Barwon and Darling Rivers, as well as the groundwater samples collected from the Gunnedah Formation (grey) and the Narrabri Formation (black). All samples show an evaporation trend (orange = surface water; grey = groundwater). The Cobar local meteoric water line (black dashed line) from Meredith et al. (2015) [20] and the Brisbane local meteoric water line (grey dashed line) are shown for reference. The symbol size for the surface water samples increases with distance from the most upstream sites.

4. Discussion

The objective of this study was to assess potential groundwater contributions to inland ephemeral streams in central east Australia. While ephemeral streams are often disconnected from regional groundwater, there are circumstances where regional aquifers intersect the river bed through up-coming strata (e.g., basement highs) or fault connectivity. More importantly, intermediate and transient storage in floodplain lenses or river banks plays an important role in sustaining flows throughout periods of declining flow. It is thought that these reservoirs also contribute to waterholes, sustaining these vital water reservoirs, which provide abundant ecosystem services in arid and semi-arid regions [11].

Here, we investigated potential groundwater contributions to waterholes along the Narran, Culgoa and Darling rivers to inform waterhole persistence modelling and provide insights on the importance of the groundwater component in the water balance. Geochemical tracers were chosen as groundwater bores are not easily accessible, have a lack of data or are often too far away from the waterholes to provide information on transient connectivity with groundwater. The focus here was primarily on isotopic tracers, such as stable isotopes and radon, as well as the electrical conductivity of water, which is a proxy for its salinity. This is based on the assumption that tracer concentrations are distinctly, if not orders of magnitude, different between the reservoirs (i.e. waterhole, rainfall, regional groundwater, freshwater lens or bank storage). When, for example, water from one reservoir, e.g., river water, contains a very low concentration of a tracer and receives water from a different reservoir, e.g., groundwater, with a high concentration of that tracer, the spatially distinct or temporally transient increase of the tracer in the first reservoirs then indicates mixing and therefore contributions from the second reservoir. This principle was applied here to qualitatively determine the extend of groundwater connectivity to the waterholes in the selected river systems.

The conditions during sampling for this study did not represent no-flow conditions and flow had not completely ceased. Higher rainfalls in early months of the year resulted in high discharge, which had declined by the time the sampling took place (Figure 2). However, it is reasonable to assume that these flow peaks lead to recharge into the river banks and the floodplains.

4.1. Characterisation of the Reservoirs

Groundwater in the regional aquifers has often high salinities from brackish to hyper saline conditions. This is relatively widespread in shallow aquifer systems in the Murray-Darling Basin [43]. Radon concentrations in the regional aquifer is generally high as well.

Freshwater lenses often develop around the river systems and form an half cylindrical freshwater reservoir on top of the saline groundwater (Figure 9) [28]. The principle is analogous to freshwater lenses in coastal and island settings, where the lens expands or contracts depending on freshwater recharge conditions [44,45]. The water chemistry is often characterised by minor salinity increases compared to the river water it was recharged from and similar stable isotope values. The slight increase in salinity can stem from minor water/rock interactions or mixing with the adjacent saline groundwater. The one distinct difference to the original river water are in general higher radon concentrations due to the fast equilibration time (secular equilibrium) between radium and radon. While no bores in the freshwater lens were sampled, examples from the region support the assumptions for this study [27,44].

River bank storage is arguably part of a freshwater lens but it may also occur in regions where freshwater lenses are absent. One could distinguish the two by modes of recharge, where river banks are possibly more often recharged directly through the upper parts of the river bed and banks during in-channel flows and floods, whereas freshwater lenses are recharged mainly through flooding of the floodplain.

The waterholes are generally filled by events but some filling from local rainfall can occur. They are depressions in river bed topography that hold larger volumes of water than the rest of the channel. The water chemistry is often close to that of rainfall. Salinity is slightly increased from minor rock-water interactions, salt dissolution during runoff processes and nutrient transport from runoff. The stable isotope composition is also similar to rainfall unless significant evaporation occurs, which is seen in the waterholes along the Culgoa, the Narran and the Darling rivers. Radon concentrations are low as rainfall contains insignificant radon, compared to groundwater, and if present, radon tends to evade quickly from surface water bodies due to the low solubility.

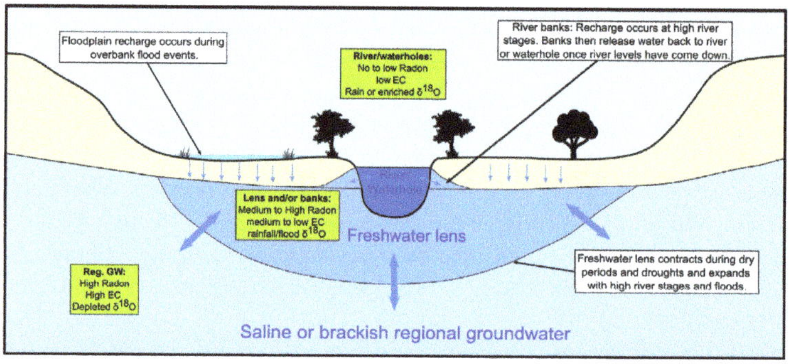

Figure 9. Concept of river and waterhole processes with groundwater reservoirs.

4.2. Groundwater Connectivity Along the Culgoa, Narran, and Darling Rivers Waterholes

Along the Narran River EC-values in Amaroo, Killarney and Narrandool were highest. While this is potentially an indication of groundwater contributions to the waterholes, radon values for the waterholes suggest that Narrandool, Narran Park, Bomali and some parts of Killarney may also have some groundwater contributions (indicated by higher radon values combined with higher EC-values), but this is not the case for Amaroo and the rest of Killarney (Figure 4A, Table 2). Higher radon concentrations but low EC may be the result of water contribution from parafluvial zone or from bank return flow where salinities are low but radon can accumulate to secular equilibrium levels [28]. However, these rivers have not developed large parafluvial zones compared to, for example, braded river systems, and bank return flow would be the most likely source. While some samples have slightly elevated radon concentrations in combination with low EC (e.g., Ballandool, Booligar, Wesmunda), groundwater fluxes on the basis of the radon concentrations would only be small (Figure 4A), considering that diffusive fluxes from the riverbed are likely to be similar to the lowest values detected for each waterhole. Higher EC values on the Culgoa River occurred at Culgoa National Park and Weilmoringle Weir with slightly higher radon concentrations ranging from 128 to 200 and 119 to 182 Bq/m^3, respectively. There is the possibility of small groundwater inflow or bank return flow to these waterholes (Figure 4B).

Groundwater is one potential cause of increasing salinities in rivers, and evaporation is another, especially when flow ceases. Stable isotopes $\delta^{18}O$ and δH fractionate during evaporation and the residual water becomes enriched in both isotope pairs. The stable isotope ratios of the water samples from the waterholes followed an evaporation trend, which progressed towards more enriched values along the flow path of the Narran and the Culgoa rivers (Figure 5A,B). The most enriched values were found at Killarney, Amaroo and Narran Park with $\delta^{18}O$ of +5.14‰ and δ^2H of 19.32‰. These waterholes also had the highest EC-values and lowest radon, which in the case of Killarney and Amaroo suggests that the salinity increase is not resulting from groundwater discharge to the river but is best explained by evaporation. The waterholes with possible groundwater contribution on the Culgoa River, Culgoa National Park and Weilmoringle Weir, also had high stable isotope ratios indicating significant evaporation and subsequent increases in salinity. Regional groundwater is believed to have much more depleted stable isotope values, which is inferred from values obtained in the Darling River groundwater bores and reported by Meredith et al. (2015) [20] ($\delta^{18}O$ of ∼−1.5‰ and δH of ∼−30‰, [20]; $\delta^{18}O$ of −3.9‰ and δ^2H of −30.4‰, data obtained during this study from bores on the Darling flood plains). Hence, groundwater discharge to the river would not only increase the EC and radon concentrations but would also mix the highly enriched stable isotope surface water values with more depleted groundwater isotope values. This should result in more depleted isotope values. The three isotope clusters in Figure 5A,B are an indication that this may happen. The middle cluster represents waterholes where the mixing between slightly higher saline and enriched water and low salinity with depleted isotope values occurs, which is also in line with the EC–$\delta^{18}O$ relationship observed in Figure 6A,B.

The results from the Darling river are very different compared to the headwater streams, the Culgoa and the Narran rivers. EC values remain more constant and radon values are significantly lower along the Darling. Here, information on groundwater chemistry from the bores is valuable for the conceptualisation of the system. B36842-1 is located close to Tilpa close to Glen Villa (Weir 19A). It's low EC value of 541 µS/cm indicates a freshwater lens or multiple fresh water lenses in the proximity of the river, which is in line with descriptions on Cooper Creek by Cendon et al. (2020) [18] and some parts of the Murray River in Cartwright et al. (2011) [27]. The extent of these lenses is only know in areas where bore data exist or geophysical surveys were conducted, such as airborne electromagnetic (AEM) data, but alluvial sediment heterogeneities and bedrock topography may be key factors influencing the geometry of a lens or lenses. Bores B36937 and B36853 are within the vicinity of the Upper Darling salt interception scheme, which is indicated as an area of preferential groundwater discharge to the Darling River [42]. D'Hautefeuille and Williams (2003) [42] reported a lens being present in the area and that the fresh lens recharges from the stream at higher flows >4500 ML/Day. It subsequently requires extended periods of low flow to deplete this freshwater lens to the extent that saline groundwater then reaches the river.

During the sampling in September 2015 river flow was not at its minimum with flow still exceeding 400 ML/Day, but much lower than the 4500ML/Day reported by D'Hautefeuille and Williams (2003) [42]. The river water during this period was still very fresh and the low radon concentrations indicate little groundwater inflow. Low radon groundwater in the Gunnedah formation is also not likely to feed the river as the salinity would increase drastically if this groundwater discharged to the river. Despite groundwater fluxes not being of great importance, the steep increase of EC and radon after Weir 19A towards Hells Gate and Louth indicates small groundwater fluxes to the river at this location whereas all the locations further downstream appear to be rather losing or neutral (Figure 7). The source of this groundwater, however, is probably not from the extended fresh water lens but could be from the regional groundwater system in this case, as described in D'Hautefeuille and Williams (2003) [42] and Meredith et al. (2015) [20]. However, the increase in stable isotopes, for example, $\delta^{18}O$, at Weir 19A towards Nangara Bend indicates that water with more enriched stable isotope ratios is discharged to the river. If regional groundwater was the main contributing reservoir, then $\delta^{18}O$ ratios would be expected to fall towards more depleted values (Figure 7). Contributions from local bank return flow or small evaporated parafluvial zones could potentially be the source of this water. For the study sites further downstream from Weir 19A, where EC values decline again, contributions from the freshwater lens or bank return flow may lead to dilution, however, radon values remain low. Another like source of additional fresh water are tributaries that would potentially add water with low EC and low radon.

5. Conclusions

In summary, radon concentrations were generally very low in all the river systems. This is in line with other studies on inland, low topography river systems. While some waterholes show evidence for possible minor groundwater contributions due to slightly increased radon concentrations and higher EC-values compared to locations upstream and downstream, some of the salinity increase needs to be attributed to evaporation. Generally, groundwater contributions appear to be very small, especially during the sampling period of this study. Given the sampling was after floods and extended in-channel flows, it may suggest that at other, drier times (e.g. during extended drought) when waterhole persistence is most vital to the ecosystems they support, that GW influence on persistence is likely less important because there will not be any bank storage or parafluvial water left to drain into waterholes.

Much of the stable isotope and major ion data suggests that the surface water chemistry is dominated by evaporation, however, some changes cannot be explained by evaporation only. The strong influence of evaporation was also previously suggested driving the increases in EC and the enrichment in $\delta^{18}O$ [46]. The radon data, however, indicated some connectivity to groundwater, for example, at the locations along the Narran River where the EC and radon increase simultaneously. This may indicate some interaction with groundwater and/or bank storage further upstream. No significant groundwater fluxes occurred along the Darling at the time of sampling apart from the area around Weir 19A, where groundwater connectivity was postulated earlier.

While groundwater was not a major contributor to the waterhole balance at the time of sampling, it may become more important during droughts. The reservoirs that are most important for ecosystem services are the freshwater lenses and the riparian river banks. They contain and provide low salinity water to the ecosystems during no-flow periods. How long the water in the banks last is still mostly is still a major knowledge gap and needs some further studies. Howcroft et al. (2019) [47] suggested residence times of river bank storage of several month if not longer for the Barwon River in Victoria, which is in a similar environment as the rivers here.

The lack of background groundwater information within the river reaches of interest places a large constraint on the ability to interpret groundwater interaction processes within these rivers. Likewise, the lack of surface water quality or gauging data restricts the water balance constraints that can be undertaken to verify the geochemical tracer work. These important factors need to be carefully considered in attempting to interpret both the inferred groundwater interaction processes presented here, as well as the confidence that can be placed in them. A further limitation to the study was the time of sampling. Sampling after prolonged dry periods would have been more beneficial, however, this was not possible due to logistical and funding constraints.

Geophysical investigations, such as AEM or electrical resistivity tomography (ERT) can be beneficial tools to monitor freshwater lenses and possibly also be used to constrain the transient nature of these system.

Waterhole persistence is an important part of ecological health surveys. These surveys feed into water policies, plans and management strategies. Ecosystems compete over the resource with anthropogenic interests, such as agriculture, town supply and other industries. Understanding the basic requirements for ecosystems to survive extended periods without significant flow is therefore of paramount importance for consideration in water management.

Author Contributions: H.H. has developed the concept of the study, collected and analysed the data. He has drafted the manuscript and prepared all figures and maps. J.M. has collected the sample, contributed to the planing and conceptual development of the study and has revised the manuscript.

Funding: Funding was provided from the Department of Natural Resources and Mines (DNRM). The project was overseen by Rose Coburn from DNRM and Adam Sluggett from Murray–Darling Basin Authority (MDBA).

Data Availability Statement: Data will be provided upon request. Contact Dr. Harald Hofmann at h.hofmann@uq.edu.au.

Acknowledgments: The project was overseen by Rose Coburn from the DNRM and Adam Sluggett from the Murray–Darling Basin Authority (MDBA). The work contained in this report represents a collaboration between the DNRM, DSITI, Neal Foster from the NSW Department of Primary Industries (DPI) Water, Ozius Spatial, and the University of Queensland (UQ). In addition, the authors would like to acknowledge the significant contributions made to the overall project by Joshua Larsen, Katrina Cousins, Jaye Lobegeiger, John Bowlen, Joanna Blessing, John Vitkovsky, Paul Harding, Myriam Raymond, Marianna Joo, Dean Holloway, Ryan Woods, Cameron Schulz, Rohan Wallace, Sara Clifford, Kate Hodges and Jessica Roffey (DSITI), Andrea Prior and Rod Dew (DNRM), William Lucardie (MDBA), Kevin Brown (Hydro Survey Services), Barry Medway (DPI Water), Nick Marsh, and Jane Bryden-Brown (Ozius). The authors also acknowledge the support of DSITI's Remote Sensing Centre and the Joint Remote Sensing Research Program in providing input data for the waterhole time-series analysis. The authors gratefully acknowledge the traditional owners of the land on which this research was conducted and the landholders who provided access to field sites on their properties.

Conflicts of Interest: The authors declare no conflicts of interest.

References

1. Shanafield, M.; Bourke, S.A.; Zimmer, M.A.; Costigan, K.H. An overview of the hydrology of non-perennial rivers and streams. *WIREs Water* **2021**, *8*, e1504. [CrossRef]
2. Godsey, S.E.; Kirchner, J.W. Dynamic, discontinuous stream networks: Hydrologically driven variations in active drainage density, flowing channels and stream order. *Hydrol. Process.* **2014**, *28*, 5791–5803. [CrossRef]
3. Schilling, O.S.; Cook, P.G.; Grierson, P.F.; Dogramaci, S.; Simmons, C.T. Controls on interactions between surface water, groundwater and riverine vegetation along intermittent rivers and ephemeral streams in arid regions. *Water Resour. Res.* **2021**, *57*, e2020WR028429. [CrossRef]
4. Messager, M.L.; Lehner, B.; Cockburn, C.; Lamouroux, N.; Pella, H.; Snelder, T.; Tockner, K.; Trautmann, T.; Watt, C.; Datry, T. Global prevalence of non-perennial rivers and streams. *Nature* **2021**, *594*, 391–397. 397A–397T. [CrossRef]
5. Scamardo, J.E.; Wohl, E. Recognizing the ephemeral stream floodplain: Identification and importance of flood zones in drylands. *Earth Surf. Process. Landforms* **2024**, *49*, 210–235. [CrossRef]
6. Keegan-Treloar, R.; Irvine, D.J.; Sol6rzano-Rivas, S.C.; Werner, A.D.; Banks, E.W.; Currell, M.J. Fault-controlled springs: A review. *Earth-Sci. Rev.* **2022**, *230*, 104058. [CrossRef]
7. Cartwright, I.; Irvine, D. The spatial extent and timescales of bank infiltration and return flows in an upland river system: Implications for water quality and volumes. *Sci. Total Environ.* **2020**, *743*, 140748. [CrossRef]
8. Cartwright, I.; Hofmann, H. Using radon to understand parafluvial flows and the changing locations of groundwater inflows in the Avon River, southeast Australia. *Hydrol. Earth Syst. Sci.* **2016**, *20*, 3581–3600. [CrossRef]
9. McJannet, D.; Marvanek, S.; Kinsey-Henderson, A.; Petheram, C.; Wallace, J. Persistence of in-stream waterholes in ephemeral rivers of tropical northern Australia and potential impacts of climate change. *Mar. Freshw. Res.* **2014**, *65*, 1131–1144. [CrossRef]
10. Cockayne, B. Climate change effects on waterhole persistence in rivers of the Lake Eyre Basin, Australia. *J. Arid. Environ.* **2021**, *187*, 104428. [CrossRef]
11. Marshall, J.C.; Sheldon, FThoms, M.; Choy, S. The macroinvertebrate fauna of an Australian dryland river: Spatial and temporal patterns and environmental relationships. *Mar. Freshw. Res.* **2006**, *57*, 61–74. [CrossRef]
12. Tibby, J.; Marshall, J.C.; Lobegeiger, J.S.; Amos, K.J.; Pickering, G.; Myburgh, T. Sedimentation from landscape clearance-induced soil erosion threatens waterhole persistence in a semi-arid river system, southern Queensland, Australia. *Mar. Freshw. Res.* **2023**, *74*, 1050–1070. [CrossRef]
13. DSITI. Waterhole refuge mapping and persistence analysis in the Lower Balonne and Barwon–Darling rivers. Department of Science, Information Technology and Innovation. Final Report—Water Planning Ecology, December 2015, Queensland

Government. Available online: https://www.mdba.gov.au/publications-and-data/publications/waterhole-refuge-mapping-and-persistence-analysis-lower-balonne (accessed on 10 December 2024)

14. Wu, H.; Jazayeri, A.; Werner, A.D.; Tang, H.; Lu, C. Riparian freshwater lens response to flooding. *Water Resour. Res.* **2023**, *59*, e2022WR033309. [CrossRef]
15. Rau, G.C.; Halloran, L.J.S.; Cuthbert, M.O.; Andersen, M.S.; Acworth, R.I.; Tellam, J.H. Characterising the dynamics of surface water-groundwater interactions in intermittent and ephemeral streams using streambed thermal signatures. *Adv. Water Resour.* **2017**, *107*, 354–369. [CrossRef]
16. Winter, T.C. The role of ground water in generating streamflow in headwater areas and in maintaining base flow. *J. Am. Water Resour. Assoc.* **2007**, 4315–4325.. [CrossRef]
17. Bunn, S.; Thoms, M.; Hamilton, S.; Capon, S. Flow variability in dryland rivers: Boom, bust and the bits in between. *River Res. Appl.* **2006**, *22*, 179–186. [CrossRef]
18. Cendón, D.I.; Larsen, J.R.; Jones, B.G.; Nanson, G.C.; Rickleman, D.; Hankin, S.I.; Pueyo, J.J.; Maroulis, J. Freshwater recharge into a shallow saline groundwater system, Cooper Creek floodplain, Queensland, Australia. *J. Hydrol.* **2010**, *392*, 150–163. [CrossRef]
19. Villeneuve, S.; Cook, P.G.; Shanafield, M.; Wood, C.; White, N. Groundwater recharge via infiltration through an ephemeral riverbed, central Australia. *J. Arid. Environ.* **2015**, *117*, 47–58. [CrossRef]
20. Meredith, K.T.; Hollins, H.E.; Hughes, C.E.; Cendón, D.I.; Chisari, R.; Griffiths, A.; Crawford, J. Evaporation and concentration gradients created by episodic river recharge in a semi-arid zone aquifer: Insights from Cl, δ^{18}O, δ^{2}H, and ^{3}H. *J. Hydrol.* **2015**, *529*, 1070–1078. [CrossRef]
21. Hofmann, H. Estimating the Role of Bank Flow to Stream Discharge Using a Combination of Baseflow Separation and Geochemistry. *Water* **2023**, *15*, 844. [CrossRef]
22. Brodie, R.; Sundarm, B.; Tottenham, R.; Hostetler, S.; Ransley, R. *An Overview of Tools for Assessing Groundwater-Surface Water Connectivity*; Technical Report; Department of Agriculture, Fisheries and Forestry. Australian Government: Canberra, Australia 2007.
23. Cook, P.G. Estimating groundwater discharge to rivers from river chemistry surveys. *Hydrol. Process.* **2012**, *27*, 3694–3707. [CrossRef]
24. Cecil, L.D.; Green, J.R. Radon-222. In *Environmental Tracers in Subsurface Hydrogeology*; Cook, P.G., Herczeg, A.L., Eds.; Kluwer Academic Books: Boston, MA, USA, 2000; pp. 175–194.
25. Ellins, K.K.; Roman-Mas, A.; Lee, R. Using ^{222}Rn to examine groundwater/surface discharge interactions in the Rio Grande De Manati. *Puerto Rico* **1990**, *115*, 319–341. [CrossRef]
26. Hoehn, E.; Van Gunten, H.R. Radon in Groundwater: A tool to access infiltration from surface water to aquifers. *Water Resour. Res.* **1989**, *25*, 1795–1803. [CrossRef]
27. Cartwright, I.; Hofmann, H.; Sirianos, M.A.; Weaver, T.R.; Simmons, C.T. Geochemical and ^{2}22Rn constraints on baseflow to the Murray River, Australia, and timescales for the decay of low-salinity groundwater lenses. *J. Hydrol.* **2011**, *405*, 333–343. [CrossRef]
28. Cartwright, I.; Hofmann, H.; Gilfedder, B.; Smyth, B. Understanding parafluvial exchange and degassing to better quantify groundwater inflows using ^{222}Rn: The King River, southeast Australia. *Chem. Geol.* **2014**, *380*, 48–60. [CrossRef]
29. Bourke, S.A.; Cook, P.G.; Shanafield, M.; Dogramaci, S.; Clark, J.F. Characterisation of hyporheic exchange in a losing stream using radon-222. *J. Hydrol.* **2014**, *519*, 94–105. [CrossRef]
30. Gilfedder, B.S.; Cartwright, I.; Hofmann, H.; Frei, S. Explicit Modeling of Radon-222 in HydroGeoSphere During Steady State and Dynamic Transient Storage. *Groundwater* **2019**, *57*, 36–47. [CrossRef]
31. Zhou, Z.; Cartwright, I. Using geochemistry to identify and quantify the sources, distribution, and fluxes of baseflow to an intermittent river impacted by climate change: The upper Wimmera River, southeast Australia. *Sci. Total Environ.* **2021**, *801*, 149725. [CrossRef]
32. Murray Darling Basin Authority (MDBA). *Assessment of Environmental Water Requirements for the Proposed Basin Plan: Lower Balonne River Floodplain, MDBA Publication No: 24/12*; Murray-Darling Basin Authority: Canberra, Australia, 2012.
33. DERM. *Refugial Waterholes Project: Research Highlights*; Queensland Department of Environment and Resource Management, Queensland Government: Brisbane, Australia 2010.
34. Waters, J.M.; Burridge, C.P.; Craw, D. The lasting biological signature of Pliocene tectonics: Reviewing the re-routing of Australia's largest river drainage system. *J. Biogeogr.* **2019**, *46*, 1494–1503. [CrossRef]
35. Meredith, K.T.; Hollins, S.E.; Hughes, C.E.; Cendon, D.I.; Hankin, S.; Stone, D.J.M. Temporal variation in stable isotopes (^{18}O and ^{2}H) and major ion concentrations within the Darling River between Bourke and Wilcannia due to variable flows, saline groundwater influx and evaporation. *J. Hydrol.* **2009**, *378*, 313–324. [CrossRef]
36. BOM. *Rainfall Data Portal*; Australian Bureau of Meteorology: Canberra, Australia 2020.
37. Woods, R.J.; Lobegeiger, J.S.; Fawcett, J.H.; Marshall, J.C. (Eds.) *Riverine and Floodplain Ecosystem Responses to Flooding in the Lower Balonne and Border Rivers—Final Report*; Department of Environment and Resource Management, Queensland Government: Brisbane, Australia 2012.

38. Danaher, T.J.; Collett, L. Development, optimisation, and multi-temporal application of a simple Landsat based Water Index. In Proceedings of the 13th Australasian Remote Sensing and Photogrammetry Conference, Canberra, Australia, 20–24 November 2006.
39. de Vries, C.; Danaher, T.; Denham, R.; Scarth, P.; Phinn, S. An operational radiometric calibration procedure for the Landsat sensors based on pseudo-invariant target sites. *Remote Sens. Environ.* **2007**, *107*, 414–429. [CrossRef]
40. Burnett, W.C.; Dulaiova, H. Radon as a tracer of submarine groundwater discharge into a boat basin in Donnalucata, Sicily. *Cont. Shelf Res.* **2006**, *26*, 862–873. [CrossRef]
41. BOM. Australian Groundwater Explorer, Australian Bureau of Meteorology. 2025. Available online: http://www.bom.gov.au/water/groundwater/explorer/map.shtml (accessed on 8 December 2024).
42. D'Hautefeuille, F.; Williams, R. *Upper Darling Salt Interception Scheme—Preliminary Investigation to May 2003*; Centre for Natural Resources New South Wales Department of Infrastructure, Planning and Natural Resources: Parramatta, Australia, 2003.
43. Rojas, R.; Gonzalez, G.; Fu, G. Resilience, stress and sustainability of alluvial aquifers in the Murray-Darling Basin, Australia: Opportunities for groundwater management. *J. Hydrol. Reg. Stud.* **2023**, *47*, 101419. [CrossRef]
44. Werner, A.D.; Laattoe, T. Terrestrial freshwater lenses in stable riverine settings: Occurrence and controlling factors. *Water Resour. Res.* **2016**, *52*, 3654–3662. [CrossRef]
45. Rotz, R.R.; Milewski, A.M. Physical modeling of inland freshwater lens formation and evolution in drylands. *Hydrogeol. J.* **2019**, *27*, 1597–1610. [CrossRef]
46. Gibson, J.J.; Sadek, M.A.; Stone, D.J.M.; Hughes, C.E.; Hankin, S.; Cendón, D.I.; Hollins, S.E. Evaporative isotope enrichment as a constraint on reach water balance along a dryland river. *Isot. Environ. Health Stud.* **2008**, *44*, 83–98. [CrossRef]
47. Howcroft, H.; Cartwright, I.; Cendón, D.I. Residence times of bank storage and return flows and the influence on river water chemistry in the upper Barwon River, Australia. *Appl. Geochem.* **2019**, *101*, 31–41. [CrossRef]

MDPI AG
Grosspeteranlage 5
4052 Basel
Switzerland
Tel.: +41 61 683 77 34

Hydrology Editorial Office
E-mail: hydrology@mdpi.com
www.mdpi.com/journal/hydrology

www.ingramcontent.com/pod-product-compliance
Lightning Source LLC
Chambersburg PA
CBHW061609210926
51850CB00009B/2270
9783725848614